Invertebrate Oxygen Carriers

Edited by Bernt Linzen

With 318 Figures and 46 Tables

Springer-Verlag
Berlin Heidelberg New York
London Paris Tokyo

Professor Dr. Bernt Linzen
Zoologisches Institut der Universität München
Luisenstr. 14, D-8000 München 2
Federal Republic of Germany

QP
99
.3
H4
I58X
1986

ISBN 3-540-16943-1 Springer-Verlag Berlin Heidelberg New York
ISBN 0-387-16943-1 Springer-Verlag New York Berlin Heidelberg

Printing: Druckhaus Beltz, 6944 Hemsbach/Bergstraße
Bookbinding: J. Schäffer, 6718 Grünstadt

2131/3140-543210

P r e f a c e

The international conference on "Invertebrate Oxygen Carriers" took place from July 29th to August 1st, 1985, at Tutzing, in a pleasant setting on Lake Starnberg. It was the 8th in a series which started in 1966 with a meeting in Naples on the "Physiology and Biochemistry of Haemocyanins" organized by F. Ghiretti. The list of contributors and participants of the Naples meeting reports thirty names among which we find that of A.C. Redfield, a pioneer of this field. In Tutzing over 100 scientists came together to discuss their research on all types of oxygen transporting proteins. Clearly the invertebrate hemoglobins have received increasing attention, and a more unified view of their complex quaternary structure is emerging. Their sequences are becoming unravelled and their functions more rationally understood. Hemerythrins have been covered, and their mechanism of binding oxygen explained lucidly. In the hemocyanin field there were exiting new results at every structural level - the X-ray structure of Panulirus hemocyanin, a wealth of sequence data for both arthropodan and molluscan types of hemocyanin, the molecular dimension of native hemocyanins provided by cryo-electron microscopy. New insights into the evolution of copper proteins have been obtained. There are new approaches to explain allosteric interaction in such complicated systems. The scope of investigations reaches from molecular genetics to organismic physiology.

The conference received major support from the Deutsche Forschungsgemeinschaft and from the Bayerisches Staatsministerium für Unterricht und Kultus. The organizers of the meeting wish to express their gratitude to both institutions. In the preparation of this book, the work and unflagging patience of Mrs. I. Krella is highly appreciated.

These Proceedings are dedicated to René Lontie who will be a professor emeritus when he registers for the next meeting. I am most grateful for the charming opening address delivered in his honour by the other "great old man", Francesco 'Ghiro' Ghiretti, who set the tone for the meeting. Mutual affection and attention has formed a wonderful basis for scientific debate, which could be felt all through the meeting. May this continue to be the common denominator in our quest for understanding nature.

B. Linzen

CONTENTS

Opening Session

Professor René Lontie

Photograph taken in 1978

HOMAGE TO RENÉ LONTIE

F. Ghiretti

C.N.R. Center of Hemocyanins and Other Metalloproteins
University of Padova, Padova (Italy)

It is a great pleasure for me to pay homage on behalf of the hemocyanin community to René Lontie on the occasion of his 65th birthday, for his scientific
career dedicated to the study of proteins, in particular of the nature and
function of hemocyanins.

The first time I met René Lontie was in 1955 at the 3rd International Congress
of Biochemistry in Brussels. Soon after I had presented my note on the catalase
activity of hemocyanin, a man younger than me, very polite, introduced himself
and looked very interested in what I was doing. I knew him by name, René Lontie
being one of the very few scientists quoted at that time in the hemocyanin literature. I was very pleased to exchange with him ideas about our common object
of love.

René Lontie graduated in Physical Chemistry at the University of Louvain in
1942. Two years later he published a study on "The influence of the wave length
on the depolarization of hemocyanin". It was the hemocyanin of the garden snail,
Helix pomatia. This is also his first scientific paper which means that on his
scientific baptism, René Lontie was sprinkled with hemocyanin and not with holy
fresh water like every common christian. This certainly left an imprinting: of
about 230 papers published until now, more than half are devoted to hemocyanin.
(I thank Prof. Linzen who kindly gave to me a curriculum vitae of R.L. from
which I have taken many biographical data).

The scientific work of René Lontie covers all aspects of the biochemistry,
physiology and physico-chemistry of proteins. Besides hemocyanin, he has studied
the proteins of milk and a number of cereals. When in 1944 Lontie was asked by
Professor Putzeys to determine the molecular weight of proteins by the method of
light scattering, it happened perhaps by chance that he used hemocyanin. Perhaps
yes, perhaps not. We don't know because we never write in a paper the reason why
that special material used has been chosen. But, whatever the reason, I like to
believe that René Lontie was enchanted by hemocyanin from the first moment.
Everybody, I guess, who looks at this protein becoming blue in the presence of
air, will never forget it. He remains bewitched and sooner or later will return
to it. And, as you know, the hemocyanin of the garden snail is one of the most
beautiful with those nuances unknown to other hemocyanins of whatever mollusc or

arthropod species.

When René Lontie published his first work on hemocyanin, papers on this respiratory protein were very scanty. Unquestionably, in the first half of the nineteenth century, with the exception of the classic work of Svedberg on the molecular weight, and of Redfield on the respiratory function, hemocyanin was almost absent in the scientific literature. Biochemists at that time were interested mainly in yeast and mammals, but hemocyanin is present neither in microorganisms nor in higher animals. In 1935 Jean Roche summarized the state of knowledge in his "Essai sur la Biochimie générale et comparée des pigments respiratoires". Of 170 pages, no more than 15 are covered by hemocyanins.

Almost contemporary with Lontie's paper was the work by Kubowitz of 1938 on "Spaltung und Resynthese der Polyphenoloxidase und des Hämocyanins", a paper by Brohult in 1940 and one by Rawlinson in 1941, both on the copper site, and in 1945 the review by Davson and Mallette "Copper proteins" in "Advances in Protein Chemistry".

The second paper on hemocyanin by René Lontie appeared 10 years later in 1954. It was entitled: "Untersuchungen über die Hämocyanine von Helix pomatia". In this paper hemocyanin is not taken as a representative of proteins as before, but it is studied on purpose as a copper protein and as an oxygen carrying pigment. "Je voudrais me limiter maintenant (I quote from the French translation) à trois propriétés des plus remarquables de l'hémocyanine d'Helix pomatia: la liaison réversible de l'oxygène, la dissociation réversible et la liaison du cuivre". Here is sketched the program of the future work of René Lontie on hemocyanin.

From this time on, René Lontie was absorbed more and more by the work on hemocyanin. We can obviously conclude that his first meeting with hemocyanin in 1944 was not an accidental infection but that he had gotten the virus and the virus had settled in his brain. Hemocyanin is an infectious disease indeed. But don't worry. In this audience we are all infected.

It would be useless, if not ridiculous, to summarize for you the contributions of René Lontie to the biochemistry and physiology of hemocyanins. To the youngest, however, I want to remember the determination of the pH stability curve of Helix pomatia hemocyanin made in 1954, which has been the basis for all future work on the subunit assembly of the molecule; the demonstration given in 1956 of the two forms, alpha and beta hemocyanin, formerly supposed by Brohult and Borgman; the suggestion that copper in hemocyanin is not bound by thiol groups as proposed by Klotz, but is chelated in a complex with the intervention of histidine. This was in 1958. All further studies have largely proved the validity of this suggestion.

Among the most recent publications I like to mention (for the benefit not only of the youngest but also of the oldest of us) the three volumes on "Copper Proteins and Copper Enzymes" René Lontie has edited last year. One chapter of 20 chapters is entitled to hemocyanin. It seems to me that specially with this work René Lontie goes back to his first study of 1944, thus closing a long journey at the end of which hemocyanin turns out again to be one protein among similar ones. However beautiful and fascinating hemocyanins are, we know indeed that their main value lies in the possibility they have to help us in the understanding of the relationships between structure and function. Our strategy is to use hemocyanin as a pilot protein. And indeed, as for the active site, it is proving to be a valuable bait for hooking other copper proteins. The same, we hope, will happen for the architecture of similar macromolecules. I believe that the decision to change the name of our periodical workshop in "Invertebrate oxygen binding proteins" or "Invertebrate respiratory proteins" or (as the present one) "Invertebrate oxygen carriers" has been profitable not only for the true content of the meeting. After a certain number of hemocyanin workshops, grant agencies started to wonder what inoffensive but expensive mental disease had affected us.

Dear René, I am sure that if, looking back, you compare the knowledge about hemocyanin we had 40 years ago with the present one, you can feel content of the work you have done. We are very grateful to you for it. Certainly, everything has not yet been discovered and many problems are still unsolved. But this is because you are not selfish and you wanted to leave some room to the young generation. To these I say: Be quiet! Hemocyanin will be a subject of study for years to come. Hemocyanin is a hard nut to crack. That's perhaps why it is also so attractive.

Consider the quaternary structure. Do you remember when in 1963, by opening the last issue of the Journal of Molecular Biology, we were breathless with astonishment looking for the first time at the hidden treasure van Bruggen had disclosed? Under the electron microscope hemocyanin molecules appear like jewels. Then a new era started in the history of this protein. One day the architecture of hemocyanin will be depicted in all its details and somebody will wonder whether it is a work of art or science. If D'Arcy Thompson had the venture to extend his analysis to the molecular level; if he could admire the geometry of the tridimensional structure of the extracellular respiratory pigments, certainly he would give us another masterpiece, perhaps a second volume of the famous "Growth and Form".

René Lontie is not only a scientist: he is also a poet. Many years ago he sent to my wife and myself a booklet of poems. Alas, they were in Flemish and, unlike scientific papers, poetry has no abstract in another language.

For his hemocyanin studies, René Lontie has used mostly the garden snail Helix pomatia, the charming pulmonate gastropod. As you know, in Saint-Petersbourg, today Leningrad, in front of the Institute of Physiology where Ivan Pavlov discovered and studied the conditioned reflexes, there is a monument to the dog. I hope that at the entrance of the Laboratory of Biochemistry in the Rue des Doyens in Louvain a monument will be erected to the snail. As the dog in the hands of Pavlov was so precious for Neurophysiology, so in the hands of René Lontie the Helix contributed so much to the knowledge of hemocyanin. Meanwhile I ask Lontie to accept a humble homage, perhaps the sketch of the monument, from the group of Padova. We wish him a long life full of intellectual and physical satisfaction.

Hemerythrins

STRUCTURE, FUNCTION AND OXIDATION LEVELS OF HEMERYTHRIN

Donald M. Kurtz, Jr.
Department of Chemistry, Iowa State University
Ames, Iowa 50011/U.S.A.

I. Introduction

The hemerythrins are a group of non-heme iron oxygen-carrying proteins found in several phyla of marine invertebrates, most notably sipunculids. The hemerythrins pose interesting evolutionary, physiological, and chemical contrasts to the more widespread hemoglobins. Perhaps the most intriguing question to a chemist is: how does nature adapt a non-heme iron site to reversibly bind molecular oxygen?

In pursuit of an answer to this question, detailed physical and chemical characterization of both the iron site and the surrounding polypeptide in hemerythrin have been undertaken during the past several years. Several reviews of this work are available (1-4). This report will summarize the most salient features of the earlier results and discuss certain advances which have occurred since the most recent reviews, mainly concerning the iron site.

II. Molecular Properties

Hemerythrin is contained within erythrocytes in the coelomic fluid of sipunculids. Although dimers, trimers and tetramers have been reported, hemerythrins from erythrocytes normally consist of octamers of identical or nearly identical subunits (3). A monomeric myohemerythrin has also been isolated from retractor muscles of the sipunculid Themiste zostericola (5,6). In all cases examined the subunit has a molecular weight of ~ 13,500. Each subunit contains two iron atoms and binds one molecule of oxygen.

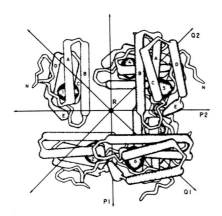

Figure 1. The quaternary and tertiary structures of Phascolopsis gouldii hemerythrin taken from Ward et al. (7). For clarity two subunits are omitted. P and Q label two-fold axes and R a four-fold axis through the octamer. The dark, numbered spheres represent iron sites.

Invertebrate Oxygen Carriers
Ed. by Bernt Linzen
© Springer-Verlag Berlin Heidelberg 1986

The quaternary and tertiary structure of octameric hemerythrin as determined from X-ray crystallography is illustrated in Figure 1. The molecule has been described as a square doughnut, 75 x 75 x 40 Å with a 20 Å central cavity. The subunits are arranged in two layers with four subunits in each layer. Each iron site is 28-30 Å away from its four nearest neighbors. Each subunit is folded into four long, approximately parallel helical regions labelled A, B, C, and D in Figure 1 with a carboxyl-terminal helical stub E, and a 20 residue amino-terminal non-helical arm. The tertiary structures of octameric hemerythrin from Themiste dyscritum (8) and myohemerythrin from Themiste zostericola (9) are quite similar to that shown in Figure 1.

Despite the large number of intersubunit contacts (7,8), cooperativity in oxygen binding and a Bohr effect have been demonstrated for purified, octameric hemerythrins only from the brachiopod Lingula (10,11). However, the lack of significant allosteric interactions in other purified hemerythrins may not reflect the situation within the erythrocyte (12).

Perchlorate is known to be an artificial heterotropic allosteric effector of hemerythrin, although its effect on oxygen binding is small (8,13,14). Using ^{31}P NMR in this laboratory, we have detected O-phosphorylethanolamine (PEA) within intact erythrocytes of P. gouldii. Furthermore, we have shown that PEA (which is much too large to bind directly to the iron site) inhibits azide binding to the iron site in purified methemerythrin (15). Thus, PEA is an allosteric effector and could be analogous to that of 2,3-diphosphoglycerate for human hemoglobin.

III. Structure of the Iron Site

The structure of the iron site as obtained from an X-ray crystal structure at 2.0 Å resolution of the azide adduct of methemerythrin is shown in Figure 2 (16). The two iron atoms are located between the four approximately parallel helical regions shown in Figure 1. Helices C and D provide three imidazole ligands from His residues to the top iron atom of Figure 2 plus a bridging carboxylate from an Asp residue. Helices A and B provide two imidazole ligands from His residues to the bottom iron atom plus an additional bridging carboxylate from the side chain of a Glu residue. The third bridging position has been assigned to an oxo ion derived from water. The Fe-Fe distance is ~ 3.3 Å. As shown in Figure 2 for metazidehemerythrin, the sixth coordination position of the bottom iron is occupied by N_3^-, which is bound in an end-on fashion. Difference electron density maps indicate that O_2 in oxyhemerythrin is bound to the same iron as is N_3^- in metazidehemerythrin (17).

Figure 3 shows that a structure analogous to the binuclear iron complex of hemerythrin can be prepared synthetically from aqueous mixtures of ferric salts, acetate and the tridendate ligand tris(1-pyrazolyl)borate (18). In addition to being a very useful model for structural and spectroscopic comparisons, the spontaneous self-assembly of this structure suggests that assembly of the iron site in hemerythrin requires little or no thermodynamic assistance of the polypeptide. However, up to now the native iron site in hemerythrin has not been successfully reconstituted after denaturation (19).

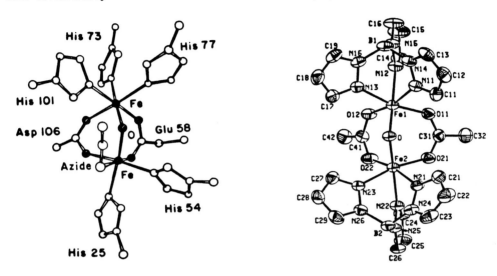

Figure 2. (Left) Structure of the iron center in metazidehemerythrin from <u>T. dyscritum</u> redrawn from Stenkamp et al. (16).

Figure 3. (Right) Structure of the synthetic compound [(HBpz$_3$)FeO(CH$_3$COO)$_2$Fe(HBpz$_3$)] from Armstrong et al. (18). HBpz$_3$ = tris(1-pyrazolyl)borate.

Probably the most definitive evidence to date for the presence of a μ-oxo bridge between the iron atoms in oxy- and methemerythrins comes from resonance Raman spectroscopy. This technique selects out vibrational modes, which are coupled to electronic transitions of the iron site. A number of anion adducts of methemerythrin have been shown to have a peak in the resonance Raman spectrum near 510 cm^{-1} which shifts downward in H$_2^{18}$O (4). This frequency is in the correct range for the symmetric stretch of a bent Fe-O-Fe system. For example, the synthetic compound whose structure is shown in Figure 3 displays an H$_2^{18}$O sensitive band at 528 cm^{-1} in its resonance Raman spectrum (18). Recently, using near-UV excitation, an H$_2^{18}$O sensitive band in the resonance Raman spectrum of oxyhemerythrin has been identified at 486 cm^{-1} (20).

This frequency together with those previously assigned to ν(O-O) at 844 cm^{-1} and ν(Fe-O) at 503 cm^{-1} of bound O$_2$ all shift in D$_2$O (20,21). Thus, protons appear to be intimately associated with the oxygen binding site. These results together with the difference electron density maps discussed above led to the proposal illustrated in Figure 4 for the oxygenation reaction. This model for oxygenation would require minimal structural reorganization of the iron site and is consistent with the high rate of oxygen binding to deoxyhemerythrin ($\sim 10^7$ M^{-1} s^{-1}) (3,14).

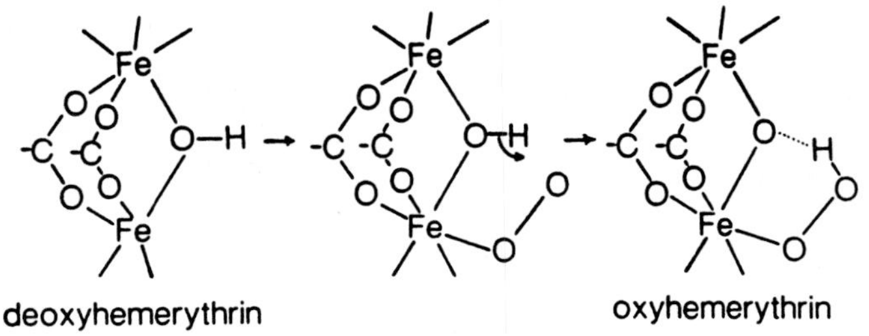

deoxyhemerythrin   oxyhemerythrin

Figure 4. Proprosed structures for the iron sites of oxy- and deoxyhemerythrin and a possible mechanism for their interconversion from Stenkamp et al. (17).

IV. Oxidation Levels

The 844 cm^{-1} frequency for $\nu(0-0)$ of bound O_2 in oxyhemerythrin is most consistent with a formal peroxide (O_2^{2-}) oxidation state (22). In addition, Mössbauer spectroscopy of deoxy- and oxyhemerythrins show essentially only high spin Fe(II) in the former and only high spin Fe(III) in the latter (1). Thus, the oxygenation reaction as written in Figure 5 is really an internal redox reaction. In addition, the autooxidation reaction:

$$[Fe(III),Fe(III)O_2^{2-}](oxy) + 2H^+ \rightarrow [Fe(III),Fe(III)](met) + H_2O_2$$

occurs with a half-time of $\sim$ 20h @ 25°C in vitro (3). For these reasons, physical and chemical characterizations of the various oxidation levels of the iron site in hemerythrin (shown in Figure 5) become important.

A. Oxy and Met

Although methemerythrin does not bind O_2, it does bind a number of small anions such as azide. These anion adducts, because of their stabilities and because the iron atoms are at the same oxidation state as in oxyhemerythrin, have been helpful in characterization of the structural and electronic properties of oxyhemerythrin. A prime example is the crystallographic studies discussed above.

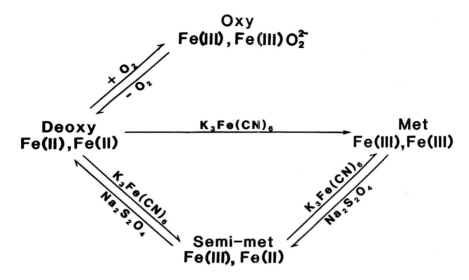

Figure 5. The oxidation levels of the iron site in hemerythrin and pathways for their interconversions.

Among the earliest and still among the best evidence for a μ-oxo bridge in oxy- and methemerythrins consists of magnetic susceptibility measurements (23). These measurements showed, despite the five unpaired electrons on each high spin Fe(III), that the ground states in oxy and met are diamagnetic. The unpaired spins on the iron atoms are antiferromagnetically coupled via a superexchange pathway through the bridging oxo ion. The magnitudes of the exchange coupling constant (-J) in oxy and met are 77 and 134 cm^{-1}, respectively. These values are similar to that measured for the "model" compound $[(HBpz_3)FeO(CH_3COO)_2Fe(HBpz_3)]$, where -J = 121 cm^{-1} (18). Thus, at ambient temperatures the iron centers in oxy and met retain some net unpaired spin, although much less than that of isolated high spin Fe(III).

B. Deoxy

Since the iron centers in oxy and met are relatively well characterized, recent attention has been focused on the deoxy and semi-met oxidation levels. On the basis of the temperature dependence of magnetic circular dichroism (MCD), Reem and Solomon have calculated that the two high spin Fe(II)'s in deoxy are antiferromagnetically coupled with -J = 13 ± 5 cm^{-1} (24). They propose a structure for the iron site of deoxyhemerythrin which is identical to that shown in Figure 4. Interestingly, in the presence of N_3^-, OCN^- or F^- the MCD of deoxyhemerythrin is consistent with a paramagnetic ground state and no antiferromagnetic coupling. This result indicates breakage of the bridging hydroxo linkage between the irons in the presence of these anions. Further evidence for this breakage is that an EPR spectrum with g ∼ 13 is observed for deoxyN$_3^-$ which is assigned to a "forbidden" M_s ± 2 transition. This type of spectrum has been observed for other isolated high spin Fe(II) systems (25). Thus, an equilibrium may exist between a coupled and uncoupled iron site in deoxyhemerythrin. Under most conditions this equilibrium probably lies overwhelmingly towards the coupled form. However, at least in the presence of N$_3^-$ or OCN$^-$, the autooxidation of oxy- to methemerythrin proceeds through a deoxy intermediate (26). In these cases it may be the uncoupled form which is actually converted to methemerythrin. The above equilibrium also explains the slow rate of exchange of the bridging oxygen with H$_2^{18}$O (20).

C. DeoxyNO

The structures shown (Fig.4) indicate that incoming dioxygen would have direct access to only one of the two iron atoms. Thus, given the formal oxidation state changes shown in Figure 5 for the oxygenation reaction, one could postulate the existence of an intermediate, which would be formulated as [Fe(II),Fe(III)O$_2^-$]. Such an intermediate has never been detected. However, a nitric oxide adduct of deoxyhemerythrin (deoxyNO) has recently been prepared and characterized in this laboratory (27). The physical and chemical properties of deoxyNO are consistent with its formulation as [Fe(II),Fe(III)NO$^-$]. Thus, deoxyNO can be viewed as an analogue of the putative intermediate in the oxygenation reaction. The iron site in deoxyNO is an odd electron system, which in the presence of antiferromagnetic coupling leads to a ground spin state S = 1/2. Evidence for this coupling in deoxyNO comes from its unique Mössbauer and EPR spectra (27), the latter of which is shown in Figure 6 (top). Anaerobic addition of excess N$_3^-$, OCN$^-$ or F$^-$ to a solution of deoxyNO results in immediate bleaching of its characteristic pine-green color and disappearance of the deoxyNO EPR signal. In the case of N$_3^-$ addition, the deoxyNO EPR signal is replaced by the deoxyN$_3^-$ EPR signal mentioned above. Excess CN$^-$ in contrast, does not bleach the deoxyNO color and has only a slight effect on the EPR spectrum. This anion specificity is that observed for bleaching of the oxyhemerythrin color (26). Thus, the NO reaction is reversible and exhibits anion specificity very similar to that of the deoxygenation reaction.

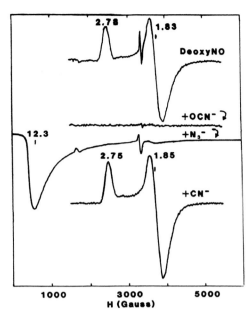

Figure 6. EPR spectra of (top to bottom) deoxyNO, and deoxyNO plus excess N_3^-, OCN^- and CN^-. Spectra were obtained at 4.2K. Numbers near each spectrum indicate positions of g values.

We propose the structure shown in Figure 7 for the iron site of deoxyNO. In this structure the proton remains bonded to the bridging oxo ion and is hydrogen bonded to the terminal oxygen of bound NO, rather than the alternative arrangement shown in Figure 4 for oxyhemerythrin. The proposed deoxyNO structure is actually close to that proposed for the intermediate between oxy- and deoxyhemerythrin shown in Figure 4.

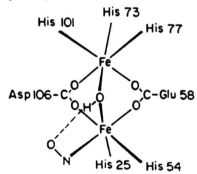

Figure 7. Proposed structure for the iron site of deoxyNO.

D. Semi-met

On the basis of the kinetics of reduction of met- to deoxyhemerythrin by dithionite, Harrington et al. in 1978 proposed an intermediate [Fe(II),Fe(III)] "semi-met" oxidation level (28). Its existence has since been confirmed by EPR (29,30), Mössbauer (31), and resonance Raman spectroscopies (32). The EPR spectra of semi-methemerythrin (Figure 8) and its anion adducts are consistent with an S = 1/2 ground state as expected for antiferromagnetic coupling of a high spin Fe(II) with a high spin Fe(III) (3,29,30).

An estimate of the magnitude of the coupling in one anion adduct, semi-metN$_3^-$, has come from the discovery of paramagnetic ^{1}H NMR spectra of hemerythrin (33). The ^{1}H NMR spectrum of semi-metN$_3^-$ at 30°C features resonances at 73 and 54 ppm downfield. On the basis of the large shifts and their disappearance in D$_2$O, these resonances are assigned to the NH protons of imidazoles coordinated to the ferric and ferrous ions, respectively of the paramagnetic iron center. The two resonances exhibit Curie temperature dependence in the range 0-50°C consistent with $-J = 10$ cm^{-1}. Although one expects an inherent decrease in exchange coupling compared to met due to the unequal oxidation states (34), the order of magnitude decrease may be more consistent with a hydroxo rather than an oxo bridge in semi-metN$_3^-$. A similar decrease in $-J$ occurs upon protonation of the μ-oxo bridge in [(HBpz$_3$)FeO(CH$_3$COO)$_2$Fe(HBpz$_3$)] (35).

E. Redox Reactions

For reasons mentioned at the beginning of this section, the redox kinetics of hemerythrins in all of the oxidation levels shown in Figure 5 have been examined with a variety of inorganic and biochemical reagents. The results of most of these studies have been thoroughly reviewed (3). Some significant new developments are discussed here.

Armstrong et al. (36) have recently examined the reduction of octameric methemerythrin from T. zostericola by several inorganic reducing agents. Their data in the presence of an excess of reducing agent are consistent with four stages in the reduction of met- to deoxyhemerythrin (37):

$$[Fe(III),Fe(III)]_8(met) \xrightarrow{k_1} [Fe(II),Fe(III)]_8(semi\text{-}met)$$

$$\downarrow k_2$$

$$[Fe(II),Fe(II)]_4[Fe(II),Fe(III)]_4$$

$$\downarrow k_3$$

$$[Fe(II),Fe(II)]_8(deoxy) \xleftarrow{k_4} [Fe(II),Fe(II)]_6[Fe(III),Fe(III)]_2$$

For a series of Co(II) complexes the rates at 25°C and pH 6.3 were 113-260 M^{-1}s^{-1} for k_1, 2.0 x 10^{-3} s^{-1} for k_2 and 1.2 x 10^{-4} s^{-1} for k_3, the second and third stages being independent of concentration or identity of the reductant. Rates for the fourth stage were not reported due to the very small absorbance changes involved and to long term instability of the reduced protein. The

kinetics of reduction of octameric methemerythrin from P. gouldii have recently
been reexamined in our laboratory (38). We also find four stages for the
absorbance changes with rate constants very similar to those reported for T.
zostericola hemerythrin. In addition, we have verified the above scheme by
examinations of the reaction solutions by EPR spectroscopy. An EPR signal can
be obtained only from [Fe(II),Fe(III)] containing subunits. As shown in Figure
8, such a signal is observed near the end of the first and second stages (1 and
30 min, respectively), but not the third stage (300 min).

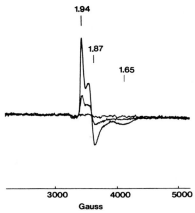

1.94

1.87

1.65

Figure 8. Semi-met EPR signals
obtained during reduction of
0.1 mM P. gouldii
methemerythrin with a 20-fold
molar excess of $Na_2S_2O_4$ at pH
7.0 and 20°C. With decreasing
intensity, spectra are of
samples frozen ~ 1, 30, and 300
minutes after mixing (38).

3000   4000   5000
Gauss

Since both of these stages are independent of concentration of reducing agent,
it is probable that conformational changes in the protein are the rate limiting
steps. Such changes have been invoked previously to explain interconversion
between two forms of semi-met obtained as kinetic products. The form obtained
upon stoichiometric oxidation of deoxy is termed (semi-met)$_0$, whereas that
obtained upon stoichiometric reduction of met is termed (semi-met)$_R$ (29,30).
These two forms can be distinguished on the basis of EPR lineshapes. The
diagram of Figure 5 can be modified to include these two forms:

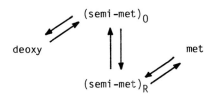

(semi-met)$_0$

deoxy   met

(semi-met)$_R$

In Figure 8 only the (semi-met)$_R$ type EPR signal is observed, as expected if a
conformational change of the type, $[(semi-met)_R]_8 \rightarrow [(semi-met)_R]_4[(semi-met)_0]_4$,
were the rate limiting step for the second stage of reduction. The
intramolecular disproportionation implicit in the third stage of reduction has
been observed previously for octameric hemerythrin from T. zostericola after
reduction of met by eight equivalents of reducing agent (3,28):

$$[Fe(III),Fe(III)]_8(met) \longrightarrow [Fe(III),Fe(II)]_8(semi\text{-}met)_R$$

$$\downarrow \text{disproportionation}$$

$$[Fe(II),Fe(II)]_4[Fe(III),Fe(III)]_4$$

Prior to the discovery of the semi-met oxidation level, Bradíc et al. (26) had demonstrated that autooxidation of oxy- to methemerythrin is accelerated in the presence of anions, nitrite among them. However, nitrite is known to function as a redox agent itself in many reactions. Also, unlike most other oxidants whose reactions with hemerythrin have been studied (3), nitrite can potentially function in an "inner sphere" mode by direct binding to iron. Doyle et al. (39) have provided evidence that oxidation of deoxy- to methemoglobin by nitrite occurs in this fashion. For these reasons reactions of deoxyhemerythrin with $NaNO_2$ were investigated in this laboratory (40). The results show that nitrite appears to be a unique oxidant of deoxyhemerythrin in that, when employed in excess, the final product is semi-met- rather than methemerythrin. Our proposed mechanism for this reaction is:

$$[Fe(II),Fe(II)] + HONO \xrightarrow{\ k_1\ } [Fe(II),Fe(III)NO] + OH^-$$
$$\text{deoxy} \qquad\qquad\qquad\qquad \text{semi-metNO}$$

$$\downarrow k_2$$

$$[Fe(II),Fe(III)NO_2^-] \xleftarrow[\ NO_2^-\ ]{\text{fast}} [Fe(II),Fe(III)]$$
$$\text{semi-metNO}_2^- \qquad\qquad\qquad \text{semi-met}$$
$$+$$
$$NO$$

A possible generalization from these results is that excesses of "outer sphere" oxidants such as $Fe(CN)_6^{3-}$ convert deoxy to semi-met, which is then further oxidized to methemerythrin. Excesses of "inner sphere" oxidants, on the other hand convert deoxy to semi-met, which is stabilized with respect to methemerythrin because the oxidant or a product of the oxidant can bind to the iron site. These results may have implications regarding the mechanism of autooxidation of hemerythrin as well as oxidation by cellular oxidants. We have shown that hemerythrin within intact erythrocytes of P. gouldii can be oxidized by extracellular addition of $NaNO_2$. A semi-metNO_2^- EPR signal was observed from these treated erythrocytes (41).

In normal, functioning human erythrocytes, hemoglobin continually undergoes oxidation. However, at any given time met accounts for less than 1% of the total hemoglobin (42,43). This low steady state level is maintained by a reductase system in which electrons are transferred to methemoglobin in the sequence: NADH$\rightarrow$cytochrome b_5 reductase$\rightarrow$cytochrome b_5-$\rightarrow$methemoglobin

(44,45). Despite the differences between mammalian and sipunculan erythrocytes, we have discovered what may be one component of a similar system in erythrocytes of P. gouldii, namely a cytochrome b_5 (46). Table I compares the properties of P. gouldii cytochrome b_5 with those of human erythrocyte cytochrome b_5. As can been seen, these properties are quite similar. Most importantly, reduced P. g. cytochrome b_5 is oxidized by P. g. methemerythrin. The second order rate constant for this reaction at 20°C and pH 7.2 is 650 ± 50 M^{-1} s^{-1}. This value is about 3 orders of magnitude faster than for oxidation of deoxymyoglobin by P. g. methemerythrin (3). The relatively high rate constant for P. g. cytochrome b_5 may reflect its specificity for reduction of methemerythrin. The low isoelectric point listed in Table I means that near pH 7, P. g. cytochrome b_5 has a net negative charge, while hemerythrin, with an isoelectric point of ~ 8 (48), has a net positive charge. In fact, a complex forms between hemerythrin and the cytochrome b_5 at < 0.3 M NaCl. This complex should provide an interesting contrast to that formed between human cytochrome b_5 and hemoglobin. In the latter case a heme to heme one-electron transfer occurs, whereas, in the case of hemerythrin a heme to non-heme iron electron transfer occurs and two electrons are required per iron site to reach deoxy- from methemerythrin.

Table I. Properties of Cytochromes b_5 from Human and P. gouldii Erythrocytes

	Human[a]	P. gouldii[b]
Molecular weight	13,700	14,000
pI	4.3	3-4
EPR g-values, oxidized	3.03,2.21,1.39	3.07,2.22,1.4
Soret maxima (nm): oxidized	412	412
reduced	423	422
α-band maximum (nm)	556	555
β-band maximum (nm)	526	526

[a] from references 45 and 47
[b] from reference 46

V. Conclusions

It is obvious for this review that a large body of knowledge has been gained about the structure and properties of the oxygen binding site of hemerythrin and of the surrounding polypeptide. On the other hand, in contrast to the case of at least vertebrate hemoglobins, relatively little is known about other cellular constituents which may affect the chemistry of hemerythrin. Superoxide dismutase and NADH diaphorase activities have been reported in sipunculan erythrocytes (49). These activities together with the discoveries mentioned above of 0-phosphorylethanolamine and of a cytochrome b_5 within erythrocytes of P. gouldii indicate that this line of research is worth pursuing.

Acknowledgements: Research on hemerythrin at Iowa State University was supported by the National Institutes of Health (GM33157) and the National Science Foundation (PCM-8216447). I acknowledge continuing collaborations of Professors Peter Debrunner at the University of Illinois, Larry Que, Jr. at the University of Minnesota, Thomas M. Loehr and Joann Sanders-Loehr at the Oregon graduate Center and their students. Special thanks go to my own students, Gudrun Lukat, Judy Nocek, Linda Pearce, Ron Utecht and Pierre Robitaille, who have all contributed to results reported in this review.

List of References

1. Kurtz, D. M., Jr., Shriver, D. F., and Klotz, I. M. Coord. Chem. Rev. 24, 145-178 (1977).
2. Loehr, J. S. and Loehr, T. M. Adv. Inorg. Biochem. 1, 235-252 (1979).
3. Wilkins, R. G., and Harrington, P. C. Adv. Inorg. Biochem. 5, 51-85 (1983).
4. Klotz, I. M., and Kurtz, D. M., Jr. Acc. Chem. Res. 17, 16-22 (1984).
5. Manwell, C. Comp. Biochem. Physiol. 1, 277-285 (1960).
6. Klippenstein, G. L., Van Riper, D. A., and Oosterom, E. A. J. Biol. Chem. 247, 5959-5963 (1972).
7. Ward, K. B., Hendrickson, W. A., and Klippenstein, G. L. Nature (London) 257, 818-821 (1975).
8. Stenkamp, R. E. , and Jensen, L. H. Adv. Inorg. Biochem. 1, 219-233 (1979).
9. Hendrickson, W. A., Klippenstein, G. L., and Ward, K. B. Proc. Nat. Acad. Sci. USA 72, 2160-2164 (1975).
10. Manwell, C. Science (Washington, D.C.) 132, 550-551 (1960).
11. Richardson, D. E., Reem, R. C., and Solomon, E. I. J. Am. Chem. Soc. 105, 7781-7783 (1983).
12. Mangum, C. P., and Kondon, M. Comp. Biochem. Physiol. 50A, 777-785 (1975).
13. Garbett, K. Darnall, D. W., and Klotz, I. M. Arch. Biochem. Biophys. 142, 455-470 (1971).
14. DeWaal, D. J. A., and Wilkins, R. G. J. Biol. Chem. 251, 2339-2343 (1976).
15. Robitaille, P.-M. and Kurtz, D. M., Jr., manuscript in preparation.
16. Stenkamp, R. E., Sieker, L. C., and Jensen, L. H. J. Am. Chem. Soc. 106, 618-622 (1984).
17. Stenkamp, R. E. Sieker, L. C., Jensen, L. H. McCallum, J. D., Sanders-Loehr, J. Proc. Nat. Acad. Sci. USA 82, 713-716 (1985).
18. Armstrong, W. H., Spool, A., Papaefthymiou, G. C., Frankel, R. B., and Lippard, S. J. J. Am. Chem. Soc. 106, 3653-3667 (1984).
19. Bradic Z., and Wilkins, R. G. Biochim. Biophys. Acta 828, 86-94 (1985).
20. Shiemke, A. K., Loehr, T. M., and Sanders-Loehr, J. J. Am. Chem. Soc. 106, 4951-4956 (1984).

21. Sanders-Loehr, J. in Bioinorganic Chemistry 85 (Xavier, A. V., eds.), VCH Publishers, Weinheim (1985), in press.
22. Dunn, J. B. R., Shriver, D. F., and Klotz, I. M. Biochemistry 14, 2689-2695 (1975).
23. Dawson, J. W., Gray, H. B., Hoening, H. E., Rossmann, G. R., Schredder, J. M., and Wang, R. H. Biochemistry 11, 461-465 (1972).
24. Reem, R. C. and Solomon, E. I. J. Am. Chem. Soc. 106, 8323-8325 (1984).
25. Hagen, W. R. Biochim. Biophys. Acta 108, 82-98 (1982).
26. Bradic Z., Conrad, R., and Wilkins, R. G. J. Biol. Chem. 252, 6069-6075 (1977).
27. Nocek, J. M., Kurtz, D. M., Jr., Sage, J. T., Debrunner, P. G., Maroney, M. J., Que, L., Jr. J. Am. Chem. Soc. 107, 3382-3384 (1985).
28. Harrington, P.C., DeWaal, D. J. A., and Wilkins, R. G. Arch. Biochem. Biophys. 191, 444-451 (1978).
29. Babcock, L. M., Bradic Z., Harrington, P. C., Wilkins, R. G., Yoneda, G. S. J. Am. Chem. Soc. 102, 2849-2850 (1980).
30. Muhoberac, B. B., Wharton D. C., Babcock, L. M., Harrington, P. C., and Wilkins, R. G. Biochem. Biophys. Acta 626, 337-345 (1980).
31. Kurtz, D. M., Jr., Sage, J. T., Hendrich, M., Debrunner, P. G., Lukat, G. S. J. Biol. Chem. 258, 2115-2117 (1983).
32. Irwin, M. G., Duff, L. L., Shriver, D. F., Klotz, I. M. Arch. Biochem. Biophys. 224, 473-478 (1983).
33. Maroney, M. J., Lauffer, R. B., Que, L., Jr., and Kurtz, D. M., Jr. J. Am. Chem. Soc. 106, 6445-6446 (1984).
34.a) Papaefthymiou, G. C., Laskowski, E. J., Frota-Pessoa, S., Frankel, R. B., and Holm, R. H. Chem. 21, 1723-1728 (1982) and references therein.
   b) Noodleman, L., and Baerends, E. J. J. Am. Chem. Soc. 106, 2316-2327 (1984).
35. Armstrong, W. H., and Lippard, S. J. J. Am. Chem. Soc. 106, 4632-4633 (1984).
36. Armstrong, G. D., Ramasami, T., and Sykes, A. G. J. C. S. Chem. Comm. 1017-1019 (1984).
37. Armstrong, G. D., and Sykes, A. G., personal communication.
38. Pearce, L., and Kurtz, D. M., Jr., manuscript in preparation.
39. Doyle, M. P., Pickering, R. A., De Weert, T. M., Hoeckstra, J. W., and Pater, D. J. Biol. Chem. 256, 12393-12398 (1981).
40. Nocek, J. M., Kurtz, D. M. Jr., Pickering, R. A., and Doyle M. P. J. Biol. Chem. 259, 12334-12338 (1984).
41. Nocek, J. M., and Kurtz, D. M., Jr., unpublished results.
42. Hsieh, H. S., Jaffe, E. R. in The Red Blood Cell, Second Edition, (Surgenor, D. M., ed.), Academic Press, New York, Vol. 2, pp. 799-824 (1975).
43. Rodkey, F. L., O'Neal, J. D. Biochem. Med. 9, 261-270 (1974).
44. Sannes, L. J., Hultquist, D. E. Biochem. Biophys. Acta 544, 547-554 (1978).
45.a) Abe, K, Sugita, Y. Eur. J. Biochem. 101, 423-428 (1979). (b) Kuma, F. J. Biol. Chem. 256, 5518-5523 (1981).
46. Utecht, R. E., and Kurtz, D. M., Jr., Inorg. Chem. 1985, in press.
47. Passon, P. G., and Hultquist, D. E. Biochim. Biophys. Acta 275, 51-61 (1972).
48. Keresztes-Nagy, S., and Klotz, I. M. Biochem. 4, 919-931 (1965).
49. Manwell, C. Comp. Biochem. Physiol. 58B, 331-338 (1977).

Structure and function of invertebrate hemoglobins

THE DISSOCIATION OF THE EXTRACELLULAR HEMOGLOBIN OF LUMBRICUS TERRESTRIS: A MODEL OF ITS SUBUNIT STRUCTURE

Serge N. Vinogradov
Biochemistry Department
Wayne State University
Detroit, MI 48201

INTRODUCTION

The invertebrate extracellular hemoglobins can be divided into four distinct groups based on their subunit structure (1). One of these groups is comprised of the annelid extracellular hemoglobins and chlorocruorins. These 60S molecules possess a characteristic appearance of a two-tiered hexagonal array of twelve pentagonally-shaped subunits (2) and, unlike the preponderant majority of all other hemoglobins and myoglobins, have a low iron content of 0.24 $\pm$ 0.03 % (3-5). The hemoglobin of Lumbricus terrestris, the common earthworm, is the best studied of the annelid hemoglobins: it has a molecular mass of 3.9×10^6, a diameter of 30nm, a height of 20nm and an iron content of 0.22% (4-6). SDS PAGE of its reduced form shows that it consists of at least six polypeptide chains: I through IV (M_r 16-19 kDa) and V (M_r 31kDa) and VI (M_r 37kDa) (7). SDS PAGE of the unreduced hemoglobin shows that it consists of four subunits: M (chain I), D1 (chain V), D2 (chain VI) and subunit T, a disulfide-bonded trimer of chains II, III and IV (4-7). Figure 1 shows the relationship between the two sets of subunits.

DISSOCIATION AT ALKALINE pH AND REASSOCIATION

Lumbricus hemoglobin dissociates at pH above 8 into three heme-containing species with sedimentation coefficients of 9-10S,, 3.5S and 2.3S (8). The presence of fragments of similar size at alkaline pH has been observed for several other earthworm hemoglobins (9-11). SDS PAGE showed that the 2.3S species consists of chain I only, the 3.5S fragment consists of chains II-VI and the 9-10S species consis-

Invertebrate Oxygen Carriers
Ed by Bernt Linzen
© Springer–Verlag Berlin Heidelberg 1986

ted of all six chains and is usually considered to represent the putative one twelfth of the whole molecule. The pattern of dissociation into a monomeric heme-containing chain and a fragment consisting of all the remaining chains has also been observed for the hemoglobins of other annelids, including oligochaetes (12, 13), polychaetes (14-17) and leeches (18,19).

The reassociation of Lumbricus hemoglobin at neutral pH following exposure to alkaline pH has been investigated in detail: we have demonstrated that both dissociation and reassociation are equilibrium processes and that there were two reassociation processes, one of which was a "dead end" pathway resulting in Ib, a fragment of the size of one twelfth of the whole molecule (20). Although the dissociations of two other earthworm hemoglobins were considered to be irreversible (9,10), evidence for the reassociation of Octalasium complanatum hemoglobin into whole molecules has been obtained recently (21). Likewise, reassociation has been demonstrated for two other oligochaete molecules: Eisenia foetida (22) and Tubifex tubifex (23) hemoglobins.

The extent of dissociation of most annelid extracellular hemoglobins at alkaline pH is decreased in the presence of divalent cations such as Ca and increased in the presence of chelators of divalent cations such as EDTA (3). This is also true of Lumbricus hemoglobin (20). The effect of ionic strength appears to be small and variable: in the case of Lumbricus hemoglobin its dissociation at alkaline pH is decreased very slightly by a large increase in ionic strength (24). In agreement with the above generalizations, the reassociation of Lumbricus hemoglobin is increased in the presence of Ca(II) and appears to be unaffected by an increase in ionic strength (20). Table 1 summarizes the conclusions obtained from the foregoing study.

DISSOCIATION AT ACID pH AND REASSOCIATION

Much less is known about the dissociation of annelid extracellular hemoglobins at acid pH relative to their dissociation at alkaline pH. Eisenia foetida hemoglobin was found to dissociate into 9S and 3S species by ultracentrifugation and into three components of 2 x 10^5, 7.9 x 10^4 amd 1.1 x 10^4 by gel filtration over the pH range 3.8 to 4.6 (11, 25). The 60S hemoglobins of Pheretima hilgendorfi and Pheretima comunissima were found to dissociate partially into 9S spe-

Table 1. Summary of the results of studies of the dissociation of
Lumbricus terrrestris hemoglobin at extremes of pH and its
reassociation upon return to neutral pH.

	Alkaline pH[a]		Acid pH[b]	
	Diss.	Reass.	Diss.	Reass.
Increase in Ca(II)[c]	Decrease	Increase	No effect	No effect
Increase in ionic strength[d]	Slight decrease	No effect	Decrease	No effect
Molecular fragments[e]	Ib 3.0×10^5	II 6.5×10^4 III 1.8×10^4	Ib 3.2×10^5	II 6.3×10^4 III 2.2×10^4
Subunit composition[f]	M,T (D1,D2)[h]	D1,D2,T M	M,T (D1,D2)[h]	D1,D2,T M
Chain composition[g]	I - IV (V, VI)[h]	II - VI I	I - IV (V, VI)[h]	II - VI I

[a] Over pH range 9 to 11.
[b] Over pH range 4 to 4.8.
[c] From 0 to 100 mM
[d] From 0 to 1.5M KCl or NaCl.
[e] Obtained by gel filtration at neutral pH on Sephacryl S-200, S-300 or Sepharose Cl-6B (20,27).
[f] Obtained by SDS PAGE of unreduced fractions.
[g] Obtained by SDS PAGE of reduced fractions.
[h] Trace amount left.

cies at pH 5.5 and completely into species of 9S and 2.5S at pH
below 4.3 (26). Tubifex tubifex hemoglobin was found to dissociate
at pH 4 into two components of 6.6×10^4 and 1.5×10^4 (23). The

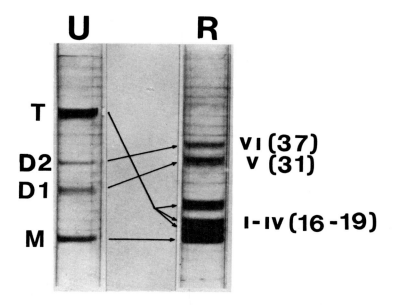

Figure 1. The relationships between the subunits of unreduced
Lumbricus terrestris hemoglobin (U) and the chains of the
reduced hemoglobin (R) obtained by SDS PAGE (7).

dissociation of Lumbricus hemoglobin at acid pH and its reassociation upon return to neutral pH has been investigated in detail (27).
Dissociation occurred at pH below 5.5 and was accompanied by time-
dependent and irreversible alterations in the absorption spectra of
both the oxy and carbonmonoxy forms to the aquomet form. The extent
of dissociation examined at pH 4.0 to 4.8 was not affected by the
presence of divalent cations but was strongly decreased by an increase in ionic strength. The reassociation at neutral pH of the
supernatant fractions, i.e. the dissociated fractions, obtained by
centrifugation in 0.1M sodium acetate buffers at pH 4.0 to 4.8, was
possible only if the supernataant was thoroughly dialyzed against
distilled water prior to dialysis against 0.1M sodium phosphate,
Tris.Cl, sodium cacodylate and imidazole.Cl buffers pH 7.0. The
extent of reassociation was not affected by either the presence of
Ca(II) or by an increase in ionic strength. Gel filtration of the
supernatant obtained by centrifugation of the hemoglobin in 0.1M

sodium acetate buffer pH 4 to 4.6, on Sephacryl S-300 at neutral pH, showed the presence of three components, Ib, II and III, whose molecular masses and subunit compositions were found to be very similar to those obtained for the three corresponding fragments observed in our studies at alkaline pH (20). The extent of reassociation varied between 40 and 80% and the SDS PAGE pattern of the reassociated molecule was identical to that of the native hemoglobin. The results of this study are also summarized in Table 1.

DISSOCIATION AT NEUTRAL pH IN THE PRESENCE OF ANIONS

The dissociation of Lumbricus hemoglobin in 0.1M sodium phosphate buffer pH 7.0, 1mM EDTA was investigated in the presence of phosphotungstate, phosphomolybdate, tungstate and molybdate anions at neutral pH (28). Of these only phosphotungstate dissociated the hemoglobin in the 1 to 10 mM concentration range. Gel filtration of the dissociated fraction, i.e. of the supernatant obtained by centrifugation, on Sephacryl S-200 at neutral pH, provided three fragments Ib, II and III, whose elution volumes corresponded closely to those of the three fragments observed in our dissociation studies at alkaline and acid pH. SDS PAGE of the three peaks showed that their subunit compositions were also similar to those of the corresponding fragments obtained earlier: peak Ib consisted of subunits M and T (chains I - IV) with traces of D1 and D2, peak II consisted of subunits D1, D2 and T (chains II - VI) and peak III was the monomer subunit M (chain I).

The effect of "chaotropic" salts on the subunit structure of Lumbricus hemoglobin at neutral pH had been investigated by Harrington and Herskovits using light scattering molecular mass measurements: the data were fitted to equilibria involving the dodecamer (whole molecule), the hexamer (half molecule) and monomer (one twelfth) without any evidence being adduced for the existence of these species by other type of measurements (29). They concluded that all three species of dodecamers, hexamers and monomers were present in 1 M sodium perchlorate. We subjected the supernatant, i.e. obtained by the centrifugation of Lumbricus hemoglobin at neutral pH in 1M sodium perchlorate, to gel filtration on Sephacryl S-200 at neutral pH (28). Only peaks II and III were observed to be present in the elution profiles of the supernatant, i.e. the dissociated fraction. Again, the elution volumes of the two peaks and their SDS PAGE patterns were similar to those observed previously. No evidence for any species intermediate in size between the whole molecule and the two fragments was observed.

DISSOCIATION AT NEUTRAL pH IN THE PRESENCE OF UREA

Herskovits and Harrington also investigated the dissociation of Lumbricus hemoglobin in the presence of urea and its alkyl derivatives at neutral pH by light scattering molecular mass measurements (30). The results were again interpreted in terms of dodecamer-hexamer-monomer equilibria. We examined the dissociation of Lumbricus hemoglobin in 0.1M sodium phosphate buffer pH 7.0, 1mM EDTA, in the presence of 1M, 2M, 3M and 4M urea by subjecting the supernatant or dissociated fraction obtained by centrifugation, to gel filtration on Sephacryl S-200 at neutral pH. No dissociation was observed in 1M urea, i.e. there was only one peak Ia, eluting at the same volume as the native hemoglobin. Two peaks, Ia and Ib, were observed in the elution profiles of the supernatants obtained in the presence of 2M and 3M urea. Gel filtration of the supernatant obtained in the presence of 4M urea provided peaks Ib, II and III and sometimes peak Ia. The relative proportion of Ib increased with the increase in urea concentration. The elution volumes and the SDS PAGE patterns of all four peaks were very similar to those of the corresponding species obtained previously. Again it was evident that peak Ib was deficient in subunits D1 and D2, i.e. chains V and VI, relative to the native hemoglobin (28). Herskovits and Harrington interpreted their results in terms of only a dodecamer-hexamer equilibrium in 2M and 3M urea and in terms of a hexamer-monomer equilibrium only in 4M urea (30). Interestingly, optical rotatory dispersion measurements at these urea concentrations showed that there was little or no alteration in the secondary structure of Lumbricus hemoglobin (30).

STRUCTURAL INTERPRETATION OF LUMBRICUS HEMOGLOBIN DISSOCIATION

The conclusions of the studies of the dissociation of Lumbricus hemoglobin at extremes of pH and at neutral pH in the presence of anions and of urea are the following: (1) the smallest fragments into which the molecule dissociated were always the same, namely the monomer M and a complex of the remaining subunits D1, D2 and T; (2)

Table 2. Possible stoichiometries of the constituent subunits
of _Lumbricus_ _terrestris_ hemoglobin.

Chains[a]	Unred. subunits	M_r, kDa	No. copies				Calctd. total M_r, kDa			
			(A)	(B)	(C)	(D)	(A)	(B)	(C)	(D)
I (H)	M	16	48	48	36	36	770	770	580	580
II (H) III IV (H)	T	50	48	48	36	36	2400	2400	1800	1800
V	D1	31	16^b	12	12^b	18	500	370	370	560
VI (H)	D2	37	16^b	12	12^b	18	590	440	440	660
Total no. of chains			224	216	168	180	4260	3980	3190	3600
No. heme groups ($0.22M_r/5585$)[c]							169	156	127	143
No. heme groups accounted for							144	144	108	108
No. heme groups left (for chain VI)							25	12	19	35

[a] All six bands in reduced hemoglobin SDS PAGE patterns stain with
approximately equal intensities (7, 31). (H) indicates whether
the chain contains heme.

[b] Assumed to be 1/3 of the number of copies of the M and T subunits
based on the proportionality of accessible surface area to $M^{2/3}$;
$((16)^{2/3}/(34)^{2/3}) = 0.31$.

[c] The iron content of the hemoglobin is 0.22 wt% (7).

the only other species observed was Ib, a fragment whose molecular size was approximately one twelfth of the whole molecule; (3) Ib was found to contain much less of subunits D1 and D2 relative to the native hemoglobin and thus could not be an exact one twelfth of the native molecule.

I have attempted to reconcile the foregoing conclusions with a model of the quaternary structure of Lumbricus hemoglobin by first abandoning the seductive concept of this molecule consisting of twelve identical subunits based on its electron microscopic appearance. Next, I assumed that the number of copies of subunits M and T but not of subunits D1 and D2, should be multiples of 12. In order to calculate some plausible stoichiometries of the four subunits, I assumed that the staining intensities of all six bands in the SDS PAGE patterns of reduced Lumbricus hemoglobin were all equal (7,31). That meant that the number of copies of the M and T subunits had to be equal. Furthermore, I took the number of the D1 and D2 subunits to be as a first approximation, one third of the number of the M and T subunits. The latter assumption was based on the known proportionality of the solvent accessible areas of proteins to their molecular masses (A_s (molecular mass)$^{2/3}$) (32).

Based on the foregoing considerations several possible stoichiometries of the four subunits of Lumbricus hemoglobin are presented in Table 2. The two principal alternatives are based on 48 and 36 as the number of copies of the M and T subunits; each of these was further subdivided based on the number of the D1 and D2 subunits. There does not appear to be any good reason for the number of copies of the two subunits to be the same. In the bottom portion of the Table I have added up the heme distribution in the whole molecule and provided estimates of the number of heme groups left for assignment to subunit D2. Although we do not know the heme content of subunit D2 (chain VI), we know that polypeptide chains I, II, and IV but not III and V, contain heme (8, 27, 33).

Examination of Table 2 does not reveal any ideal fit with what appears to be the best estimate of the molecular mass of Lumbricus hemoglobin: 3.9×10^6 (8,34). Alternatives (B) and (D) or some modification thereof are the most appealing. This exercise in numerology should not obscure the possibility that none of the four subunits occur in multiples of 12.

POSSIBLE STRUCTURAL MODEL OF LUMBRICUS HEMOGLOBIN

The assumption that the number of copies of subunits D1 and D2 is not a multiple of 12 was the first step towards a new look at the subunit structure of Lumbricus hemoglobin. The second step was to consider a plausible model to account for the facts that the molecule dissociated under a variety of conditions into the same two smallest fragments and that the intermediate size fragment was not a one twelfth of the whole molecule. The model that I would like to propose is based on the concept that subunits D1 and D2 are tightly complexed together and form a "collar" to which twelve groups of M and T subunits are attached. The latter when dissociated from the whole molecule together with some D1 and D2, provide the fragment Ib, having a molecular mass of ca. 3×10^6 and deficient in chains V and VI: they can be considered as "pseudo protomers". Next, a strong association between D1 and D2 on one hand and the trimer subunit T on the other, must be postulated since under all conditions of dissociation examined so far, they always stay together to form the fragment II. This fragment can be broken appart only at acid pH (27). The monomer subunit M is the easiest to remove from the molecule by any means; hence, its association with the other subunits must be relatively slight. A schematic diagram of the proposed model is shown in Figure 2.

The proposed model accounts for the existence of the fragment Ib having the molecular size of a one twelfth but lacking chains V and VI relative to the native Lumbricus hemoglobin, under a variety of dissociating conditions. A further possible property of this model is that upon disruption of the native quaternary structure, the fragment Ib produced may contain variable amounts of subunits D1 and D2. In addition, predictions based on the model can be tested experimentally. In particular, the model predicts that the absence of subunits D1 and D2 would prevent the reassociation of the two other subunits M and T into a whole molecule. Some of our early results on the dissociation and reassociation of Ib obtained at alkaline pH (20) clearly support this possibility. Furthermore, the Ib obtained by gel filtration of the supernatant obtained by centrifugation at neutral pH in the presence of phosphotungstate and urea, could not reassociate by itself into a whole molecule. However, when Ib was exposed to the dissociating agent and subjected to gel filtration at neutral pH on Sephacryl S-200, small amounts of peak Ia

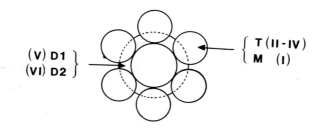

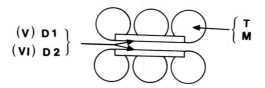

Figure 2. Diagrammatic representation of the proposed model
of *Lumbricus terrestris* hemoglobin.

were observed, i.e. reassociation into whole molecules occurred inso-
far as allowed by the content of subunits D1 and D2; at the same
time, the second generation peak Ib contained less D1 and D2 than
the starting Ib or none at all (28). Presumably the fragments Ib can
have a variable content of subunits D1 and D2. An obvious corollary
to the presence of fragments Ib is that peak Ia, which consists of
whole molecules, should also contain molecules lacking in one or
more of the "pseudo protomers". Such "incomplete" hemoglobin molecules would not be separated by gel filtration from the complete
molecules. The presence of "incomplete" molecules in the Ia fraction
has been observed in the STEM (35) and further detailed electron
microscopic evidence is being sought at present.

The model provides a structural basis for explaining the con-
flicting results obtained in different laboratories concerning the
dissociation of annelid extracellular hemoglobins at extremes of pH,
their reassociation upon return to neutral pH and in particular, the
often observed lack of reversibility of these processes. The model

also provides a structural basis for the concept of microheterogeneity invoked in the discussion of the dissociation and reassociation of earthworm hemoglobins (9,36).

ACKNOWLEDGEMENT

This work was supported by U.S. Public Health Grant HL 25952.

REFERENCES

1. Vinogradov, S. N., Comp. Biochem. Physiol. (in press) (1985).
2. Kapp, O. H., Vinogradov, S. N., Ohtuski, M. and Crewe, A. V., Biochim. Biophys. Acta 704, 546-548 (1982).
3. Chung, M. C. M. and Ellerton, H. D., Prog. Biophys. Mol. Biol. 35, 53-102 (1979).
4. Vinogradov, S. N., Shlom, J. M., Kapp, O. H. and Frossard, P., Comp. Biochem. Physiol. 67B, 1-16 (1980).
5. Terwilliger, R. C., Am. Zool. 20, 53-67 (1980).
6. Vinogradov, S. N., Kapp, O. H. and Ohtuski, M., in Electron Microscopy of Proteins (Harris, J., ed.), Academic Press, vol. 3, pp. 135-163 (1982).
7. Shlom, J. and Vinogradov, S. N., J. Biol. Chem. 248, 7904-7912 (1973).
8. Vinogradov, S. N., Shlom, J. M., Hall, B. C., Kapp, O. H. and Mizukami, H., Biochim. Biophys. Acta 492, 136-155 (1977).
9. Chiancone, E., Vecchini, P., Rossi-Fanelli, M. R. and Antonini, E., J. Mol. Biol. 70, 73-84 (1972).
10. David, M. M. and Daniel, E., J. Mol. Biol. 87, 89-101 (1974).
11. Ochiai, T. and Enoki, Y., Comp. Biochem. Physiol. 68B, 275-279 (1981).
12. Frossard, P., Biochim. Biophys. Acta 704, 524-534 (1982).
13. Vinogradov, S.N., Hersey, S. L., Frohman Jr., C. and Kapp, O. H., Biochim. Biophys. Acta 578, 216-221 (1979).
14. Chung, M. C. M. and Ellerton, H. D., Biochim. Biophys. Acta 702, 6-16 (1982).
15. Vinogradov, S. N., Kosinski, T. F. and Kapp, O. H., Biochim. Biophys. Acta 621, 315-323 (1980).
16. Vinogradov, S. N., Shlom, J. M. and Doyle, M., Comp. Biochem. Physiol. 65B, 145-150 (1980).
17. Gotoh, T., J. Sci. Univ. Tokushima 13, 1-7 (1980).
18. Andonian, M. R. and Vinogradov, S. N., Biochim. Biophys. Acta 400, 344-354 (1975).
19. Andonian, M. R., Barrett, A. S. and Vinogradov, S. N., Biochim. Biophys. Acta 412, 202-213 (1975).
20. Kapp, O. H., Polidori, G., Mainwaring, M. G., Crewe, A. V. and Vinogradov, S. N. J. Biol. Chem. 259, 628-639 (1984).
21. Santucci, R., Chiancone, E. and Giardina, B., J. Mol. Biol. 179, 713-727 (1984).
22. Frossard, P., Biochim. Biophys. Acta 704, 535-541 (1982).
23. Polidori, G., Mainwaring, M., Kosinski, T., Schwarz, C., Fingal, R. A. and Vinogradov, S. N., Arch. Biochem. Biophys. 233, 800-814 (1984).
24. Fingal, R. A. and Vinogradov, S. N., unpublished observations.
25. Ochiai, T. and Yamauchi, K., Sci. Rept. Niigata Univ. Ser. D. (Biology) No. 18, 5-13 (1981).
26. Ochiai, T. and Enoki, Y., Comp. Biochem. Physiol. 71B, 727-729 (1982).
27. Mainwaring, M. G., Lugo, S. D., Fingal, R. A., Kapp, O. H., Crewe, A. V. and Vinogradov, S. N. Biochemistry (submitted).

28. Vinogradov, S. N., Lugo, S. D. and Mainwaring, M. G., in preparation.
29. Harrington, J. P. and Herskovitz, T. T., Biochemistry 14, 4972-4976 (1975).
30. Herskovitz, T. T. and Harrington, J. P., Biochemistry 14, 4966-4971 (1975).
31. Kapp, O. H., Zetye, L. A., Henry, R. and Vinogradov, S. N., Biochem. Biophys. Res. Comm. 101, 509-516 (1981).
32. Teller, D. C. Nature 260, 729-731 (1976).
33. Shishikura, F., Mainwaring, M., Lightbody, J. L., Yurewicz, E. C., Walz, D. and Vinogradov, S. N., Biochim. Biophys. Acta (submitted).
34. Pilz, I.,Schwartz, E. and Vinogradov, S. N., Int. J. Biol. Macromol. 2, 279-283 (1980).
35. Kapp, O. H. and Crewe, A. V., unpublished observations.
36. Antonini, E. and Chiancone, E., Ann. Rev. Biophys. Bioeng. 6, 239-271 (1977).

SUBUNIT STRUCTURE OF EARTHWORM ERYTHROCRUORIN

Y. Tsfadia and E. Daniel

Department of Biochemistry, Tel-Aviv University, Tel-Aviv 69978
Israel

Introduction

Electron microscopy of earthworm erythrocruorin shows two hexagonal rings stacked
on one another (1-5). Existing models of the molecule involve 12 identical subunits
arranged in dihedral symmetry (6-10). In the present report, the adequacy of this
model to account for the dissociation pattern of earthworm erythrocruorin is examined.
A new model is proposed.

Results

Fig. 1 presents the dissociation pattern of earthworm erythrocruorin brought

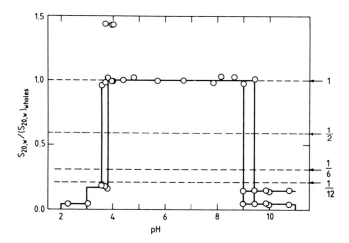

Fig. 1. Dissociation of earthworm erythrocruorin by pH. $S_{20,w}$ is the observed sedimentation coefficient (taken from David & Daniel (7)), and $(S_{20,w})_{wholes}$ is the value
(61.2 S, average of 14 determinations shown) for whole, undissociated molecules.
Dashed lines indicate values predicted for 1/2, 1/6 and 1/12 molecules from the
Kirkwood theory (11) assuming compact spherical subunits and dihedral symmetry.

Invertebrate Oxygen Carriers
Ed. by Bernt Linzen
© Springer-Verlag Berlin Heidelberg 1986

about by pH.   Dissociation into species with $S_{20,w}/(S_{20,w})_{wholes}$ equal to 0.15 and
0.04 is seen to occur.   By comparison with corresponding values predicted on the
assumption of dihedral symmetry, we can immediately exclude dissociation into 1/2 or
1/6 molecules.   Further, the observed values of $S_{20,w}/(S_{20,w})_{wholes}$ are significantly smaller than the values expected for 1/12 molecules.

The dissociation patterns in NaI, urea and acetic acid are shown in Fig. 2.
Dissociation into 1/2 or 1/6 molecules can again be excluded.

The molecular weight of native 60 S earthworm erythrocruorin has been redeter-
mined (12).   A value of 4.4×10^6 was obtained.

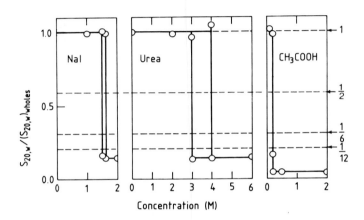

Fig. 2.   Dissociation of earthworm erythrocruorin by NaI, urea and acetic acid.
Values for $(S_{20,w})_{wholes}$ used: 66.2 (NaI), 60.2 (urea) and 54.6 (acetic acid.

Discussion

Existing models for earthworm erythrocruorin assume dihedral symmetry.   Non-
identity of the intersubunit bonds in the dihedral model opens the possibility of
obtaining 1/2 and 1/6 molecules as intermediates in the dissociation to 1/12 molecules (Fig. 3).   Dissociation by pH does not involve neither 1/2 nor 1/6 molecules
as is clearly seen in Fig. 1.   Attempts to detect intermediates with sedimentation
coefficients corresponding to 1/2 or 1/6 molecules by dissociation with NaI, urea
or acetic acid were not successful (Fig. 2)   These findings do not support the
dihedral model.

Additional difficulties concerning the dihedral model exist.   Dissociation by
pH yields species with sedimentation coefficient of 9-10 S $(S_{20,w}$ $(S_{20,w})_{wholes}$
equal to 0.15 in Fig. 1), significantly smaller than the value of 12.6 S predicted
according to the dihedral model.   Further, on a molecular weight basis, it seems

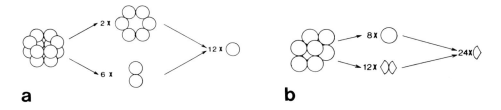

a   b

Fig. 3. Modes of dissociation predicted by the dihedral (a) and the cubic (b) models.

highly improbable that the 9-10 S species represents a 1/12 molecule. The molecular weight of the isolated 9-10 S species has been previously determined to be 163 000 (7). The molecular weight of the whole molecule is 4.4×10^6. These findings are more consistent with a 24-subunit structure as proposed by David & Daniel (7).

    In the light of our findings, we propose a new model for earthworm erythro-cruorin. The model consists of 24 subunits organized in groups. Each group consists of 3 subunits and is assumed to be spherical in shape. The 8 groups occupy the corners of a cube. Support for the cubic model comes from electron microscopy, dissociation pattern and small angle X-ray scattering.

Electron microscopy. Projections of the model (Fig. 4) are consistent with profiles seen in electron microscopy. The projections along the 4- and 2-fold axes account for the square and rectangular profiles (see, for example, Figs. 1b, 1c of ref. (4)). The projection along the 3-fold axis is compatible with the hexagonal profile (see Fig. 1a of ref (4)). Failure to observe more than 6 subunits in the hexagonal pro-file is probably a result of accumulation of stain in the central cavity of the cubic molecule. The presence of a central subunit in hexagonal profiles of earthworm erythrocruorin has been reported by Ohtsuki & Crewe (13).

Dissociation pattern. According to the cubic model for earthworm erythrocruorin, dissociation into 1/8, 1/12 and 1/24 is possible (Fig. 3). The sedimentation coef-ficient predicted for the 1/24 molecule, assuming compact spherical subunits, is 7.8 S, significantly lower than the observed sedimentation coefficient of the 9-10 S

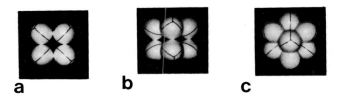

a   b   c

Fig. 4. Projections of the cubic model along the 4- (a), 2- (b) and 3-fold (c) axes.

species. An improved calculation of predicted sedimentation coefficients, assuming an array of smaller spheres for each subunit, yields sedimentation coefficients of 17.5, 13.3 and 9.5 S for 1/8, 1/22 and 1/24 molecules (14). The observed sedimentation coefficient of the 9-10 S species is thus consistent with the predicted value for a 1/24 molecule. It should be remarked that dissociation into 1/2 molecules is not expected according to the cubic model, consistent with observation.

Small angle X-ray scattering. A model consisting of 12 spherical subunits arranged in two superimposed hexagonal rings did not fit the scattering data satisfactorily (9). The best fit with such a model was obtained by assuming the existence of protein mass between the subunits and in the center of the molecule. As is seen in Fig. 4, the cubic model is equivalent to a hexagonal distribution of mass plus mass along the central axis. The advantage of the cubic model is that the central mass is an integral part of the model.

References

1. Roche, J., Bessis, M., and Thiery, J.P., Biochim. Biophys. Acta 41, 182-184 (1960).

2. Levine, O., J. Mol. Biol. 6, 95-101 (1963).

3. Royer, W.E., Braden, B.C., Jacobs, H.C., and Love, W.E., in Invertebrate Oxygen Binding Proteins (Lamy, J., and Lamy J., eds.), pp. 337-342, Marcel Dekker, New York (1980).

4. Kapp, O.H., Vinogradov, S.N., Ohtsuki, M., and Crewe, A.V., Biochim. Biophys. Acta 704, 546-548 (1982).

5. Kapp, O., Polidori, G., Mainwaring, M., Crewe, A.V., and Vinogradov, S., J. Biol. Chem. 259, 628-640 (1984).

6. Chiancone, E., Vecchini, P., Rossi Fanelli, M.R., and Antonini, E., J. Mol.Biol. 70, 73-84 (1972).

7. David, M.M., and Daniel, E., J. Mol. Biol. 87, 89-101 (1974).

8. Vinogradov, S.N., Shlom, J.M., Hall, B.C., Kapp, O.H., and Mizukami, H., Biochim. Biophys. Acta 492, 136-155 (1977).

9. Pilz, I., Schwarz, E., and Vinogradov, S.N., Int. J. Biol. Macromol. 2, 279-283 (1980).

10. Hendrickson, W.A., Life Chem. Rep. Suppl. 1, 167-185 (1983).

11. Van Holde, K.E., in The Proteins, 3rd ed. (Neurath, H., and Hill, R., eds.), vol. 1, pp. 225-291, Academic Press, London and New York. 1975.

12. Tsfadia, Y., and Daniel, E., in preparation.

13. Ohtsuki, M., and Crewe, A.V., J. Ultrastr. Res. 83, 312-318 (1983).

14. Garcia Bernal, J.M., and Garcia de la Torre, J., Biopolymers 20 129-139 (1981).

CHARACTERIZATION OF TRIMERIC HEMOGLOBIN COMPONENT FROM THE EARTH WORM, LUMBRICUS TERRESTRIS

F. SHISHIKURA[*], M. G. MAINWARING[#], E. C. YUREWICZ[@], J. J. LIGHTBODY[#], D. A. WALZ[*], and S. N. VINOGRADOV[#]
Departments of Physiology[*], Biochemistry[#] and Obstetrics and Gynecology[@] Wayne State University School of Medicine, Detroit, MI 48201

The extracellular hemoglobin of Lumbricus terrestris has a molecular mass of ca. 4×10^6, contains 156 ± 5 heme groups and exhibits at least six different polypeptide chains (I through VI) on SDS PAGE in the presence of 2-mercaptoethanol (1). The 50 KDa hemoglobin subunit consists of a disulfide-bonded trimer of chains II, III and IV, which allow it to retain heme (2).

Separation of the individual chains of the Lumbricus hemoglobin is of interest because in contrast to all other hemoglobins, not all of the constituent polypeptide chains of Lumbricus hemoglobin and the other known annelid hemoglobins bear heme (1, 3, 4). The purified hemoglobin chains could also be useful in clarifying the molecular evolution of hemoglobin in invertebrates.

We report here the chromatofocusing of chains II, III and IV from the 50 KDa subunit of Lumbricus hemoglobin, and their partial characterization. Our results indicate that only chains II and IV carry heme.

MATERIALS AND METHODS

Lumbricus terrestris hemoglobin was prepared as described previously (2). The 50 KDa (trimer) subunit with heme was isolated by the gel filtration on Sephadex G-200 in 0.1 M sodium acetate buffer, pH 4.0 (5) and by SDS gel filtration (2).

The 50 KDa subunit of Lumbricus hemoglobin was converted to the cyanferriform and reduced with 1 mM DTT in 25 mM imidazole buffer, pH 7.0 at 4°C (6) for 3 hrs. The reduced material was subjected to chromatofocusing on a 0.9 x 50 cm column PBE 94 (Pharmacia) and eluted with 10% polybuffer 74 (Pharmacia) pH 4.5. The absorbance of fractions (2 ml) was monitored at 280 nm and at 410 nm. To remove the polybuffer, the pooled fractions were subjected to chromato-

Invertebrate Oxygen Carriers
Ed. by Bernt Linzen
© Springer–Verlag Berlin Heidelberg 1986

graphy on Phenyl-Sepharose CL-4B column (Pharmacia, 0.7 x 2.5 cm).

SDS PAGE was carried out using 10% disc gels by the method of Weber and Osborn (7).

Protein samples were hydrolyzed in 6 N HCl at 110°C for 22 hrs, dried and derivatized to the PTC-form immediately before chromatographic analysis on Waters Pico-Tag system (8). Three nmoles of protein, previously denatured by performic acid oxidation (9) was degraded sequentially by a Beckman 890 M sequencer (10) and the resulting PTH derivatives were separated by a column of Ultrasphere C18 using a mixture of 0.1 M sodium acetate pH 4.9 and acetonitrile (62: 38, v/v) (11) and a Waters HPCL system.

Lyophilized protein samples were analyzed by gas chromatography of their trimethylsilyl methylglycosides following the procedure of Bhatti et al. (12).

Hybridomas were obtained by using a modification of the procedure of Kohler and Milstein (13). BALB/c mice were immunized with Lumbricus hemoglobin (2 mg) and one month later the spleen cells were removed and fused to azaguanin-resistant NS1 cells (14) obtained from Dr. M. D. Poulik (Beaumont Hospital Detroit) in the presence of 50% polyethyleneglycol 4000. After 15 days the supernatants of clones were tested for antihemoglobin activity using a double antibody binding assay.

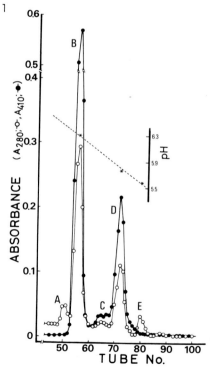

Fig. 1. Chromatofocusing of Lumbricus 50 KDa subunit.

RESULTS AND DISCUSSION

Figure 1 shows the result of chromatofocusing of the Lumbricus hemoglobin 50 KDa subunit obtained by gel filtration at pH 4.0. Three main peaks (B, D and E) and additional two peaks (A and C) were eluted, whose SDS PAGE patterns are shown in Fig. 2. Peaks B and D, identified as chains II and IV, respectively, have an absorbance both at 410 nm and at 280 nm indicative of the presence of heme. This finding presents the first evidence for the existence of a non-heme-binding poly-

peptide chain of an extracellular
Lumbricus hemoglobin.

The amino acid composition
(Table 1) and the N-terminal
sequence of chain II (K-K-Q-C-
G-V-L-E-G-L-K-V-K-S-) are identical to chain A-III isolated
and sequenced by Garlick and
Riggs (15). A comparison of the
amino acid compositions and acetic
acid-urea PAGE patterns (data not
shown) indicate that our chains
III and IV correspond to chains
A-II and A-I, respectively, isolated by Garlick and Riggs (15).

Table 2 shows the result of
carbohydrate analysis of
Lumbricus hemoglobin, the 50
KDa subunit and its isolated
chains, indicating that
Lumbricus hemoglobin is a gly-
coprotein with a high mannose
content. In the 50 KDa subunit,
it appears that most of the
carbohydrate side chain groups
reside on polypeptide chain IV.

Of the monoclonal antibodies,
hybridoma EHb 1 reacts only with
the subunit obtained by the nondenaturing method, however, EHb
9 reacts with the 50 KDa subunit
obtained by denaturing and nondenaturing methods (Table 3);
thus, it is clear that SDS
alters the conformation of a
portion of the 50 KDa subunit.
Both EHb 1 and EHb 9 react
weak against chains II, III and
IV. This indicates that these
chains are a structural building
block of the native Lumbricus

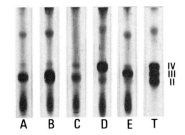

Fig. 2. SDS PAGE of fractions
from Fig. 1. A; peak A, B; peak
B, C; peak C, D; peak D, E; peak
E, T; 50 KDa subunit (trimer).

Table 1. Amino acid compositions
of three polypeptide chains.

Amino acid	II[a]	III[b]	IV[c]	50 KDa[d]	A-II[e]
Aspartic acid	10.3	8.1	10.6	54.6	13
Glutamic acid	18.8	14.6	16.1	46.0	20
Serine	9.8	15.8	19.3	20.7	8
Glycine	14.9	17.9	19.8	27.0	13
Histidine	7.0	6.9	7.4	27.2	10
Arginine	8.7	7.9	15.2	31.8	10
Threonine	5.3	6.0	6.9	13.5	5
Alanine	16.7	17.5	18.4	38.4	17
Proline	3.8	5.8	6.0	13.7	5
Tyrosine	2.4	2.4	4.6	6.8	3
Valine	7.1	8.6	15.2	26.5	10
Methionine	0	0	0	0	0
Cystine	1.0	0.6	1.0	3.5	3
Isoleucine	7.7	10.3	9.7	27.0	9
Leucine	11.3	21.5	20.1	50.0	13
Phenylalanine	5.7	6.2	10.6	20.2	7
Lysine	7.3	9.8	8.7	29.1	9
Tryptophane	N.D.	N.D.	N.D.	N.D.	2

[a]Residues/14,000, [b] residues/16,000
[c]residues/19,000, [d] residues/49,500
[e]residues from the sequence (15).

Table 2. Carbohydrate analysis of
three chains.

Material	Carbohydrate content (wt%)	Mannose (nanomoles/mg)	GlcNAc (nanomoles/mg)	Mole ratio (Man/GlcNAc)
Hemoglobin	2.0±0.5	113±29	12±3	9.4±0.5
50 KDa	1.8±0.5	92±28	11±4	8.4±1.0
Chain II	N.D.	N.D.	N.D.	--
Chain III	N.D.	N.D.	N.D.	--
Chain IV	4.8	17.4	2.4	7.4

N.D., not detected.

hemoglobin molecule and are not artifacts of SDS PAGE or of the method used in its purification.

ACKNOWLEDGMENTS

We are indebted to the National Institutes of Health for support of this research through Grants AM 30382 and HL 35953.

Table 3. ELISA of Lumbricus hemoglobin, the 50 KDa subunit and its constituent polypepdie chains.

Material	Monoclonal antibody			NS[b]
	EHb-1	EHb-6	EHb-9	
Hemoglobin	1.7	1.7	1.7	0
50 KDa (I)	1.6	0	1.6	0
50 KDa (II)	0	0	1.6	0
Chain II	0.16	0	0.15	0
Chain III	0.16	0	0.15	0
Chain IV	0.16	0	0.15	0

[a]Absorbance at 405 nm after 20 min reaction time using undiluted tissue culture supernatant
[b]Tissue culture supernatant.

REFFERENCES

1. Vinogradov, S.N., Shlom, J.M., Hall, B.C., Kapp, O.H., and Mizukami, H., Biochim. Biophys. Acta 492, 136-155 (1977).
2. Shlom, J.M., and Vinogradov, S. N., J. Biol. Chem. 248, 7904-7912 (1973).
3. Fushitani, K., Morimoto, H., and Ochi, O., Arch. Biochem. Biophys. 218, 540-547 (1982).
4. Vinogradov, S.N., Comp. Biochem. Physiol., in press (1985).
5. Mainwaring, M.G., Lugo, S.D., Fingal, R., Kapp, O.H., Crewe, A.V. and Vinogradov, S.N., in preparation (1985).
6. Gotoh, T., J. Sci. Univ. Tokushima XV, 1-10 (1982).
7. Weber, K., and Osborn, M., J. Biol. Chem. 244, 4406-4412 (1969).
8. Henrickson, R.L., and Meredith, S.C., Anal. Biochem. 136, 65-74 (1984).
9. Hirs, C.H.W., J. Biol. Chem. 219, 611-617 (1956).
10. Edman, P., and Begg, G., Eur. J. Biochem. 1, 80-91 (1967).
11. Chan, M.M.S., Beckman Chromatogram 5, 2-5 (1984).
12. Bhatti, R., Chambers, R.E., and Clamp, J.R., Biochim. Biophys. Acta 222, 339-347 (1979).
13. Kohler, G., and Milstein, C., Nature 256, 495 (1975).
14. Kohler, G., and Milstein, C., Eur. J. Immunol. 6, 511-519 (1976).
15. Garlick, R.L., and Riggs, A., J. Biol. Chem. 257, 9005-9015 (1982).

THE STRUCTURE OF THE EXTRACELLULAR HEMOGLOBIN OF ANNELIDS

A. Ghiretti Magaldi, F. Ghiretti, G. Tognon and G. Zanotti[*]
Dept. of Biology, C.N.R. Center for Hemocyanins and other Metallo-
Proteins, *Dept. of Organic Chemistry, C.N.R. Center for Biopolymers,
University of Padova, Padova, Italy.

Annelid extracellular hemoglobins (also named erythrocruorins and chlorocruorins after the nature of the porphyrin group) are very similar proteins as indicated by their functional, structural and chemical properties. We believe that a single model can elucidate how the subunits and the polypeptide chains are arranged in the double hexagonal prism of these respiratory proteins.

Erythrocruorins and chlorocruorins from many annelid species have been studied by several authors in many laboratories using different techniques (1-5). Due to the complexity and to the high dimensions of the molecules, a general agreement on their structure has not been attained and different models have been proposed (2-8).

With the aim of finding a unitary model we have performed two-dimensional image analysis and reconstruction by optical and computed methods on electron micrographs of crystalline monolayers of Spiro-graphis spallanzanii and Ophelia bicornis extracellular hemoglobins.

The results are presented in this work together with a preliminary three-dimensional analysis and reconstruction of S. spallanzanii chlorocruorin.

Methods

Highly purified (5,7) protein solutions were diluted down to 70-100 μg/ml with distilled water containing 0.05% Polyethylene Glycol (PEG) -1000. Equal volumes of protein and unbuffered 1% ammonium molybdate were rapidly mixed on a Vortex shaker and brought to pH 6.2-6.4. After at least 15 min this mixture was applied onto a freshly cleaved mica sheet. The excess liquid was drained, the mica was dried over-night in a closed container and the preparation was finally transfer-red to the evaporation chamber of an Edwards metallizer implemented with a liquid nitrogen trap. A carbon membrane, less than 10 nm thick, was evaporated and the carbon and protein were detached by floatation

Invertebrate Oxygen Carriers
Ed. by Bernt Linzen
© Springer-Verlag Berlin Heidelberg 1986

over 1% unbuffered uranyl acetate. Observations were made in a Hitachi
H-600 electron microscope; Kodak SO-163 planar films were used with
a constant magnification of 40000 x.

Large crystalline monolayers can be obtained from S. spallanzanii
chlorocruorin when 0.025 mM $ZnCl_2$ or $TbCl_3$ are added to the protein
before dilution. Zn^{2+} binding to chlorocruorin was followed by equilibrium dialysis and estimated with a Perkin-Elmer atomic absorption
spectrophotometer.

A preliminary selection of images of bidimensional crystals was
performed by optical diffraction using a Polaron optical bench.

Selected regions of the micrographs (15 x 15 mm) corresponding to
areas of 0.4 x 0.4 μm in the crystal, were digitized using a photo-
scan P1000 Optronics automatic microdensitometer controlled by a PDP
11/34 microcomputer. The scanning aperture was 50 x 50 μm wide and
the optical densities were collected at intervals of 1.2 nm in the
crystal.

Two- and three-dimensional reconstructions were obtained using the
SPIDER software package (9) running on the Vax 780/158 of the University of Padova.

Areas of 256 x 256 data points, corresponding to about 250 unit
cells, were windowed out for every projection and padded into a larger
area (512 x 512 points) with a background constant equal to the mean
value of the examined region. The Fourier transform was computed using
the FFT algorithm (10) and, after indexing of the transform, a mask
was applied corresponding to the theoretical position of the lattice
points. The resolution of the best crystal extended to about 3 nm.
Inverse Fourier transform of the masked diffraction spectra were then
calculated.

For the three-dimensional reconstruction, selected smaller areas
(64 x 64 data points) from the filtered two-dimensional reconstructions of the tilted images were chosen and aligned by using the cross-
correlation algorithm (11). Subsequent back-projection into the three-
dimensional volume was performed (12) after appropriate weighing.
Projections titled around a single axis were used.

Results and Discussion

The monolayered crystals of S. spallanzanii and O. bicornis hemoglobins
are shown in Plate I Figs. 1 and 3 and Plate II Figs. 5 and 7 respectively. Monolayered arrays are not obtained if the original protein
solution contains either sulfate or calcium ions. A low concentration
of PEG-1000 is sufficient to induce crystal formation in O. bicornis
erythrocruorin, while, in order to obtain sufficiently large foils
from chlorocruorin, the addition of Zn^{2+} or Tb^{2+} is necessary.

Zn^{2+} binding to chlorocruorin is of the order of 170 M/M of protein
(3×10^6D), i.e. approximately one Zn^{2+} per polypeptide chain. The
role of zinc in the crystalline organization of this protein seems to
be similar to that observed in the formation of tubulin foils (13).

Both, erythrocruorin and chlorocruorin, give foils in the hexagonal
(axial) and in the rectangular (lateral) projections. Often both types
of foils are seen in the same region. We have also observed that the
hexagonal foils of O. bicornis erythrocruorin can be deprived of the
central subunit (8). When it is present, however, the final contrast
is poor, probably due to a lower penetration of uranyl acetate. In
the lateral projections the central subunit is always seen.

When, as it happens sometimes, the foils of polymers are multi-
layered, the top layer is not exactly superposed to the lower ones,
but it looks as inserted in the funnel between two underlying polymers.

Two-dimensional analysis and reconstruction studies give the same
results for S. spallanzanii chlorocruorin and O. bicornis erythrocruorin:

1) The diffraction spectra of the axial projections reveal a hexagonal
   lattice with a real cell parameter of 24-25 nm. The computed reconstructions of the axial projections (Plate I Fig. 2, Plate II Fig.
   6) give highly symmetrical structures: hexagons made by six nearly
   triangular units with a hole in the center and related by a six
   fold axis. Each of the units appears to be formed by three globular
   structures.

2) The filtered computed images of the lateral projections (Plate I
   Fig. 4, Plate II Fig. 8) show two layers, each made by two units

separated by a channel. These units have the same general appearance
as the corresponding ones of the top projection, being made by
three globular interconnected structures around an empty space.
The presence of a channel in the latter reconstructions needs an explanation. If, as it appears, the polymers lie on a side of the hexagon,
three units per layer should be seen. In the computed and also in the
optical reconstructions (8) the third unit is not visible. This could
be due either to masking by the stain or to a low degree of order in
this part of the polymer.

The results of the two-dimensional reconstructions are consistent
with the physicochemical properties reported in Table 1. They show a
basic similarity and allow to propose a model which is valid at least
for the two extracellular hemoglobins we have studied.

TABLE 1

MAIN SIMILARITIES BETWEEN CHLOROCRUORIN (S. spallanzanii)
AND ERYTHROCRUORIN (O. bicornis).

	CHLOROCRUORIN		ERYTHROCRUORIN
MOLECULAR DIMENSIONS (1)			
Height	18.6 nm		19.8 nm
Width	23.9		26.1
MOLECULAR WEIGHTS			
Whole Molecules (2)	2.8×10^6 D		2.7×10^6 D
(3)	3.0×10^6		3.2×10^6
Main subunit (1/12th)(4)	257000	(5)	242000
Minimal subunit (4)	62000	(5)	58000 + 63000
Polypeptide chains (average)	15000		15200

1 - Optical diffraction
2 - Sedimentation and diffusion (5,7)
3 - Gel filtration (5,7)
4 - Equilibrium sedimentation (7)
5 - Calculated from the relative concentration of the dissociation
    products (7)

This model based, for the moment, only on two-dimensional reconstructions, indicates that the main subunit contains three globular
units both in the top and the lateral projection. The simplest way of
representing this subunit is a tetrahedron which in both projections
shows only three of its components, the fourth being masked by the
others. A tetrameric structure has been recently suggested also for
the main subunit of the chlorocruorin of <u>Mixycola infundibulum</u> (14).

A preliminary three-D analysis and reconstruction has been made on
hexagonal foils of <u>S. spallanzanii</u> hemoglobin. Tilted images up to
60° along one axis have been analysed and from the computed sections
a three-D image has been obtained (Fig. 9). Being the image reconstructed from a limited number of tilts along one axis only, some
features of the molecule have been preserved while others are altered.

References

1. Chung, M.G.M., and Ellerton, H.D., Prog.Biophys.Mol.Biol. <u>35</u>,
   53-102 (1979).
2. Chiancone, E., Vecchini, P., Rossi Fanelli, M.R.,and Antonini,
   E., J.Mol.Biol. <u>70</u>, 73-84 (1972).
3. Hendrickson, W.A., <u>in</u>Structure and function of Invertebrate respiratory proteins (Wood, E.J., ed.), Life Chem.Rep., Suppl. 1,
   pp. 167-185, Harwood Acad.Publ., London (1983).
4. Messerschmidt, U., Wilhelm, P., Pilz, I., Kapp, O.H., and Vinogradov,
   S.N., Biochim.Biophys.Acta <u>742</u>, 366-373 (1984).
5. Mezzasalma, V., Di Stefano, L., Piazzese, S., Zagra, M., Ghiretti
   Magaldi, A., Carbone, R., and Salvato, B., <u>in</u>Structure and function
   of Invertebrate respiratory proteins (Wood, E.J., ed.), Life Chem.
   Rep., Suppl. 1, pp. 187-191, Harwood Acad.Publ., London (1983).
6. Ghiretti Magaldi, A., Zanotti, G., Salvato, B., Tognon, G.,
   Mezzasalma, V., and Di Stefano, L., <u>in</u>Structure and function of
   Invertebrate respiratory proteins (Wood, E.J., ed.) Life Chem.Rep.
   Suppl. 1, pp. 193-196 (1983).
7. Mezzasalma, V., Di Stefano, L., Piazzese, S., Zagra, M., Salvato,
   B., Tognon, G., and Ghiretti Magaldi, A., Biochim.Biophys.Acta
   <u>829</u>, 135-143 (1985).
8. Ghiretti Magaldi, A., Zanotti, G., Tognon, G., and Mezzasalma, V.
   Biochim.Biophys.Acta <u>829</u>, 144-149 (1985).
9. Kranck, J., Shimkin, B., and Dowse, H., Ultramicroscopy <u>6</u>, 343-358
   (1981).
10. Ten Eyck, L.F., Acta Cryst. <u>A33</u>, 486-492 (1977).
11. Saxton, W.O., and Franck, J., Ultramicroscopy <u>2</u>, 219-227 (1977).
12. Gilbert, P.C.F., J.Theoret.Biol. <u>36</u>, 105-117 (1972).
13. Amos, L.A., and Baker, T.S. Nature <u>279</u>, 607-612 (1979).
14. Vinogradov,S.N.,Standley,P.R.,Mainwaring,M.G.,Kapp,O.H.,Crewe,A.V.
    B.B.A. <u>828</u>, 43-50 (1985).

50

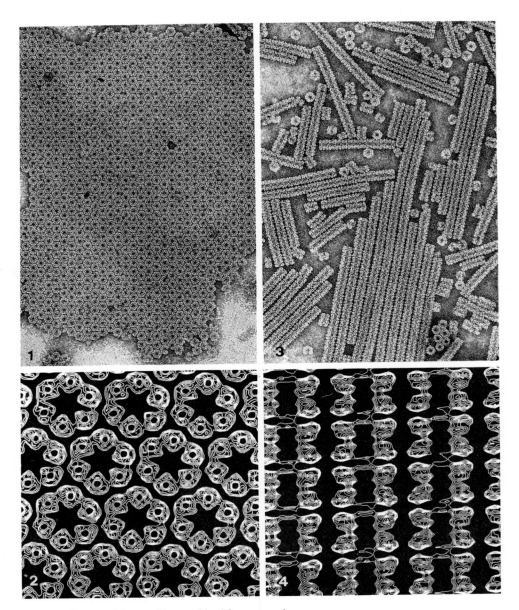

PLATE I <u>Spirographis spallanzanii</u> chlorocruorin
1 Bidimensional crystal in the axial projektion. 2 Relative computed reconstruction.
3 Bidimensional crystal in the lateral projection. 4 Relative computed reconstruction.
Magnification electron micrographs 90000 x.

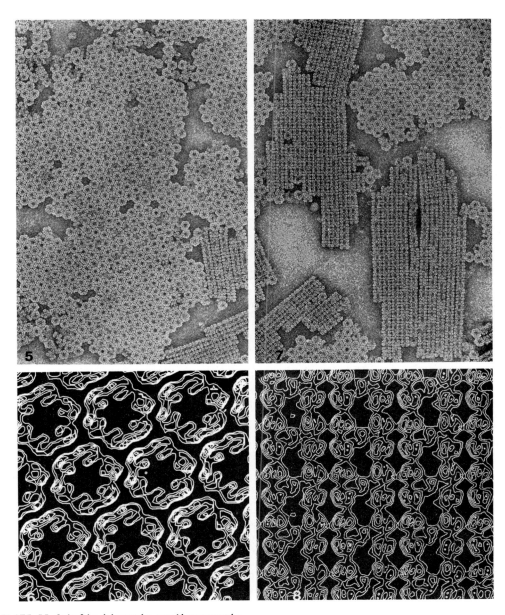

PLATE II Ophelia bicornis erythrocruorin
5 Bidimensional crystal in the axial projection. 6 Relative computed reconstruction.
7 Bidimensional crystal in the lateral projection. 8 Relative computed reconstruction.
Magnification electron micrographs 90000 x.

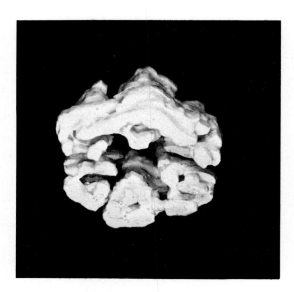

Fig. 9. Three dimensional reconstruction of Spirographis spallanzinii chlorocruorin. The preliminary model has been built from the computed section profiles obtained from tilted images along one axis up to 60°.

Ophelia bicornis ERYTHROCRUORIN. SOME PECULIAR FEATURES

A. Ghiretti Magaldi and G. Tognon
Dept. of Biology and C.N.R. Center for Hemocyanins and other Metallo-
Proteins, University of Padova, Padova, Italy.

Ophelia bicornis erythrocruorin has two distinctive structural
features as compared with the majority of the extracellular hemoglobins
from the other annelid species: (1) two fractions with different molecular weight are present in the hemolymph; (2) the central cavity of
the double hexagonal prism contains an additional subunit.

Two fractions, a major one with 50 S and a minor one with 95 S,
have been found also in the hemolymph of Euzonus mucronata (1), another
ophelid polychaete. This erythrocruorin contains also a central additional subunit.

The minor fraction (about 12% of the total hemoglobin of Ophelia)
appears to be a dimeric form by sedimentation analysis and by electron
microscopy. As indicated by ultracentrifugation performed after leaving
the solution for several days in the cold, the two fractions are not
in equilibrium. We believe that they might be stable conformers of an
identical molecular species, as shown by the aminoacid composition
and by the identity of the polypeptide chains obtained by complete
dissociation (2).

Almost complete separation of the dimeric and monomeric components
can be obtained by Biogel A-5M filtration or by sedimentation in the
preparative ultracentrifuge for 2 hours at 200000 x g over 30% sucrose.

Electron micrographs of the dimer and of the monomer are shown in
Figs. 1 and 2. In both fractions molecules with the central subunit
are present. The dimer in the axial projection (Fig. 1) can be easily
recognized because of the stronger stain deposition along its sides.
The lateral projection of the dimer appears as made by four stacked
discs of which the internal ones are rotated relative to the external
(Fig. 1 and 2). To this fact is due the apparent lack of the hexagonal
symmetry of the top projections of the dimer which is restored when
a photographic image enhancement is done with 120° and 60° rotation
(Fig. 3).

Invertebrate Oxygen Carriers
Ed. by Bernt Linzen
© Springer–Verlag Berlin Heidelberg 1986

The central subunit is present in both the monomer and the dimer. This subunit, which appears as having a three-fold axis (Figs. 2 and 3), is quite unstable and is easily lost under the conditions for obtaining negative staining and crystalline monolayers (3). Most of the crystals in the axial projection, prepared from the purified mono-meric fraction, do not have the central subunit which is present only in a few fields (Fig. 5). Foils of polymers, studied by the image analysis and reconstruction method, show always the presence of the central subunit.

Until now the additional subunit has been observed in a few other Annelid species: Nephtys incisa (4), N. Hombergi (5) and Oenone fulgida (6). The nature and structure of this peculiar subunit are still unkown; it would be interesting to investigate also its evolutionary signifi-cance.

1. Terwilliger, R.C., Terwilliger, N.B., Schabtach, E., and Dangott, L., Comp.Biochem.Physiol. 57A, 143-149 (1977).
2. Mezzasalma, V., Di Stefano, L., Piazzese, S., Zagra, M., Salvato, B., Tognon, G., and Ghiretti Magaldi, A., Biochim.Biophys.Acta 829, 135-143 (1985).
3. Ghiretti Magaldi, A., Zanotti, G., Tognon, G., and Mezzasalma, V. Biochim.Biophys.Acta 829, 144-149 (1985).
4. Wells, M.R.G., and Dales, R.P., Comp.Biochem.Physiol. 54A, 387-394 (1976).
5. Messerschmidt, U., Whilhelm, P., Pilz, I., Kapp, O.H., and Vinogradov S.N., Biochim.Biophys.Acta 742, 366-373 (1983).
6. Van Bruggen, E.F.J., and Weber, R.E., Biochim.Biophys.Acta 359, 210-212 (1974).

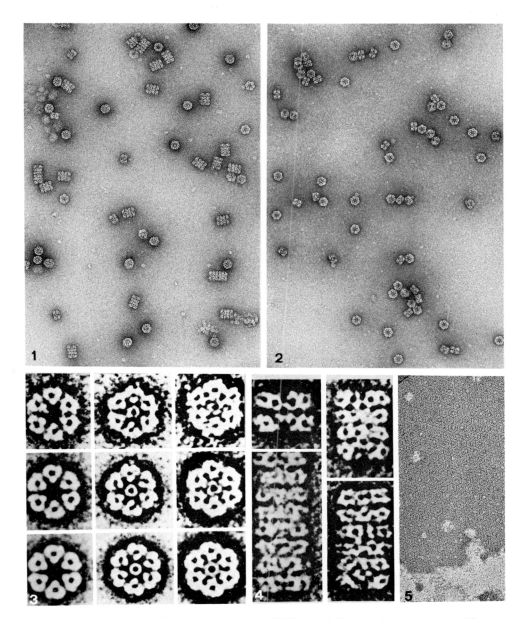

1 Dimeric component. 1% uranyl acetate. 150000 x. 2 Monomeric component. 1% uranyl acetate. 150000 x. 3 Photographic image enhancement after rotation of axial projections. (a) monomer; (b) monomer with central subunit; (c) dimer. 4 High magnification (1,080,000 x) of lateral projections: monomer, dimer, trimer and tetramer. 5 Monolayered crystal in the axial projection of molecules with the central subunit. 150000 x.

STRUCTURE OF ANNELID HEMOGLOBINS - SMALL ANGLE X-RAY STUDIES

I.Pilz, M.Theuer, P.Wilhelm and E.Schwarz
Institut für Physikalische Chemie, Karl-Franzens Universität Graz
A-8010 Graz, Austria.
S.N.Vinogradov
Biochemistry Department, Wayne State University, Detroit
MI 48201, USA.

Small angle X-ray scattering is a suitable method to get information on the quaternary structure of extracellular hemoglobins in solution. The method allows to calculate directly from the scattering curve and the distance distribution function (p(r)-function) molecular parameters as molecular weight M_r, radius of gyration R and the maximum distance D_{max}. The scattering curve reflects the conformation of the macromolecule in the reciprocal space; the distance distribution function reflects the molecule in the real space and represents the frequency of the distances combining any volume element with any other within the molecule. Models describing the conformation of the hemoglobins in solution can be deduced by comparing their scattering curves and p(r)functions with the experimental ones (1,2).
The following studies were performed:
1. Comparison of the conformation of different extracellular hemoglobins
2. Comparison of native and reassociated hemoglobins
3. Investigation of subunits of hemoglobins.

1. Comparison of the conformation of the hemoglobins of Tubifex tubifex (3), Lumbricus terrestris (4), Arenicola marina (4), Nephtys incisa (5) and Macrobdella decora (5). The dimensions of these hemoglobins are

Molecular parameters of extracellular hemoglobins of annelids

	NEPHTYS INCISA	ARENICOLA MARINA	LUMBRICUS TERRESTRIS	TUBIFEX TUBIFEX	MACROBDELLA DECORA
R [nm]	10.62 ± 0.15	11.30 ± 0.15	11.20 ± 0.15	10.66 ± 0.15	10.60 ± 0.15
D_{max} [nm]	28 ± 1	29 ± 1	29 ± 1	30 ± 1	28 ± 1
$M_r \times 10^6$		3.85 ± 0.15	3.95 ± 0.15	3.09 ± 0.15	

Invertebrate Oxygen Carriers
Ed. by Bernt Linzen
© Springer–Verlag Berlin Heidelberg 1986

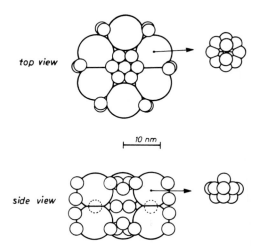

top view

10 nm

side view

Fig.1 Model T.tubifex hemoglobin

summarized in the table. The values are very similar and in good agreement with the dimensions obtained by STEM (4,5,6). Only the molecular weight of T.tubifex Hb is clearly lower (3).

The overall shape of the four extracellular hemoglobins is always best approximated by a model which consists of 12 subunits arranged in two superimposed hexagonal layers (Fig.1). A difference was found in the region of the central cavity. The p(r) function (Fig.2) and the height of the submaxima are very sensitive to the amount of protein mass present in the central cavity. The experimental results suggest that there exists more protein mass in the central cavity of T.tubifex than in those of L.terrestris,A.marina and M.decora molecule.

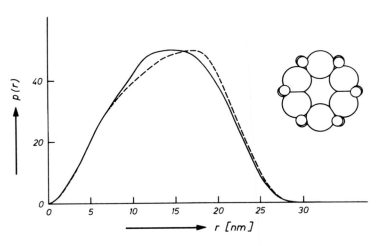

Fig.2. Distance distribution function p(r) of T.tubifex native hemoglobin (—) and the curve calculated for the model without protein mass in the central cavity shown in the illustration (---).

2. Comparison of native and reassociated hemoglobins of T.tubifex.
The native T.tubifex hemoglobin and the product of its reassociation at neutral pH subsequent to dissociation at alkaline pH (7) were examined by small angle X-ray scattering. The scattering curves and p(r) functions of both forms are very similar (Fig.3). The small differences

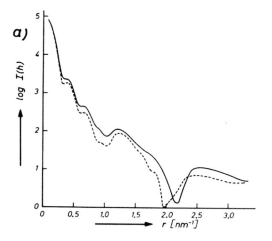

a) *log I(h)* vs *r [nm⁻¹]*

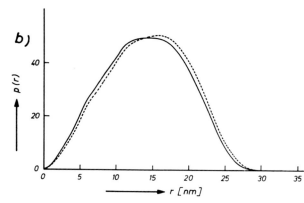

b) *p(r)* vs *r [nm]*

Fig.3. Experimental curves of T.tubifex
native (——) and reassociated (....).
a) Scattering function, b) p(r) function

indicate that the dimensions
of the reassociated molecule
are slightly increased relative to those of the native
molecule by an amount slightly greater than the error in
measurement. (In some cases
no difference was found.)

3. Subunits of Lumbricus terrestris hemoglobin. L. terrestris hemoglobin dissociates at
alkaline pH and in the presence of denaturants into
different subunits (8). Up
to now we did not succeed
in getting homodisperse solutions of the subunits. A
relatively homogenous fraction (Fraction II) could be
characterized by the following molecular parameters:
molecular weight about
200000, radius of gyration:
5.4 nm and maximum dimen-
sion 17.5 nm. The shape of
this fragment is best ap-
proximated by a trimer (dimensions 10.8 x 17.5 x 2.7nm)
shown in Fig.4. Another fraction (Fraction III) consists mainly of a
small particle with a molecular weight of about 13000, a radius of gyration of 2 nm and a maximum dimension of 2.7 nm, and is a rather iso-
tropic particle.

References

1. Pilz,I., in Small angle X-ray scattering (Glatter,O. and Kratky,O.eds)
   pp 239-294,Academic Press (1982).

2. Pilz,I., Glatter,O.and Kratky,O. in Methods in Enzymology, Vol.61,
   pp 148-249, Academic Press (1979).

3. Theuer,M., Pilz,I., Schwarz,E., Wilhelm,P., Mainwaring,M.G. and
   Vinogradov,S.N., Int.J.Biol.Macromol. 7, 25 (1985).

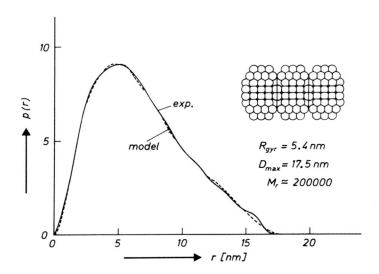

Fig.4. Fraction II of L.terrrestris hemoglobin; distance distribution functions p(r) of the experimental (——) and model (....) curve.

4. Kapp,O.H., Vinogradov,S.N.,Ohtsuki,M.and Crewe, A.V., Biochim.Biophys.Acta 704, 546, (1982).

5. Messerschmidt,U., Wilhelm,P., Pilz,I., Kapp,O.H. and Vinogradov, S.N., Biochim.Biophys.Acta 742, 366 (1983).

6. Kapp,O.H., and Crewe,A.V., Biochim.Biophys.Acta (submitted).

7. Polidori,G., Mainwaring,M., Kosinski,T.,Schwarz,C., Fingal,R. and Vinogradov,S.N., Arch.Biochem.Biophys.(submitted).

8. Vinogradov,S.N., Skolm,J.M., Hall,B.C., Kapp,O.H. and Mizukami,H. Biochim.Biophys.Acta 492, 136 (1977).

STRUCTURAL ANALOGIES IN ANNELID ERYTHROCRUORINS

Michela Zagra, Vincenza Mezzasalma and Benedetto Salvato[*]

Institute of Histology and Embryology, University of Palermo, Italy
[*]Department of Biology, University of Padova, Italy

INTRODUCTION

Erythrocruorins from many annelid species have been studied by many authors. In spite of the great similarity of most physicochemical properties and of the electron-micrographs of these proteins, some debated experimental data divide the scientific community, e.g. the Mr of the whole molecule (ranging from 2,500 to 4,1000 kDa) and, in consequence, the Mr of the 1/12 molecules, the Mr and number of the smallest covalently bound subunits and the Mr and number of the globin-like monomers (1).

In this paper, we present the results of the dissociation, by means of citraconic anhydride, of the erythrocruorins from four annelid species (Lumbricus terrestris, Allolobophora caliginosa, Eisenia foetida and Octodrilus complanatus). The data show great similarity to those obtained for Ophelia bicornis erythrocruorin, for which we have proposed a four-tetramer structure (2).

MATERIALS AND METHODS

Erythrocruorins from the four annelid species were isolated according to the method described for O. bicornis (2). In L. terrestris and O. complanatus, the erythrocruorin was collected by perfusion from the cut dorsal blood vessel.

Alkaline dissociation of erythrocruorins was done by dialyzing for 9 days against 0.1 M carbonate-bicarbonate buffer, pH 9.6, in the cold.

Citraconic anhydride dissociation was performed as described (2). The dissociation products were analyzed on a Sephacryl S300 column (1.8x150 cm), equilibrated with 0.1 ionic strength Tris-HCl buffer, pH 8, containing 0.4 M NaCl. A Sephadex G100 column (4.0x 200 cm), equilibrated with the same buffer, was used to achieve a better separation.

SDS-PAGE was carried out according to Laemmli (3).

RESULTS AND DISCUSSION

As already found for O. bicornis erythrocruorin, alkaline pH dissociation is very slow and often incomplete.

Under the conditions described, E. foetida and L. terrestris erythrocruorins dissociate into subunits having molecular masses of about 15 kDa, 30-40 kDa, 60-70 kDa and multiples of the latter (mainly 120-130 kDa). A. caliginosa erythrocruorin dissociates into a 120-130 kDa fraction and a fraction of 60-70 kDa. O. complanatus erythrocruorin has not been subjected to alkaline dissociation (Fig. 1).

In contrast, if dissociated by citraconic anhydride, the four annelid erythrocruorins yield very similar products, viz. components of three Mr classes, 60-70 kDa, 30-40 kDa and ca. 15 kDa (Fig. 1).

Invertebrate Oxygen Carriers
Ed. by Bernt Linzen
© Springer–Verlag Berlin Heidelberg 1986

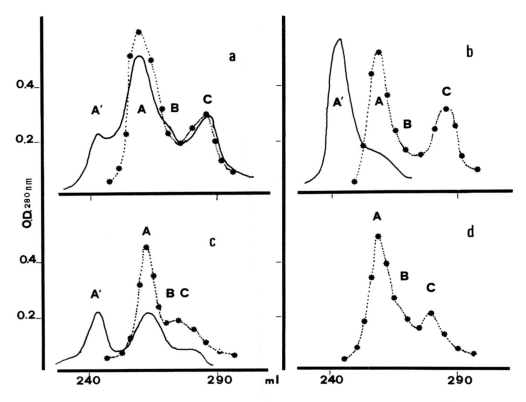

Fig. 1   Dissociation profiles of hemoglobins from a) L. terrestris, b) A. caliginosa, c) E. foetida and d) O. complanatus. The full profiles refer to alkaline pH dissociation, the star profiles to citraconic anhydride dissociation. The capital letters indicate the Mr classes: A' (120-130 kDa); A (60-70 kDa); B (30-40 kDa); C (15 kDa).

The 30-40 kDa component is resolved only on G100 gel chromatography. In Fig. 2, the elution pattern of citraconylated O. bicornis erythrocruorin is shown.

SDS-PAGE of citraconylated erythrocruorins is shown in Fig.3.The three Mr fractions are clearly distinct and the 30-40 kDa fraction seems to be composed of more than one component. Similar patterns are obtained also after S-S reduction: the 60-70 kDa fraction completely dissociates into three main components of 18-15 kDa, whereas the 30-40 kDa class seems to be unaffected by the reducing agent (Fig. 3).

To these similarities in chromatographic profiles and in SDS-PAGE patterns of dissociated erythrocruorins, we have to add that we have determined similar values for molecular parameters of the whole molecules, such as diffusion constant, sedimentation coefficient and elution volumes (2). All these data support the idea of a common (or largely comparable) structure for all annelid erythrocruorins.

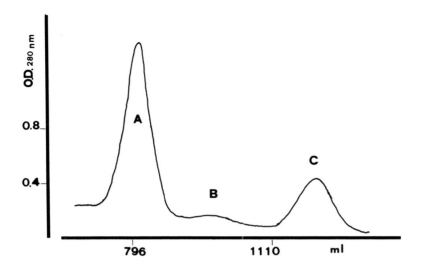

Fig. 2   G100 chromatographic profile of citraconylated O. bicornis Hb.

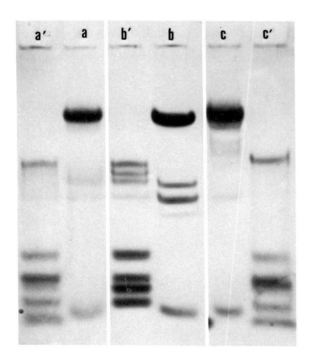

Fig. 3
SDS-PAGE of a,a') L. terrestris; b,b') A. caliginosa
and c,c') E. foetida citraconylated erythrocruorin.
a,b,c) unreduced samples,
a',b',c') reduced samples.

This work was supported by contribution of Ministero della Pubblica Istruzione.

REFERENCES

1. Chung, M.C.M., and Ellerton, H.D., Prog. Biophys. Mol. Biol. 35, 53-102 (1979).
2. Mezzasalma, V., Di Stefano, L., Piazzese, S., Zagra, M., Salvato, B., Tognon, G. and Ghiretti-Magaldi, A., Biochim. Biophys. Acta 829, 135-143 (1985).
3. Laemmli, L.K., Nature 227, 680-685 (1970).

ROLE OF SH-GROUPS AND S-S BRIDGES IN THE MAIN SUBUNIT (1/12 MOLECULE) OF SPIROGRAPHIS SPALLANZANII CHLOROCRUORIN

Vicenza Mezzasalma, Michela Zagra, Lucio Di Stefano and Benedetto Salvato[*]

Institute of Histology and Embryology, University of Palermo, Italy
[*]Department of Biology, University of Padova, Italy

INTRODUCTION

In all species of annelids described, the extracellular hemoglobins show (by electron microscopy) the same quaternary structure constituted by twelve identical subunits. The fine structure of the 1/12 subunit is not well resolved and several models have been proposed for it (2).

On electron micrographs of S. spallanzanii chlorocruorin, the main subunit seems to be constituted of four components placed in two different levels (2). After dissociation it appears to be constituted of subunits of Mr 60, 30 and 15 kDa, held together by disulfide bridges (3).

In this paper we report the number of SH groups and S-S bridges in each isolated subunit and propose a likely intra- and inter-chain distribution.

MATERIALS AND METHODS

Samples of native, apo- and alkylated chlorocruorin were prepared as already described (4). Acetylation of native chlorocruorin was performed according to Fraenkel-Conrat (5). Sulfitolysis was performed according to Chan (6) both on functional chlorocruorin and on apo-chlorocruorin. After the reaction, the mixture was dialyzed exhaustively and chromatographed on Sephacryl S-300.

^{35}S labeled sodium sulfite was purchased from Amersham International. Initial specific acitivity was 29.3 mCi/mmol. It was diluted by adding carrier until the specific activity was close to 34,000 cpm/μmole. Radioactivity was determined by counting 1 ml of each fraction in vials containing 10 ml of Instagel (Packard). Counting efficiency was about 60%.

Protein concentration was assayed either spectrophotometrically from the absorption at 280 nm in 8M urea ($E_{280}^{1\%}$ = 11) or by the method of Read et al. (7). BSA, native and apo-chlorocruorin were used for calibration. SDS-PAGE was performed according to Laemmli (8).

RESULTS AND DISCUSSION

When acylated with acetic anhydride, the whole molecule of chlorocruorin dissociated completely into subunits having an apparent Mr of about 240 kDa (1/12) (Fig. 1). The acetylated chlorocruorin has the same absorption bands as the whole molecule and the same pattern in SDS-PAGE (inset in Fig. 1): three components of Mr 30, 21 and 15 kDa. The same dissociation pattern is given by the apo-chlorocruorin in denaturing agents (SDS, 8M urea, 6M guanidine) (Fig. 2).

Sulfitolysis, performed on acetylated chlorocruorin at 40 °C for 24 hrs, produces subunits of Mr 30 and 15 kDa (Fig. 3). At shorter reaction-times an intermediate subunit of 60 kDa is obtained. By SDS-PAGE the 30 and 15 kDa subunits appear to be constituted of monomers of 15 kDa only, while the 60 kDa subunit yields monomers and dimers of 30 kDa (inset in Fig. 3). Sulfitolysis performed on apo-chlorocruorin, results in complete dissociation into monomers of 15 kDa, only a small quantity of the 30 kDa subunit remaining (Fig. 4).

Invertebrate Oxygen Carriers
Ed. by Bernt Linzen
© Springer–Verlag Berlin Heidelberg 1986

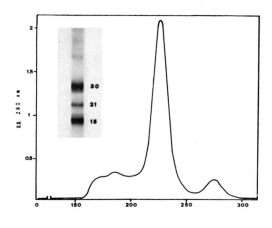

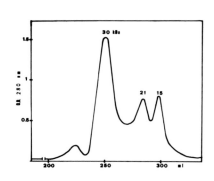

Fig.1   Fig.2

Fig.1 - Chromatography on Sephacryl S-300 of acetylated chlorocruorin
        (column 1.9 x 140 cm) in Tris buffer I=0.1 pH8 and NaCl 0.4M.
Fig.2   Elution profile of apo-chlorocruorin on Sephacryl S-200
        (column 2.5 x 100 cm) in 6M guanidina pH7.

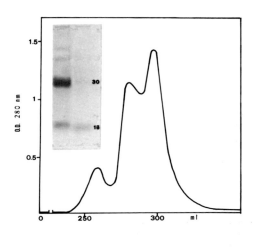

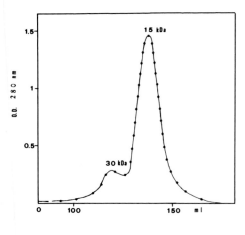

Fig.3   Fig.4

Fig.3   Elution profile of acetylated chlorocruorin after sulfito-
        lysis,on Sephacryl S-300 (see fig.1).
Fig.4   Elution profile of apo-chlorocruorin after sulfitolysis on
        Sephacryl S-300 (column 1.5 x 150) in 6M guanidine pH8.

The same procedure was performed on alkylated apo-chlorocruorin, in order to distinguish between free and bound SH groups.

From the radioactivity incorporated into each component, the number of SH groups per globin unit was calculated, assuming a Mr of 15 kDa (table).

On the basis of the present results it is concluded:
1. By acetylation it is possible to obtain a monodisperse solution of 1/12 subunits.
2. Sulfitolysis confirms that the 1/12 subunits are constituted of tetramers of 60 kDa made up, in turn, of monomers which form dimers via S-S bridges, and of monomers with free -SH, which form dimers not covalently bound, but detectable by gel chromatography (Fig. 3). The ratio between dimers (S-S) and dimers (-SH) being about 3:1, they should be assembled at least in two different tetrameric forms. Moreover, the 21 kDa component is a dimer, whose anomalous apparent Mr is most likely ascribed to disulfide bridges (both inter- and intra-chain) which are, however, hardly accessible.
3. According to our results (table), we propose one of the possible distributions of the SH groups and the S-S bridges in the two tetramers (Fig. 5).
The mixture of the two tetramers should contain a total of 16 SH groups:
4 free -SH located in the dimer not covalently bound;
4 involved in two disulfide bonds (one S-S/each dimer of 30 kDa);
8 involved in two inter- and two intra-chain bonds in the 21 kDa dimer which is resistant to reduction and in which up to 3 -SH/polypeptide chain can be titrated.

Table. Number of SH groups determined in Spirographis chlorocruorin and its dissociation products.

No. SH groups	Subunit	Protein
1.85*(2)		Whole mol.
1.54 (1.5)		" alkyl.
0.55 (0.5)	60 kDa	
2.24 (2)	30 kDa	Chl
1.09 (1)	15 kDa	
3.00 (3)	30 kDa	Apo-Chl
2.33 (2)	15 kDa	
1.22 2.9 (≤3)	30 kDa	Alkyl. apo-chl
1.46 (1.5)	15 kDa	

*experimental values
()expected values for the proposed model

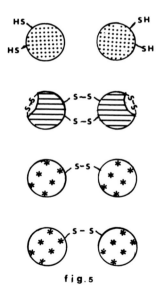

fig.5

Fig. 5. Proposed model of Spirographis chlorocruorin showing distribution of SH groups and disulfide bridges.

This work was supported by contributions of Ministero della Pubblica Istruzione.

REFERENCES

1. Chung,M.C.M. and Ellerton,H.D.,Prog.Biophys.Mol.Biol. 35,53-102 (1979).
2. Ghiretti-Magaldi,A.,Zanotti,G.,Salvato,B.,Tognon,G.,Mezzasalma,V. and Di Stefano,L.,Life Chem.Rep.,1 suppl.1,193-196 (1983).
3. Mezzasalma,V., Di Stefano,L.,Piazzese,S.,Zagra,M.,Ghiretti-Magaldi, A.,Carbone,R.and Salvato,B.,Life Chem.Rep.,1 suppl.1,187-191(1983).
4. Di Stefano,L.,Mezzasalma,V.,Piazzese,S.,Russo,G.and Salvato,B., FEBS Lett. 79,337-339 (1977).
5. Fraenkel-Conrat,H. in Methods in Enzymology(Colowick,S.P. and Kaplan,N.O. eds.)vol.4 p.247 (1957).
6. Chan,W.W.C.,Biochem. 7,4247-4254 (1968).
7. Read,S.M. and Northcote,D.H.,Anal.Biochem. 116,53-64 (1981).
8. Laemmli,L.K.,Nature 227,680-685 (1970).

SUBUNIT STRUCTURE AND AMINO ACID SEQUENCE OF THE EXTRACELLULAR
HEMOGLOBIN FROM THE POLYCHAETE TYLORRHYNCHUS HETEROCHAETUS

Tomohiko Suzuki[1] and Toshio Gotoh[2]

1 Department of Biology, Faculty of Science, Kochi University, Kochi
780, JAPAN ; 2 Department of Biology, College of General Education,
University of Tokushima, Tokushima 770, JAPAN

Introduction

The extracellular hemoglobin from the polychaete Tylorrhynchus
heterochaetus has a molecular weight of about 3×10^6 and consists
of two types of subunits: a "monomeric" subunit (chain I) and a
disulfide-bonded trimer of chains IIA, IIB, and IIC(1-4). Each chain
is linked to one heme(1,5). As the first step for construction of a
model of the molecular assembly of a giant multisubunit hemoglobin,
we determined the amino acid sequences of all the constituent polypeptide chains and their molecular masses exactly. The dissociation
of Tylorrhynchus hemoglobin in the presence of SDS can be summarized
as follows:

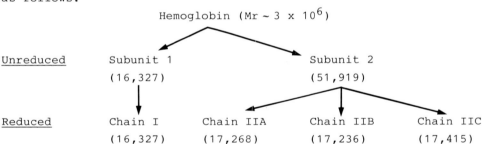

Hemoglobin (Mr ~ 3×10^6)

Unreduced	Subunit 1 (16,327)		Subunit 2 (51,919)	
Reduced	Chain I (16,327)	Chain IIA (17,268)	Chain IIB (17,236)	Chain IIC (17,415)

In this report we describe the sequence homologies among the
four constituent chains of Tylorrhynchus hemoglobin and the chain
AIII of Lumbricus hemoglobin (6).

Results and Discussion

The amino acid sequences of the four Tylorrhynchus chains I(3),
IIA (Suzuki and Gotoh; manuscript in preparation), IIB(2) and IIC(1)

Invertebrate Oxygen Carriers
Ed. by Bernt Linzen
© Springer–Verlag Berlin Heidelberg 1986

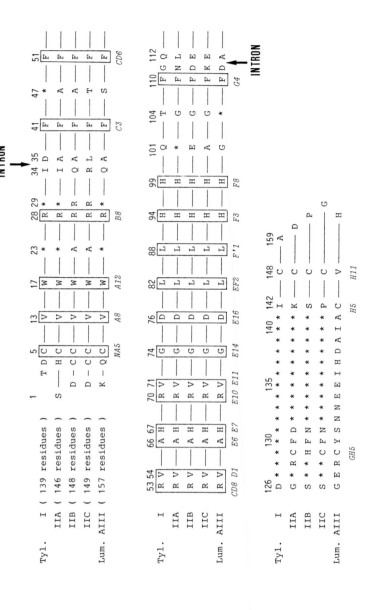

Fig.1 Alignment of the amino acid sequences of _Tylorrhynchus_ chains (I, IIA, IIB and IIC) with that of _Lumbricus_ chain AIII(6). Intronic positions (⟶) are based on the assumption described in the text. Only the following residues are indicated; the residues conserved in five globins (boxed), the positions containing deletions (*), the positions containing half-cystines, the residues neighbouring to the intronic positions, and the N- and C-terminal residues.

are aligned with that of <u>Lumbricus</u> chain AIII(6) in Fig.1. The tentative alignment is based on the assumption that the helical segments present in most other globins are also present in the chains of giant hemoglobins(2,6). In the five chains so far sequenced, 19 residues, including distal(E7) His, distal(E11) Val and proximal(F8) His which have a large influence on the oxygen-binding properties of mammalian hemoglobins and myoglobins, appear to be invariant.

The invariant 19 residues are distributed from the NA segment to the beginning of the G segment as shown in Fig.1. Here a con-sideration of the relation between the structural units of the polypeptide chain and the corresponding DNA is of particular inte-rest. It is well known that there are three exons separated by two introns in the DNA sequence of vertebrate hemoglobins (7,8). According to our alignment, the C-terminal exonic region starts at position G6. Surprisingly, there is no invariant residue in this region. In addition it cannot be avoided to take many deletions when <u>Tylorrhynchus</u> chains and <u>Lumbricus</u> chain are aligned. A possible explanation of these big differences is that the whole of the C-terminal exon has been replaced as a result of "exon-shuffling"(9) in the hemoglobin gene during the evolution of annelids. An alternative explanation is that heavy point mutations and insertions in the C-terminal exon have resulted in the sequence of <u>Tylorrhynchus</u> chains and <u>Lumbricus</u> chain.

It should be emphasized that the half-cystinyl residues have a key role in the unique molecular architecture of annelid extra-cellular hemoglobins, since these giant hemoglobins differ from other hemoglobins in possessing disulfide-bonded trimers or tetra-mers (1,10,11). In fact, <u>Tylorrhynchus</u> hemoglobin is dissociated into monomers in the presence of a reducing agent without protein-denaturant(1,5). In Fig.1, all the five chains have conserved the half-cystinyl residues in the segments NA, GH and H. In contrast, the intracellular monomeric hemoglobin from the polychaete <u>Glycera dibranchiata</u> has no half-cystinyl residue in the corresponding positions (12), although the C-terminal sequence of its chain shows strong homology with that of <u>Tylorrhynchus</u> chain I (3).

Most of the invariant residues in <u>Tylorrhynchus</u> chains and <u>Lumbricus</u> chain are located in the central exonic region as shown in Fig.1. This might reflect that the central exonic region is the fundamental entity of binding a heme group. Consistent with this idea, Craik et al. show that the product of the central exon of the human beta globin gene is a complete functional domain capable of binding heme tightly (13). On the other hand, all the half-cystinyl residues of Tylorrhynchus and Lumbricus chains are located in the side exonic regions. This suggests that the side exonic regions have an important role of building up the giant hemoglobins. Anyway, it is awaited to determine the sites where the interchain-disulfide bonds are formed. This would reveal the contact region between each chain in the "trimer".

Acknowledgement

We are grateful to Drs. K.Konishi, K.Shikama, T.Takagi, K.Nakamura and T.Furukohri for their invaluable suggestions.

References

1. Suzuki,T., Furukohri,T., and Gotoh,T. (1985) J.Biol.Chem. 260, 3145-3154.
2. Suzuki,T., Yasunaga,H., Furukohri,T., Nakamura,K., and Gotoh,T. (1985) J.Biol.Chem. in press.
3. Suzuki,T., Takagi,T., and Gotoh,T. (1982) Biochim.Biophys.Acta 708, 253-258.
4. Gotoh,T., and Kamada,S. (1980) J.Biochem.(Tokyo) 87, 557-562.
5. Suzuki,T., Takagi,T., Furukohri,T., and Gotoh,T. (1983) Comp. Biochem.Physiol. 75B, 567-570.
6. Garlick,R.L., and Riggs,A.F. (1982) J.Biol.Chem. 257, 9005-9015.
7. Nishioka,Y., and Leder,P. (1979) Cell 18, 875-882.
8. Go,M. (1981) Nature 291, 90-94.
9. Gilbert,W. (1978) Nature 271, 501.
10.Waxman,L. (1975) J.Biol.Chem. 250, 3790-3795.
11.Vinogradov,S.N., Shlom,J.M., Hall,B.C., Kapp,O.H., and Mizukami, H. (1977) Biochim.Biophys.Acta 492, 136-155.
12.Imamura,T., Baldwin,T.O., and Riggs,A. (1972) J.Biol.Chem. 247, 2785-2797.
13.Craik,C.S., Buchman,S.R., and Beychok,S. (1980) Pro.Natl.Acad. Sci.U.S.A. 77, 1384-1388.

A PRELIMINARY STUDY OF THE HEMOGLOBIN OF ARENICOLA MARINA

J.Sgouros, T.Kleinschmidt and G.Braunitzer
Max-Planck-Institut für Biochemie
D-8033 Martinsried (Munich), FRG.

The extracellular hemoglobin of Arenicola marina, having a molecular weight of 3.9 MDa, consists of twelve identical subunits arranged in two superimposed hexagonal rings. Investigation of the molecule has revealed the existence of only two types of polypeptide chains, I and II, with molecular weights of approximately 17 kDa.They were separated (Fig.1) by reversed-phase high performance liquid chromatography and isolated in pure form. Reconstitution experiments showed that both are capable of binding heme (Fig. 2). Their amino acid composition was determined(p.74), as well as the N-terminal sequence of chain I. Chain II has a blocked N-terminus. Separation of the cyanogen bromide peptides is shown in Fig. 3. As can be seen, the N-terminal region of chain I shows remarkable homology with the corresponding region of the small chain of the hemoglobin from the polychaete annelid Tylorrhynchus heterochaetus. Electron microscopic examination of the molecule is being carried out, to allow the construction of a precise three-dimensional model.

The N-terminal sequence of chain I compared to that of the small chain from Tylorrhynchus (from the latter,only differences against Arenicola are shown).

Arenicola: - Asp-Cys-Gly-Pro-Leu-Gln-Arg-Leu-Lys-Val-Lys-
Tylorrhynchus: Thr   Ile   Ile
Arenicola: His-Gln-Trp-Val-Gln-Val-Tyr-Ser-Gly-His-Gly-Tyr-
Tylorrhynchus: Gln   Ala   Val-Gly-Glu-Ser-
Arenicola: Glu-Arg-
Tylorrhynchus: Arg-Thr-

Invertebrate Oxygen Carriers
Ed. by Bernt Linzen
© Springer–Verlag Berlin Heidelberg 1986

Amino acid composition of chains

	Chain I	Chain II
Asp	17.60	17.73
Thr	3.48	5.20
Ser	13.77	9.01
Glu	14.71	15.57
Pro	5.02	5.33
Gly	9.59	10.30
Ala	11.53	10.42
Cys	1.96	1.30
Val	12.64	9.10
Met	2.86	1.81
Ile	4.58	5.35
Leu	13.28	14.33
Tyr	3.65	3.27
Phe	8.24	9.16
His	7.72	6.11
Trp	2.18	2.65
Lys	7.95	7.92
Arg	7.74	8.08

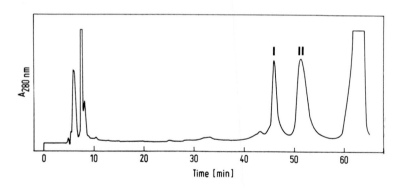

Fig.1 Separation of the chains of <u>Arenicola</u> <u>marina</u> hemoglobin by RP-HPLC. Column: 0.8 cm x 25 cm Merck LiChrosorb RP2. Eluent: 30%-60% acetonitrile in 0.1M ammonium acetate/5% formic acid, pH 3.0. Flow rate: 1.0 ml/min. Absorbance measured at 280 nm.

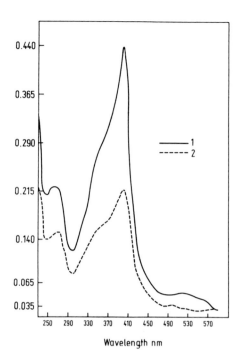

Fig. 2 Spectra of both chains after reconstitution with hemin chloride. The extinction maximum at 410 nm (Soret band) indicates that both chains are capable of binding heme.

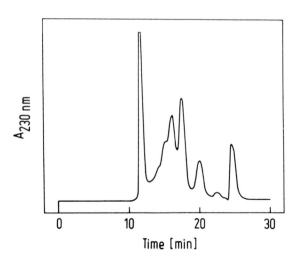

Fig. 3 Separation of the cyanogen bromide peptides of chain I by HPSEC. Column: 0.3 cm x 60 cm TSK G2000SW. Eluent: 0.3% trifluoroacetic acid. Flow rate: 1.0 ml/min. Absorbance measured at 230 nm.

References

1. Waxman, L., J.Biol.Chem. 246(23), 7318-7327 (1971).

2. Vinogradov, S.N., Shlom, J.M., Kapp, O.H., and Frossard, P.,
   Comp.Biochem.Physiol. 67(B), 1-16 (1980).

3. Van Bruggen, E.F.J., and Weber, R.E., Biochim.Biophys.Acta 359,
   210-214 (1974).

4. Suzuki, T., Takagi, T., Gotoh, T., Biochim.Biophys.Acta 708,
   253-258 (1982).

5. Suzuki, T., Furukohri, T., Gotoh, T., J.Biol.Chem. 260(5),
   3145-3154 (1985).

OXYGEN EQUILIBRIA OF WHOLE MOLECULES, ISOLATED SUBUNITS AND THE SUBUNIT
TRIMER OF HEMOGLOBIN FROM LUMBRICUS TERRESTRIS.

Fushitani, K., Imai, K. and Riggs, A.F.
Department of Zoology, University of Texas, Austin, Texas 78712, USA and Department
of Physicochemical Physiology, Medical School of Osaka University, Osaka, Japan.

Abstract. The extracellular hemoglobin of Lumbricus terrestris is composed of four
kinds of heme-containing chains in equimolar proportions. Each combines reversibly
with oxygen. Three of the chains form a disulfide linked trimer. Oxygen equilibria of the intact molecule are consistent with previous reports (1,2) but prior
incubation at low or high pH does not result in the irreversible changes reported
for the same or similar hemoglobins (3,4). Oxygen equilibria of the isolated subunits, trimer, and the reconstructed 4 chain product show that the trimer retains
both the Bohr and the Ca^{2+} effects. Addition of the 4th chain to the trimer results in regeneration of much of the cooperativity present in the native molecule.

Introduction. Both the stoichiometry and the number of different chains are uncertain in the hemoglobin of Lumbricus terrestris. The minimum molecular weight
based on heme differs from that of the polypeptide chains on the basis of polyacrylamide gel electrophoresis in SDS. Garlick and Riggs (5) reported 3 chains one
of which lacked heme. Kapp et al. (4) proposed 6 different chains. We have approached this problem by isolating the constituent polypeptides in as native a
form as possible.

Materials and Methods. Hemoglobin was extracted as the CO-form and purified by
the following sequence of procedures at $4°C$: precipitation by polyethylene glycol,
ultracentrifugation in the presence of phenylmethylsulfonylfluoride, and gel filtration on Sepharose CL-6B. Trimer and monomer were isolated on Ultragel ACA44
with 0.1 M borate buffer, pH 9.3, 1mM EDTA. The 3 constituent chains of the
trimer were isolated by chromatography on DE52 cellulose in the presence of dithiothreitol. The NH_2-terminal sequences were determined with the Beckman sequencer.
Oxygen equilibria were done either with the Imai machine at $25°C$ or the Gill cell
at $15°C$ (6,7). Gel electrophoresis in SDS was carried out with 15% polyacrylamide
as described (8).

Results and Discussion. Gel filtration on Ultragel ACA44 showed only the presence
of the trimer (T) and the monomer (M) in a stoichiometric molar ratio close to 1:1
on the basis of absorbance in the visible and UV regions. Gel electrophoresis
without mercaptoethanol showed that M and T have sizes of 15 and 49 K_d, respectively. The component chains of the trimer, a, b and c, were isolated on DE52. Chain d,

was obtained from the monomer fraction. No non-heme peak was obtained in either chromatography. All four chains, a, b, c and d, have heme and are 16.5-17.4 Kd in size.

NH_2-terminal sequences of the isolated chains are: a: His-Ile-Trp-, b: Lys-Lys-Gln-, c: Asp-Glu-His-, d: Glu-?-Leu-. However, the isolated chain a is evidently a cleaved product of the original chain a. The mobility of the isolated chain a did not correspond to that of any of the bands revealed by electrophoresis of the trimer or of the whole molecule. NH_2-terminal sequence analysis of the trimer showed three types of chains with NH_2-terminal sequences of Lys-Lys-Gln-, Asp-Glu-His-, and Ala-Asp-Glu-. We conclude that the latter is the NH_2-terminal sequence of chain a and that the His-Ile-Trp- sequence is the result of proteolysis. Chains a, b and c correspond to chains AII, AIII and AI of Garlick and Riggs (5) and bands IV, III and II of Kapp et al. (4), respectively. Chain d corresponds to band I (4). We conclude that the four chains are in equimolar proportions, because (i) the molar ratio of monomer to trimer is close to 1:1, (ii) isoelectric focusing of the trimer in 8 M urea shows only one major band and some minor bands, (iii) the assumed trimer produces three different chains in approximately equal amounts.

Oxygen binding by the whole molecule (9) showed a rather small free energy of cooperativity (1.6-2.8 Kcal/mol) but large, pH dependent, Hill coefficients (n_{max}, ca. 2.5-7.9). Protons lower the oxygen affinity primarily by shifting the upper asymptote in the Hill plot to the right. Ca^{2+} increases the oxygen affinity by shifting the upper asymptote to the left; an enhanced Bohr effect is associated with Ca^{2+} binding. Ca^{2+} shifts the pH of the maximal Hill coefficient from 8.1 to 7.6 as observed in chlorocruorin (10). Oxygenation of the intact hemoglobin at pH 7.8 was not significantly affected by prior incubation at either pH 6.2 or pH 9.0. Neither the Hill coefficient nor the oxygen affinity varied significantly between 6 and 600 μM (heme). Fig. 1 shows the pH dependence of the Hill coefficient, n, and log P_{50} of chain d, the mixture of chain d and the trimer (presumably the 1/12 subunits) and the native molecule. The trimer has a Bohr effect close to that of the native Hb. The trimer-chain d complex has a greatly enhanced cooperativity over that of the trimer alone. The effect of Ca^{2+} on the oxygen affinity of both the trimer and the reconstructed product is very similar to that of the whole molecule. The trimer has the functionally linked Ca^{2+} binding site(s). Fig. 2 shows oxygen equilibria corresponding to the data at pH 7.7 in Fig. 1.

The emerging picture of Lumbricus hemoglobin is that conformational changes accompany both the formation of the trimer and the trimer-chain d complex because each formation is associated with the appearance of new properties. (Supported by grants from NSF, PCM 8202760, Welch Foundation, F-213, and NIH, GM28410, to A.F.R.)

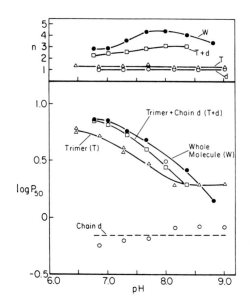

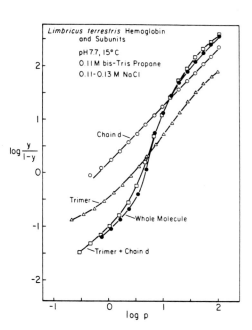

Fig. 1 pH dependence of the Hill coefficient, n, and the oxygen affinity. Buffer: 0.11 M bis-Tris-propane with 0.11-0.13 M NaCl at 15° C by Gill Cell.

Fig. 2 Oxygen equilibrium curves in Hill plot, taken from Fig. 1 at pH 7.7.

References

1. Vinogradov, S.N., Shalom, J.M., Hall, B.C., Kapp, O.H. and Mizukami, H., Biochim. Biophys. Acta 492, 136-155 (1977).
2. Weber, R.E., Nature 292, 386-387 (1981).
3. Giardina, B., Chiancone, E., and Antonini, E., J. Mol. Biol. 93, 1-10 (1975).
4. Kapp, O.H., Polidor, G., Mainwaring, M.G., Crewe, A.V., and Vinogradov, S.N., J. Biol. Chem. 259, 628-639 (1984).
5. Garlick, R.L. and Riggs, A., Arch. Bhichem. Biophys. 208, 563-575 (1981).
6. Imai, K. Methods in Enzymology 76, 438-449 (1981).
7. Gill, S.J., Methods in Enzymology 76, 427-438 (1981).
8. Laemmli, U.K., Nature 277, 680-685 (1970).
9. Fushitani, K., Imai, K. and Riggs, A., in preparation (1985).
10. Imai, K. and Yodhikawa, S., Eur. J. Biochem. 147, 453-463 (1985).

THE AMINO ACID SEQUENCE OF A STRUCTURAL UNIT ISOLATED FROM THE HIGH MOLECULAR WEIGHT GLOBIN CHAINS OF <u>ARTEMIA</u> SP.

L. Moens[1], M.L. Van Hauwaert[1], D. Geelen[1], G. Verpooten[1] and J. Van Beeumen[2].

(1) Departement Biochemie, Universiteit Antwerpen (UIA), B-2610 Wilrijk, Belgium.

(2) Laboratorium voor Mikrobiologie, Rijksuniversiteit Gent (RUG), B-9000 Gent, Belgium.

The primary structure of invertebrate globins (myoglobins and haemoglobins) is only available for a limited number of species. All the sequenced polypeptides belong to the low Mr globin chain types having a Mr of about 16 000 and binding one haem group. Their sequence alignment shows a definite relationship with their vertebrate counterparts (1,2).

In contrast, no structural information is available on the invertebrate high Mr globin-chain types.

These globin chains, with Mr 32 000 to 300 000 and binding two to sixteen haem groups are found in molluscs and arthropod extracellular haemoglobins. They are, like the haemocyanins, built up as polymers of structural and functional units resembling the low Mr globin-chain types (Mr ± 16 000) (3,4).

In the present communication, the primary structure of a single structural unit isolated from the high Mr globin chains of the brine shrimp <u>Artemia</u> <u>sp</u>. is described (5,6,7).

Materials and Methods

A fraction (E), containing a collection of structural units (Mr ± 16 000) was isolated from subtilisin digested <u>Artemia</u> haemoglobins by gel filtration as described previously (6). From this a single type of structural unit (E$_1$) was purified to homogeneity by isoelectric focusing (7).

Invertebrate Oxygen Carriers
Ed. by Bernt Linzen
© Springer–Verlag Berlin Heidelberg 1986

Tryptic, chromotryptic, thermolytic and S. aureus V-8 protease digestion was carried out on denatured samples (8).
The resulting peptides were separated by HPLC and TLC.
Amino acid sequence was determined either automatically using a gas phase sequencer (Applied Biosystems : 470 A) (9) or manually using the DABITC-PITC double coupling method (10).

Results and Discussion

The isolated structural unit E_1 is homogeneous on O'Farrell analysis. It has an apparent Mr of 15 800 ± 800 (n = 6) and a pI = 4.8 and is still able to bind dioxygen reversibly (P_{50} = 0,357 ± 0,046 mm Hg; n = 10) (7).
The amino acid composition of E_1 is very similar with that of the intact globin chain (7).
The primary structure was determined from tryptic, chymotryptic, S. aureus V8 and thermolytic peptides and the proposed sequence aligned with the human ß chain (Fig. 1).
Compared with the latter there is an extension of 6 residues on the amino terminus and a shortening with two residues on the carboxy-terminus. Most likely this is the result of the proteolytic cleavage of the E_1 structural unit from the intact globin chains.
Two insertions (residues : 17-18; 81-84) and four deletions (residues : 49; 97-101; 119; 124-125) are necessary to obtain maximal homology (38 residues : 26%).
The alignment suggests that the helical segments present in all other globin chains are also present in the E_1 chain. There are three histidines. Residues 63 and 92 seem to be the distal and proximal histidine respectively and thus the haem ligands. Histidine at position 35 which is also present in myoglobin and in Glycera haemoglobin may be a relict from a chytochrome b like origin (2).
The majority of residues which are identical in vertebrate myoglobin and haemoglobin are also identical in the Artemia E_1 chain (NA_3, A_{12}, A_{14}, A_{15}, B_6, C_2, CD_1, CD_4, E_7, E_{11}, F_4, F_8, G_{16}, GH_5) or they are kept conservative (A_8, B_{10}, B_{12}, C_4, CD_7, E_8, FG_1) (11,14).
Moreover the hydrophobicity profiles of the human ß and Artemia E_1 chain show with exception of the G-helix a good overall similarity.

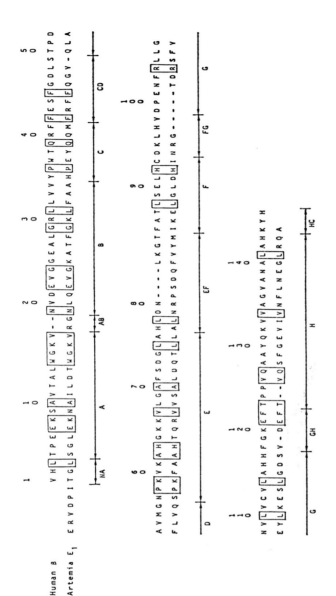

Fig. 1 : Comparison of the proposed sequence for Artemia E1 with that of the human β chain. The helix designations are those for the human chain.

As the packing of the structural units in the _Artemia_ haemoglobin must be different from that in the vertebrate tetramer the major differences can be expected in the subunit contact (packing and sliding) zones and this includes the G helix. Protein families that have the same pattern of residue hydrophobicity along their amino acid sequences tend to share a same type of folding (13).

This suggests that the E_1 structural unit may show the myoglobin fold.

The _Artemia_ globin chains may be considered as covalent polymers of eight "myoglobin folded" structural units which are definitely related to the classical globin family.

Vertebrate and invertebrate globins with low or high Mr thus all seem to be structurally related, confirming their monophyletic origin (1,2).

References

1. Goodman, M., Prog. Biophys. Mol. Biol. 37, 105-164 (1981).
2. Runnegar, B., J. Mol. Evol. 21, 33-41 (1984).
3. Chung, M.C.M. & Ellerton, M.D., Prog. Biophys. Mol. Biol. 35, 53-102 (1979).
4. Wood, E.J., Essays Biochem. 16, 1-47 (1980).
5. Wood, E.J., Barker, C., Moens, L., Jacob, W., Heip, J. & Kondo, M., Biochem. J. 193, 353-359 (1981).
6. Moens, L., Geelen, D., Van Hauwaert, M.L., Wolf, G., Blust, R., Witters, R. & Lontie, R., Biochem. J. 223, 861-869 (1984).
7. Moens, L., Van Hauwaert, M.L. & Wolf, G., Biochem. J. 227, 917-924 (1985).
8. Allen, G., in "Laboratory techniques in biochemistry and molecular biology" (Work, T.S. & Burdon, R.M., Eds) vol. 9, Elsevier North Holland, 1981.
9. Hewick, R.M., Hunkapiller, M.W., Hood, L.E. & Dreyer, W.J., J. Biol. Chem. 256, 7990-7997 (1981).
10. Chang, J.Y., Bauer, D., Wittman-Liebold, B., FEBS Lett. 93, 205-214 (1978).
11. Dickerson, R.E. & Geis, I., in "Hemoglobin : Structure, function, evolution and pathology" The Benjamin Cummings publishing Co. Inc., London, 1983.
12. Kyte, J. & Doolittle, R.F., J. Mol. Biol. 157, 105-132 (1982).
13. Eisenberg, D., Annu. Rev. Biochem. 53, 595-623 (1984).
14. Lesk, A.M. & Chothia, C., J. Mol. Biol. 136, 225-270 (1980).

QUATERNARY STRUCTURE AND MODULATION OF FUNCTION
IN EARTHWORM ERYTHROCRUORIN

E. Chiancone, F. Ascoli, B. Giardina, R. Santucci,
P. Vecchini, and D. Verzili

CNR Center of Molecular Biology, Institutes of Chemistry and
Biochemistry, University of Rome 'La Sapienza', and
Department of Cell Biology, University of Camerino, Italy

In the study of the structural and functional properties of erythrocruorins one of the most interesting aspects regards the elucidation of the reciprocal effects arising from the interplay of protons,
monovalent and divalent cations. It may be recalled that erythrocruorins are characterized by a complex hierarchical quaternary structure. Thus, in the electron microscope the native molecule appears as
an hexagonal bilayer with an apparently empty central cavity, each
layer consisting of six subunits; in turn the 1/12 subunits are formed by three or four lower molecular weight ones (1,2). However, the
detailed arrangement of the constituent polypeptide chains in the
subunits appearing in the electron microscope is still elusive.

Earthworm erythrocruorin binds both anions and cations, but only
cations affect its structural and functional properties. Ion binding
measurements performed by quadrupole relaxation NMR show that monovalent cations are bound with a roughly 10^3 lower affinity than divalent ones and that there is competition between Na^+ and Ca^{2+} for some
of the binding sites in the whole molecule, but not in the 1/12 subunits. The picture of the calcium binding sites proposed on this basis is that of carboxyl groups anchored at the interface of neighbouring 1/12 subunits and forming intersubunit chelates. Such binding
sites, which are saturated at concentrations below 1 mM Ca^{2+} and
whose number can be calculated to be ~0.3/heme, can be anticipated to
play an important role in the stabilization of the native quaternary
structure (3). These expectations are confirmed by the effect of Ca
on the alkaline dissociation of erythrocruorin from the earthworm Oc-
tolasium complanatum and on the reassembly of the subunits thus obtained. Indeed the presence of Ca^{2+} at concentrations around 0.1-1 mM

Invertebrate Oxygen Carriers
Ed. by Bernt Linzen
© Springer–Verlag Berlin Heidelberg 1986

markedly inhibits the alkaline dissociation of the erythrocruorin molecule, and a large excess of Na^+ diminishes the stabilizing effect of Ca^{2+} (4). Moreover, addition of Ca^{2+} to the buffer used to neutralize freshly dissociated subunits enhances their reassembly into the native molecule, and Na^+ (in roughly tenfold molar excess over Ca^{2+}) abolishes the effect of the divalent cation. The study of the pH dependence of the assembly process in the absence and presence of Ca^{2+} provides further information on the structurally relevant Ca^{2+} binding sites. In the absence of Ca^{2+} assembly is at a maximum around pH 6 and decreases significantly at pH ~8, whereas in the presence of Ca^{2+} the process is essentially pH independent, since the effect of the divalent cations is small at pH 6, but large at pH 8. It may be envisaged that some of the relevant carboxyl groups in these binding sites display abnormally high pK values, being protonated at pH 6 and deprotonated at pH 8 as in polyelectrolytes or small organic molecules involving clusters of carboxylates (5). Hence, at acid pH values, there are no electrostatic effects to counteract the specific recognition between the 1/12 subunits and Ca^{2+} is not required for assembly; in contrast, at alkaline pH values, Ca^{2+} is necessary to neutralize and chelate the relevant negatively charged groups brought into juxtaposition by the recognition process (6).

The effect of cations on the functional properties of earthworm erythrocruorin has been recognized only recently (7), although it was well known for the erythrocruorins of marine species (8-10). Qualitatively the linkage pattern of cations is similar in all erythrocruorins: cations bind more strongly to the oxygenated derivative thereby increasing the oxygen affinity; a marked increase in cooperativity parallels the change in oxygen affinity; divalent cations are effective at lower concentrations than are monovalent ones; the influence of cations depends on pH and, in a reciprocal way, the shape and position of the Bohr effect depend on the concentration of cations. Within this common pattern each erythrocruorin displays distinctive features. In the case of O. complanatum erythrocruorin (7) the most unusual one concerns the interplay between the oxygen-linked binding of protons and cations. Thus, the Bohr effect shows the expected

shift as a function of the concentration of cations, but is completely abolished at low cation concentrations (Fig.1). It follows that in
O. complanatum erythrocruorin at such low cation concentrations the
intrinsic dissociation constant of the Bohr protons is the same in
the oxygenated and deoxygenated derivative and hence that the Bohr
effect is totally due to the oxygen-linked binding of cations. In
their competition with protons, monovalent and divalent cations differ only by a scale factor. In fact, the oxygen affinity and cooperativity (measured by the Hill coefficient, $n_{\frac{1}{2}}$) of O. complanatum erythrocruorin correspond to log $p_{\frac{1}{2}}$ = 0.76 torr and $n_{\frac{1}{2}}$ = 3.0 in water or
at sufficiently low cation concentration and shift to log $p_{\frac{1}{2}}$ = -0.15
torr and $n_{\frac{1}{2}}$ = 5.5 in the cation saturated protein irrespective of pH.
Moreover, the oxygen binding curves that yield a given set of log $p_{\frac{1}{2}}$
and $n_{\frac{1}{2}}$ values are superimposable, again irrespective of the combination of proton and cation concentration. The analysis of these data
in terms of a modified two-state model brings out that cations influence the ligand binding properties of the state which they stabilize
and that cooperativity arises within functional constellations whose
dimensions depend on the physicochemical conditions.

The question arises as to whether there is a correspondence be-

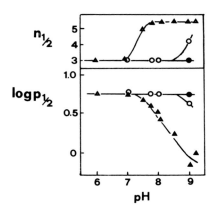

Fig.1 Effect of Tris and bis-Tris cations on the oxygen affinity and
degree of cooperativity of O. complanatum erythrocruorin. Cation
concentration: (●) 1 mM, (O) 2 mM, (Δ) 100 mM; temperature: 20°C.

tween the divalent cation binding sites relevant to the stabilization of the quaternary structure and those relevant to the regulation of function. The results on O. complanatum erythrocruorin, like those obtained on the erythrocruorin from the marine species Amphitrite ornata (9), indicate that the two types of site may be distinguished on the basis of their affinity for divalent cations. Those with a high affinity are always saturated at physiological concentrations of these ionic species (10 mM in the earthworm) and have a structural role; on the other hand, those involved in the modulation of oxygen transport have a degree of saturation that is sensitive to changes in the concentration of divalent cations and of other allosteric effectors within their physiological range.

REFERENCES

1. Antonini, E., and Chiancone, E., Annu.Rev.Biophys.Bioeng. 6, 239-271 (1977).
2. Shlom, J.M., and Vinogradov, S.N., J.Biol.Chem., 248, 7904-7912.
3. Chiancone, E., Bull, T.E., Norne, J.-E., Forsén, S., and Antonini, E., J.Mol.Biol. 107, 25-34 (1976).
4. Chiancone, E., Vecchini, P., Ascoli, F., and Antonini, E., Life Chem.Rep.Suppl. 1, 205-209 (1982).
5. Tanford, C., in Physical Chemistry of Macromolecules, pp. 548-562 (1961), J. Wiley, Inc., New York.
6. Chiancone, E., Vecchini, P., Verzili, D., and Ascoli, F., J.Mol. Biol., 172, 545-558 (1984).
7. Santucci, R., Chiancone, E., and Giardina, B., J.Mol.Biol., 179, 713-727 (1984).
8. Everaarts, J.M., and Weber, R.E., Comp.Biochem.Physiol., 48A, 507-520 (1974).
9. Chiancone, E., Ferruzzi, G., Bonaventura, C., and Bonaventura, J., Biochim.Biophys.Acta, 670, 84-92 (1981).
10. Weber, R.E., Nature (London), 292, 386-387 (1981).

CALCIUM DEPENDENT ALLOSTERIC MODULATION AND ASSEMBLY OF THE GIANT HEMOGLOBIN FROM THE EARTHWORM, EISENIA FOETIDA

Y. Igarashi, K. Kimura*, and A. Kajita
Department of Biochemistry and *Laboratory of Medical Sciences
Dokkyo University School of Medicine Mibu, Tochigi 321-02, Japan

INTRODUCTION

Extracellular hemoglobins(Hb) from annelids have been known to show no heterotropic interaction with the organic phosphate, such as 2,3-diphosphoglycerate. In an attempt to investigate allosteric effectors of annelid Hb, we estimated the oxygen equilibrium of Eisenia Hb and found that Ca plays an important role on the modulation of the oxygen affinity and cooperativity, as a result of the binding to the oxy-form of the Hb, under a physiological range of Ca concentration(1).

Present study deals with the effect of Ca on the allosteric parameters and the direct estimation of Ca binding. The role of Ca in the assembly of the molecule is also proposed.

MATERIALS AND METHODS

Preparation of Hb from Eisenia foetida and determination of oxygen equilibrium curves with an automatic oxygenation apparatus were performed as described previously(1). The oxygen equilibrium data were analyzed according to the two-state model for allosteric transition(2). K_T and K_R are defined as ligand association constants for the binding of the first and the last oxygen molecules to the Hb. L and N denote the allosteric constant and the number of interacting sites, respectively. The four parameters were obtained by perturbation method of the least-squares minimization procedure(3) which was fitted to the Hill equation with the aid of a micro-computer(NEC PC-9801). The cooperativity of oxygen binding is designated by n_{max} and the maximum slope of the Hill plot was computed by the differential method.

Reassociation of the Hb was performed with the CO form of freshly prepared sample, incubating with 50 mM bis-Tris propane pH 9.3 for 3 h at room temperature and then bringing back to about pH 8 by dialysis against 0.5 M NaCl–50 mM Tris buffer pH 8.0.

Metal chelate affinity chromatography(MCAC)(4) was carried out using a column(18 x 1.6 cm) of Chelating Sepharose 6B(Pharmacia Fine Chemicals), which charged with 5 mg/ml $ZnCl_2$ by passing through. After charging, the column was equilibrated with 0.5 M NaCl–50 mM Tris buffer. An aliquot (5 ml) of Hb at 1 mM

Invertebrate Oxygen Carriers
Ed. by Bernt Linzen
© Springer–Verlag Berlin Heidelberg 1986

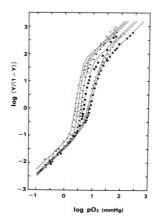

Fig. 1. Hill plots of the oxygen equilibria of <u>Eisenia</u> hemoglobin in 50 mM bis-Tris propane containing 0.1 M Cl⁻ at pH 7.5. The solid lines were constructed from the best-fit MWC parameters. 1 mM EDTA [●], no additive [○], 5 mM [■], 20 mM [▲], 50 mM [□] or 100 mM CaCl₂ [△].

heme was applied, washed with the same buffer and eluted with a histidine linear density gradient(0-50 mM) in the same buffer at pH 7.5. The reassembled Hb was analyzed by HPLC, using TSKgel G3000SW and G4000SW column(Toyo Soda).

RESULTS AND DISCUSSION

Fig. 1 shows Hill plots of the oxygen equilibria of Eisenia Hb at various Ca concentrations at pH 7.5 and 25 °C. It is evident from the figure that CaCl₂ increased the oxygen affinity and cooperativity of <u>Eisenia</u> Hb. Further it may be pointed out that the curves converge to a very close asymptote at low saturation range, while they diverge at high saturation range.

Table I summarizes the effect of Ca ion on the allosteric parameters describing the oxygen equilibrium by <u>Eisenia</u> Hb at pH 7.5 and 25 °C. As seen in the table, p50 was lowered with increasing Ca, indicating an increase of the oxygen affinity. The n_{max} value attained a maximum value of 9.76 in the presence of 20 mM Ca which is a two-fold increase as compared with that in the absence of Ca. The fact that K_T is insensitive to the change of Ca concentration whereas K_R depends upon Ca concentration, indicates the binding of Ca to the oxy-form in accordance with the previous report(1).

Table I also indicates that the number of interacting oxygen binding sites (N) increased from about six to ten in the presence of 20 mM Ca at pH 7.5, and then decreased gradually with further increase in Ca. This fact shows that the two-

Table I. Effect of calcium on the oxygen equilibrium parameters of
Eisenia hemoglobin. Conditions are the same as described in Fig. 1.

addition	p50 mmHg	$Pm^{a)}$ mmHg	Hill's n	$\Delta G(R,T)$ Kcal/mol	$K_R^{b)}$ $mmHg^{-1}$	$K_T^{c)}$ $mmHg^{-1}$	$log^{d)}$ L	$N^{e)}$
EDTA 1 mM	7.81	6.67	4.68	2.48	2.44	.0373	8.35	6.87
none	6.73	5.83	5.32	2.82	4.42	.0380	8.44	5.99
$CaCl_2$								
5 mM	4.72	4.32	7.86	3.03	5.16	.0311	11.09	8.21
20 mM	3.82	3.45	9.76	3.29	8.72	.0335	14.17	9.55
50 mM	2.96	2.75	9.59	3.33	10.30	.0371	13.46	9.23
100 mM	2.40	2.23	8.80	3.17	11.91	.0562	12.77	8.98

a) Median oxygen pressure.
b) Oxygen association constant of the R state.
c) Oxygen association constant of the T state.
d) Allosteric equilibrium constant.
e) Number of interacting oxygen binding sites.

state model is plausible to describe oxygen binding with a functional entity
(allosteric unit), having a variable number of the binding sites, although the
number never exceeds the oxygen binding site of a submultiple, which is a putative
functional unit, containing 12 hemes.   A similar phenomenon has been observed for
the hemocyanins of Limulus polyphemus(3) and Helix pomatia(5).

According   to   the   linkage   theory(6),the slope of   the   log   Pm   versus   log
concentration   of   Ca which stands for the number of bound ligands   per   heme,   was
calculated to be -0.22.   Consequently,   about 3 molecules of Ca are considered   to
bind   to   a   submultiple.   Direct   estimation of the Ca   binding   sites   by   using
an ultrafiltration method and Scatchard plots, revealed the presence of two kinds   of
binding sites.One kind is a high affinity site (Ka = 4.1 x 10^3 /M) and the other is
a low   affinity site (Ka = 39 /M).The oxy-form of the Hb contained four high affinity
sites per submultiple, which was compatible with the number as above mentioned.

The   effect   of Ca on the reassembly of the Eisenia Hb was studied of   alkali
treated   molecules of 60 s Hb.   The reconstructed Hbs were fractionated into   two
major peaks by a zinc chelate affinity column.   The first peak(peak I) was composed
of 10 s component only, while the second(peak II) included 10 s and 60 s components
in a ratio of 3/2 which was confirmed by HPLC(G4000SW) and ultracentrifuge.   Each
peak collected was dialyzed against 50 mM bis-Tris propane pH 7.5,   in the presence
and   absence   of 100 mM $CaCl_2$.   Dialyzates were subjected to HPLC   using   G4000SW
column.   The   lower   curves(Fig.   2) demonstrate the   controls,   where   Hbs   were
reassociated   in   the   absence   of Ca.   Peak I(left) consisted of 60 s and   10   s
components   in a ratio of 1/9 while peak II(right) consisted of 60 s and 10 s in a
ratio of about 4/6.   The upper curves(Fig. 2) illustrate the effect of Ca on the
reassociation.   Composition of the peak I was essentially the same as the control.
On the other hand,   composition of the peak II greatly changed as compared with the

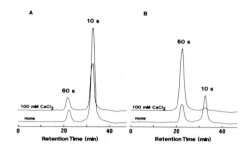

Fig. 2.   High performance liquid chromatogram of the reconstructed
hemoglobin on TSKgel G4000SW column.   A: peak I, B: peak II.

control.   The   fact   that   60 s/10 s ratio increased remarkably from 4/6   to   96/4
indicates clearly the involvement of Ca in the reassociation.

Presence of two kinds of 10 s components in the Peak I and II implies that one
is   native   which is able to reassemble to 60 s,   and that the other   is   denatured
which is not capable of reassociation.   Ca appears to enhance the former   process.
Consequently, both 10 s components were isolated by HPLC (G3000SW) and subjected to
two-dimensional   SDS   polyacrylamide gel electrophoresis (the first: SDS   and   the
second: SDS + mercaptoethanol) to investigate their subunit compositions.

The   results   evidently showed that 10 s component from the peak II   gave   the
same subunit composition as that of native 60 s Hb(i.e.,   Ia, Ib, Ic, II, III, IV),
whereas the 10 s component from the peak I had lost the subunit II.Therefore, it is
probable that this subunit(M.W.: 54K) is indispensable for the reassociation of the
60 s Hb and may be involved in the binding of Ca.

All   these results lead us to conclude that Ca plays an important role in   the
reassociation of the native Hb, as well as in the allosteric modulation.

REFERENCES

1. Igarashi, Y., Kimura, K., and Kajita, A., Biochem. Int. 10, 611–618 (1985).
2. Monod, J., Wyman, J., and Changeux, J. P., J. Mol. Biol. 12, 88–118 (1965).
3. Brouwer, M., Bonaventura, C., and Bonaventura, J., Biochemistry 22, 4713–4723
   (1983).
4. Fanou-Ayi, L., and Vijayalakshmi, M., Ann. N. Y. Acad. Sci. 413, 300–306(1983).
5. Zolla, L., Kuiper, H. A., Vecchini, P., Antonini, E., and Brunori, M., Eur. J.
   Biochem. 87, 467–473 (1978).
6. Wyman, J., Adv. Protein Chem. 19, 223–286 (1964).

QUATERNARY STRUCTURE OF ERYTHROCRUORIN FROM THE AQUATIC SNAIL Helisoma trivolvis

E. Ilan[1], I. Hammel[2], M.M. David and E. Daniel
Department of Biochemistry, Tel-Aviv University, Tel-Aviv 69978
[1]Department of Bio-medical Engineering, Technion, Haifa 32000
[2]Department of Pathology, Tel-Aviv University, Tel-Aviv 69978
Israel

Introduction

Among the gastropods, planorbid snails are the only species that make use of heme-containing erythrocruorin, rather than the copper containing hemocyanin, as oxygen carrier. The first structural study of a planorbid erythrocruorin was carried out by Svedberg & Eriksson-Quensel (1). Recent studies include those on erythrocruorin from Biomphalaria glabrata (2), Planorbis corneus (3) and Helisoma trivolvis (4). The present report is concerned with the quaternary structure of Helisoma trivolvis erythrocruorin. Our main goal was the determination of the number of polypeptide chains in the molecule.

Materials and Methods

Erythrocruorin from Helisoma trivolvis was isolated by preparative ultracentrifugation. SDS gel electrophoresis was performed on 5% polyacrylamide gels (5). Molecular weights were determined by meniscus depletion sedimentation equilibrium as described elsewhere (6).

Results and Discussion

Fig. 1 presents SDS polyacrylamide gel electrophoresis of Helisoma erythrocruorin. In the absence of 2-mercaptoethanol Helisoma erythrocruorin migrates as a single band with a mobility slower than that of thyroglobulin which has a molecular weight of 3.3×10^5. In the presence of 2-mercaptoethanol, one band corresponding to a molecular weight of 1.9×10^5 is observed. The same pattern was obtained with hemolymph drawn from a live animal. These results indicate that Helisoma erythrocruorin is composed of identical polypeptide chains, and that inter-chain disulfide bonds play a role in the organization of the native molecule.

The determination of the exact number of chains in Helisoma erythrocruorin, like

Invertebrate Oxygen Carriers
Ed. by Bernt Linzen
© Springer-Verlag Berlin Heidelberg 1986

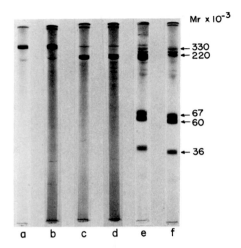

Fig. 1. SDS polyacrylamide gel electrophoresis of <u>Helisoma</u> erythrocruorin.
(a) Same as (c) except for omission of 2-mercaptoethanol; (b) Same as (d) except
for omission of 2-mercaptoethanol; (c) Hemolymph from live animal; (d) Purified
erythrocruorin; (e) Purified erythrocruorin + protein markers; (f) Protein markers

in other multi (8 or more) polypeptide chain proteins, depends critically on the
precise measurement of the molecular weight of the native molecule and that of the
polypeptide chain.   Special care was therefore placed on molecular weight determinations.   The native protein, $s^0_{20,w}$ = 34.7 S, was found to have a molecular weight
of 2.25×10^6 as determined by sedimentation equilibrium (Fig. 2).   This value is

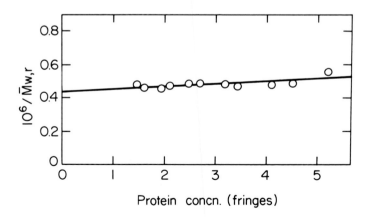

Fig. 2. Reciprocal weight-average molecular weight of <u>Helisoma</u> erythrocruorin as
a function of protein concentration.   Solvent: 0.1 M phosphate buffer, pH 6.7 containing 0.01 M $MgCl_2$.

considerably higher than the values reported for erythrocruorin from the same
$(1.7x10^6$ (4)) and from other (<u>Planorbis corneus</u>, $1.65x10^6$ (3); <u>Biomphalaria</u>
<u>glabrata</u>, $1.75x10^6$ (2)) planorbid snails.   There are, in our opinion, two possible
reasons for this discrepancy.   First, we made use of an experimentally measured
value for the partial specific volume 0.740 rather than the value 0.725-0.733 ml/g
calculated from the amino acid composition.   Second, our results show a definite
dependence of molecular weight on protein concentration.   To obtain a correct
molecular weight, we have therefore extrapolated the data to zero protein concentration rather than using a slope of ln c <u>versus</u> r^2 plot.   Sedimentation equilibrium
in 6 M guanidine hydrochloride in the presence and absence of 2-mercaptoethanol
gave molecular weights of $1.87x10^5$ and $3.82x10^5$ respectively.   These values are in
agreement with the corresponding values reported by Terwilliger <u>et al.</u> (4).   We
conclude that a <u>Helisoma</u> erythrocruorin molecule is composed of 12 single polypeptide chain subunits arranged in pairs.   The two subunits in each pair are held
together by disulfide bonds.

A model for the arrangement of the 12 subunits is presented in Fig. 3.   The 12
subunits occupy equivalent positions on the surface of a sphere.   A possible
symmetry is a tetrahedral shell symmetry (7).   It is easy to see that an electron
micrograph of a molecule oriented as in Fig. 3 will consist of 10 spots arranged
in a ring with an additional spot at the centre.   This prediction is consistent
with the results of electron microscopy of <u>Helisoma</u> erythrocruorin reported by
Terwilliger <u>et al.</u> (4).

<u>Fig. 3.</u> Proposed model for <u>Helisoma</u> erythrocruorin.   The model is composed of
12 identical spheres evenly distributed on the surface of a sphere.

References

1. Svedberg, T., and Eriksson-Quensel, I.B., J. Am. Chem. Soc. 56, 1700-1705 (1934).
2. Almeida, A.P., and Neves, A.G.A., Biochim. Biophys. Acta 371, 140-146 (1974).
3. Wood, E.J., and Mosby, L.J., Biochem. J. 149, 437-445 (1975).
4. Terwilliger, N.B., Terwilliger, R.C., and Schabtach, E., Biochim. Biophys. Acta 453, 101-110 (1976).
5. Weber, K., Pringle, J.R., and Osborn, M., Methods Enzymol. 26, 3-27 (1972).
6. Ilan, E., and Daniel, E., Biochem. J. 183, 325-330 (1979).
7. Haschemeyer, R.H., Adv. Enzymol. 33, 71-118 (1970).

IMMUNOCHEMICAL PROPERTIES OF MOLLUSCAN HEMOGLOBINS

D. Verzili, G. Citro, F. Ascoli, and E. Chiancone

CNR Center of Molecular Biology, Institutes of Chemistry
and Biochemistry, University of Rome 'La Sapienza',
Regina Elena Institute for Cancer Research, Rome, and
Department of Cell Biology, University of Camerino, Italy

The immunochemical study of the two hemoglobin components from the
Arcid mollusc Scapharca inaequivalvis has been approached as a means
to probe the quaternary structure of Arcid hemoglobins on the basis
of the following considerations. Sequence changes on the surface of
the molecule exceeding 40% usually suffice to eliminate immunological
crossreactivity between genetically homologous proteins (1). In Arcid
hemoglobins the three polypeptide chains that constitute the dimeric
HbI and the tetrameric HbII display a characteristic pattern of variable and invariant regions; in the latter, notably the E and F helices, the homology reaches 75% and 100%, respectively (2-4). Thus, the
presence or absence of crossreactivity between the two hemoglobins
can be related in a simple way to the location of the homologous regions on the surface or in the interior of the molecule. The lack of
crossreactivity between S. inaequivalvis HbI and HbII indicated that,
in contrast to expectations based on the assembly of vertebrate hemoglobins, the E and F helices are not exposed to solvent (5) as shown
by the low resolution X ray data of Royer et al. (6, and this volume)

This paper reports an extension of the previous study. Specific
anti-HbI and anti-HbII Fab fragments have been used to obtain information on the stoichiometry of the immunocomplexes and to study the
effect of antibody binding on the functional properties of the two
molluscan hemoglobins.

S. inaequivalvis HbI and HbII were prepared and purified as described in (7). The Fab fragments were obtained by papain digestion
of specific immunoglobulins (8) and were purified by means of ion exchange chromatography on the Mono S column of the FPLC system (Pharmacia). Five well separated fractions were obtained; two of them were
capable of binding the respective antigen as judged by immunodiffusion experiments. The active Fab fragments were titrated in fluore-

scence quenching experiments with the specific antigen; they yielded a stoichiometry of the reaction corresponding to ~ 3 Fab/Hb chain for both HbI and HbII, but differed slightly in the fraction of the fluorescence quenched (Fig. 1). The number of antigenic sites (or regions) therefore is similar in the three polypeptide chains that constitute the dimeric and tetrameric hemoglobin.

The knowledge of the stoichiometry of the immunocomplex has been exploited to study the effect of increasing quantities of specific Fab fragments on the oxygen binding properties of the two hemoglobins. Similar experiments using the specific IgG fractions had shown that both anti-HbI and anti-HbII are sensitive to the conformational changes that accompany ligand binding. Thus, an effect on the oxygen binding parameters was observed only when the antibody was mixed with the oxygenated protein. However, a decrease in oxygen affinity was observed, in contrast with the linkage theory, given the preferential binding of the antibody to the oxygenated protein. As a possible explanation a polymerization of the immunocomplex was invoked, although only a slight excess of IgG over Hb chain had been employed (molar ratio 2:1) (5). The use of Fab fragments allowed us to clarify this specific point. The results obtained with HbII, where the de-

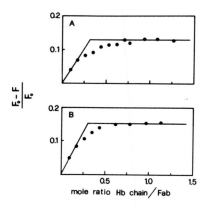

Fig.1 Fluorescence quenching titration of anti-HbI (A) and anti-HbII (B) Fab fragments with the respective antigen. Conditions: excitation wavelength, 285 nm; emission wavelength, 330 nm; 3.5×10^{-6} M Fab fragments in 0.05 M phosphate buffer, pH 7.0; 20°C.

crease in oxygen affinity produced by IgG binding was greatest, are shown in Fig. 2. The addition of a stoichiometric amount of Fab fragments (3 Fab/Hb chain) to the oxygenated protein results in an increase of oxygen affinity, as expected on the basis of linkage relationships, with no significant effect on cooperativity. The presence of Fab fragments in a substoichiometric amount does not affect the oxygen binding properties of HbII (data not shown). Similar results have been obtained with HbI.

The relatively modest effect of antibody binding on the functional properties of the S. inaequivalvis hemoglobins has to be contrasted with the behaviour of HbA. In HbA, the presence of a single Fab fragment per $\alpha\beta$ unit suffices to alter the oxygen equilibrium significantly. Thus, the Hill plot is heterogeneous and displays two distinct phases. Cooperativity, as measured by the Hill coefficient, is less than 1 in the first phase, but greater than 1 in the second. Both the relative proportion of protein in the first phase and its oxygen affinity are increased in the presence of greater amounts of antibody; on the other hand the second phase is essentially unaf-

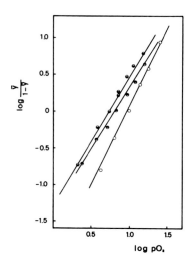

Fig.2 Hill plots of oxygen equilibria of HbII combined with the specific Fab fragments. Native protein (o), native protein mixed with Fab fragments fraction 2 (●) and 3 (◉) in stoichiometric amounts. Conditions: 0.1 M phosphate buffer at pH 7.0, 20°C.

fected (9).

In the attempt to rationalize these differences between S. inae-quivalvis hemoglobins and HbA, one can assume that in S. inaequi-valvis HbI and HbII the antigenic regions have the same location along the polypeptide chains as in human hemoglobin. This assumption is corroborated by the recent appreciation that there is a general link between high mobility of protein segments and antigenic activity (10). In particular, in globins the high helix content and compact folding of the molecule confines the regions with high relative fle-xibility essentially to the chain termini and to the bends of the chain . The different mode of assembly of Arcid hemoglobins and HbA places the antigenic determinants in a different position with re-spect to the subunit interfaces; one can speculate that this fact results in a different sensitivity of the bound antibody to the qua-ternary conformational changes that accompany oxygenation.

REFERENCES

1. Prager, E. M., and Wilson, A. C., J.Biol.Chem. 246, 7010-7017 (1971).
2. Furuta, H., and Kajita, A., Biochemistry 22, 917-922 (1983).
3. Petruzzelli, R., Goffredo, B. M., Barra, D., Bossa, F., Boffi, A., Verzili, D., Ascoli, F., and Chiancone, E., FEBS Lett. 184, 328-332 (1985).
4. Gilbert, A. T., and Thompson, E. O. P., Aust.J.Biol.Sci. in press.
5. Verzili, D., Citro, G., Ascoli, F., and Chiancone, E., FEBS Lett. 181, 347-352 (1985).
6. Royer, W. E., Love, W. E., and Fenderson, F. F., Nature in press.
7. Chiancone, E., Vecchini, P., Verzili, D., Ascoli, F., and Anto-nini, E., J.Mol.Biol. 152, 577-592 (1981).
8. Porter, R.R., Biochem.J. 73, 119-126 (1959).
9. Reichlin, M., Udem, L., and Ranney, H. M., Biochim.Biophys.Acta 175, 49-54 (1969).
10. Westhof, E., Altschuh, D., Moras, D., Bloomer, A. C., Mondragon, A., Klug, A., and Van Regenmortel, M. H. V., Nature 311, 123-126 (1984).

STUDIES OF THE HEME ENVIRONMENT IN MOLLUSCAN HEMOGLOBINS BY EPR

C. Spagnuolo, A. Desideri, E. Chiancone, and F. Ascoli

CNR Center of Molecular Biology, Institutes of Chemistry and
Biochemistry, University of Rome 'La Sapienza', and
Department of Cell Biology, University of Camerino, Italy

The red cells of the mollusc Scapharca inaequivalvis, as those of other Arcid molluscs, contain a dimeric (HbI) and a tetrameric (HbII) hemoglobin component constructed from three myoglobin-like chains assembled into a homodimer and an $\alpha_2\beta_2$ type of structure, respectively. From a structural viewpoint, the heme carrying helices, E and F, are essentially invariant in all three polypeptide chains, but differ in sequence from the corresponding regions of vertebrate hemoglobins and myoglobins. The E and F helices of the Arcid hemoglobins are particularly rich in hydrophobic residues, a feature that has been ascribed to their involvement in the dimeric intersubunit contact (1, 2). This characteristic topology is accompanied by a number of spectroscopic features, i.e. the absorption shoulder and the unusually high ellipticity at 590 nm in both deoxy-HbI and HbII, which indicate that in the deoxygenated derivative the structure at the heme site is constrained (3). Moreover, the anisotropy of the EPR spectrum of deoxy-CoHbI is indicative of a distorted coordination of the hindered proximal histidine (4). As a model for diamagnetic O_2 and CO complexes the ferrous NO derivatives of the S.inaequivalvis hemoglobins have been prepared. Since NO contains one unpaired electron the ferrous NO complexes are easily amenable to EPR studies.

The NO-derivative of hemoproteins at neutral pH values usually display a rhombic spectrum diagnostic of an hexacoordinated iron atom (5). This EPR spectrum is characterized by a 9-line superhyperfine pattern in the g_z region that is due to the concomitant contribution of the N_ε atom of the proximal histidine and of the NO nitrogen. The various hemoproteins differ in the degree of resolution of the superhyperfine structure in the g_z region (6). A second type of EPR spectrum, characterized by a triplet in the high field region, may also be observed; on the basis of comparisons with model compounds

Invertebrate Oxygen Carriers
Ed. by Bernt Linzen
© Springer–Verlag Berlin Heidelberg 1986

this latter spectrum has been ascribed to NO-complexes in which the iron is pentacoordinated (7). A transition from the hexacoordinated to the pentacoordinated species has been observed in both myoglobins and hemoglobins at low pH values under conditions that induce the protonation of the proximal histidine and hence the cleavage (or the severe weakening) of the Fe-N_ε bond (8, 9). In addition, the penta-coordinated EPR spectrum has been observed in human hemoglobin at pH values around neutrality upon addition of inositolhexaphosphate. This allosteric effector appears to switch the quaternary conformation of HbNO from R to T and to induce cleavage of the Fe-N_ε bond in the subunits (7). It may be added that the transition from the hexa- to the pentacoordinated structure can be followed also spectrophotome-trically in the Soret region, where a small decrease in absorbance and a slight shift of the absorption band towards lower wavelengths takes place (9,10).

The NO derivatives of S. inaequivalvis HbI and HbII were prepared by adding a small excess of dithionite to a carefully degassed pro-tein solution equilibrated with pure N_2. After addition of a proper volume of a KNO_2 solution the protein was mixed with buffer at the desired pH. EPR spectra were recorded at 110°K with a Varian E-9 spectrometer, absorption spectra at 25°C with a Cary 219 spectropho-tometer.

The EPR spectra of S. inaequivalvis HbI-NO and HbII-NO are illu-strated in Fig. 1A and B, respectively. At pH values around neutra-lity both HbI-NO and HbII-NO have the expected rhombic symmetry (see also Table I). In both spectra there is no 9-line superhyperfine structure for the g_z signal. The degree of resolution in this spectral region has been related to the strength of the interactions of the unpaired electron spin density with the iron bound proximal nitrogen or to differences in the electronic relaxation times of the various ferrous-NO complexes (11).

Upon lowering the pH below 6, HbI-NO and HbII-NO undergo a rever-sible transition into a pentacoordinated structure, characterized by the 3-line splitting in the high magnetic field region, whose EPR parameters are reported in Table I. The fraction of hexacoordinated

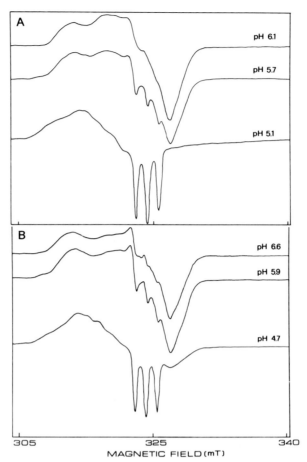

Fig. 1 pH dependence of X-band EPR spectra of HbI-NO (A) and HbII-NO
(B). Spectra were recorded in 0.25 M acetate buffer or sodium
acetate. Setting conditions: 9.14 GHz microwave frequency, 20
mW microwave power, 0.1 mT modulation amplitude. T = 110°K.

TABLE I

X-BAND EPR PARAMETERS OF THE FERROUS NITRIC OXIDE DERIVATIVES
OF S. INAEQUIVALVIS DIMERIC AND TETRAMERIC HEMOGLOBINS

Protein	pH	g_x	g_y	g_z	A_xNO (mT)	A_yNO (mT)	A_zNO (mT)
HbI	6.1	2.080	1.990	2.009	-	-	2.2
HbI	5.2	2.100	2.060	2.010	-	1.8	1.6
HbII	6.6	2.080	1.990	2.008	-	-	2.1
HbII	4.7	2.100	2.060	2.010	-	1.7	1.6

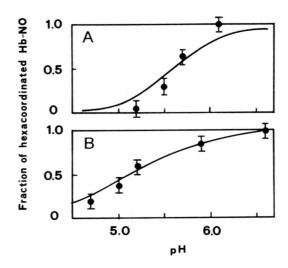

Fig. 2 pH dependence of the fraction of hexacoordinated HbI-NO (A)
and HbII-NO (B). Buffers and other conditions as in Fig. 1.

heme as a function of pH in the range 4.7-6.6 for both HbI-NO and
HbII-NO is shown in Fig. 2. It appears that all three types of chain
undergo essentially a complete hexa-to pentacoordinate transition in
the range studied; in HbII-NO the transition conforms to the titration of a single group with a $pK_a \sim 5.1$, in HbI-NO the pK_a value is
slightly higher (~ 5.6). These pK values are only 1 to 1.5 pH units
lower than that of a histidyl residue exposed to solvent ($pK_a \sim 6.8$)
and fall in the upper range of the values measured in a number of
monomeric hemoproteins (pK_a values from 3.9 to 5.4) (9,12) pointing
to an overall similarity of the free energy change in the binding of
the proximal imidazole (~ 2 Kcal/mole) to the ferrous NO-porphyrin.
However, HbI-NO displays a unique and most intriguing feature in that
the hexa- to pentacoordinate transition is cooperative, as confirmed
also by spectrophotometric titrations in the Soret region (data not
shown). The molecular basis for this cooperativity is unknown, although it may reside in the unusual assembly of Arcid hemoglobins in
which the heme carrying E and F helices are involved in the dimer
contact (2). If this is the case, one may speculate that cooperativity is obscured in HbII by heterogeneity arising from small pK_a.

differences in the hexa- to pentacoordinate transition of the two
types of chain.

REFERENCES

1. Petruzzelli, R., Goffredo, B.M., Barra, D., Bossa, F., Boffi, A., Verzili, D., Ascoli, F., and Chiancone, E., FEBS Lett. 184, 328-332 (1985).
2. Royer, W.E., Love, W.E., and Fenderson, F.F., Nature, in press.
3. Chiancone, E., Vecchini, P., Verzili, D., Ascoli, F., and Antonini, E., J.Mol.Biol. 152, 577-592 (1981).
4. Verzili, D., Santucci, R., Ikeda-Saito, M., Chiancone, E., Ascoli, F., Yonetani, T., and Antonini, E., Biochim.Biophys.Acta 704, 215-220 (1982).
5. Rein, H., Ristau, O., and Scheler, W., FEBS Lett. 24, 24-26 (1972).
6. Trittelvitz, E., Gersonde, K., and Winterhalter, K.H., Eur.J. Biochem. 51, 33-42 (1975).
7. Szabo, A., and Perutz, M.F., Biochemistry 15, 4427-4428 (1976).
8. Hille, R., Olson, J.S., and Palmer, G., J.Biol.Chem. 254, 12110-12120 (1979).
9. Ascenzi, P., Giacometti, G.M., Antonini, E., Rotilio, G., and Brunori, M., J.Biol.Chem. 256, 5383-5386 (1981).
10. Salhany, J.M., Ogawa, S., and Schulman, R.G., Biochemistry 14, 2180-2190 (1975).
11. O'Keeffe, D.H., Ebel, R.E., and Peterson, J.A., J.Biol.Chem. 253, 3509-3516 (1978).
12. Ascenzi, P., Coletta, M., Desideri, A., and Brunori, M., Biochim.Biophys.Acta 829, 299-302 (1985).

THE COOPERATIVE DIMERIC HEMOGLOBIN
FROM THE MOLLUSC SCAPHARCA INAEQUIVALVIS:
STRUCTURAL AND FUNCTIONAL CHARACTERIZATION

F. Ascoli, D. Verzili, and E. Chiancone

CNR Center of Molecular Biology, Institute of Chemistry
and Biochemistry, University of Rome 'La Sapienza', and
Department of Cell Biology, University of Camerino, Italy

The minor hemoglobin component (HbI) present in the red cells of
the bivalve mollusc Scapharca inaequivalvis, an indopacific species
of the Arcid family that has settled in the Adriatic sea in recent
years, is a dimer constructed of two identical chains. S. inaequivalvis HbI does not undergo ligand-linked association-dissociation
reactions, binds oxygen in a cooperative fashion ($n_{\frac{1}{2}}$ = 1.5) with a
fairly low affinity ($p_{\frac{1}{2}}$ ~10 torr), lacks the alkaline Bohr effect and
is not sensitive to the allosteric effectors of vertebrate hemoglobins [1]. These properties, which are shared by all the dimeric
Arcid hemoglobins [2-4], render S. inaequivalvis HbI an ideally simple system for the study of cooperative phenomena in hemoproteins.
Hence, the equilibrium and kinetic aspects of ligand binding to S.
inaequivalvis HbI have been analyzed in detail and compared with the
behaviour of human hemoglobin (HbA). At equilibrium the main contribution to cooperativity in oxygen binding to HbI is primarily entro-
pic as in HbA. However in HbA the ΔH values are uniform after correction for the contribution of the release of heterotropic ligands
[5], while in HbI the ΔH values are not uniform [6]. Moreover, the
decrease of ΔH in the second oxygenation step indicates that the
stabilization of the oxy-quaternary structure in HbI is linked to an
endothermic process [6]. Kinetically, cooperativity in ligand binding
manifests itself as in HbA. Thus, the oxygen dissociation rate constant decreases in the second step of the reaction, and the carbon
monoxide combination rate constant increases while the reaction proceeds [7,8].

The appreciation founded on these data that the presence of unlike
chains is not a prerequisite for cooperative ligand binding leads to
the conclusion that in HbI cooperativity must have a different stru-

Invertebrate Oxygen Carriers
Ed. by Bernt Linzen
© Springer–Verlag Berlin Heidelberg 1986

ctural basis from that operating in HbA and tetrameric vertebrate hemoglobins. In this connection it is important to recall that the dimeric structure of HbI is very stable towards dissociation in high salt concentrations and towards pH changes in the range 5-9 (1,9). Hence, the breakage of salt bridges, which is an important factor contributing to cooperativity in HbA, is not likely to play a major role in HbI.

In the attempt to characterize factors governing subunit interactions and involved in the ligand linked quaternary conformational changes, the distribution of polar and apolar residues along the HbI sequence (10) has been analyzed in comparison with that occurring in the myoglobin and hemoglobin chains. A preliminary analysis revealed significant differences at the level of the regions corresponding to the $\alpha_1\beta_1$ and $\alpha_1\beta_2$ contacts of HbA (11). The comparison was then refined by means of the OMH (Optimal Matching Hydrophobicity) scale proposed by Eisenberg (12). The latter analysis allowed us to identify hydrophobic residues that substitute polar side chains in the regions of the HbI sequence corresponding to the E and F helices of HbA. Such additional hydrophobic residues provide sticky patches for intersubunit contacts in these helices which are exposed to solvent in HbA and in myoglobins. These data, taken together with the immunological properties of HbI and of the major tetrameric S. inaequivalvis component HbII (13), are in agreement with the unusual assembly of the two myoglobin-like subunits demonstrated by recent low resolution X ray data: the two chains are assembled "back to front" with respect to vertebrate hemoglobins, thus involving the E and F helices in the intersubunit contacts (14, and this volume). This radically different arrangement of the subunits provides a rationale for the conservation of the E and F helices in all the polypeptide chains of Arcid hemoglobins sequenced thus far (15-18). The X ray work also brought out that in the two S. inaequivalvis hemoglobins there is an additional helix called N that extends from the A helix towards the amino terminus of each chain and renders the A helix shorter than in vertebrate hemoglobins and myoglobins. We have therefore attempted to predict the helical regions in HbI making use of both the "central

residue" initiation mechanism of Chou and Fasman (19) and the "terminal" initiation mechanism proposed subsequently by Blagdon and Goodman (20). Two restrictions have been imposed in the analysis: the
presence of Phe CD1, which is common to all globin sequences, and of
the two histidines in the proximal and distal position on the basis
of the results given in (21). The predicted helical regions are shown
in Fig. 1. In accordance with the X ray data the N helix is clearly
defined as is the short A helix and the only cysteine residue is situated at the beginning of the F helix. There is no evidence of a D
helix and in the region of the G one two short helices are deline-
ated. The overall α-helical content is 70%, in agreement with circular dichroism data (1). The E and F helices seem to differ somewhat
in length with respect to the vertebrate hemoglobin ones; in particular the F helix appears to comprise 17 residues. A long F helix is
present also in Aplysia limacina myoglobin, the only molluscan globin whose structure is known at 2 Å resolution (22). A further similarity between the latter protein and HbI resides in the location of
the two tryptophan residues which are in the A and H helices. Lastly,
the prediction of the helical regions just reported confirms that
the additional hydrophobic residues are positioned in the E and F helices. In conclusion, the present analysis strengthens the hypothesis
(10) that such hydrophobic residues may constitute a topological cha-

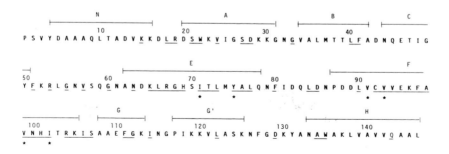

Fig.1 Primary structure and predicted helical regions of S. inaequivalvis HbI. The residues common to all dimeric and tetrameric
Arcid hemoglobin chains sequenced thus far are underlined.
Asterisks indicate the hydrophobic residues that have been proposed to be involved in the dimeric contact (10).

racteristic of Arcid hemoglobins; it will be of interest to establish
how this topology is related to the molecular basis of cooperativity
in these hemoproteins.

REFERENCES

1. Chiancone, E., Vecchini, P., Verzili, D., Ascoli, F., and Antoni-
    ni, E., J.Mol.Biol. 152, 577-592 (1981).
2. Ohnoki, S., Mitomi, Y., Hata, R., and Satake, K., J.Biochem.
    (Tokyo) 73, 717-725 (1973).
3. Furuta, H., Ohe, M., and Kajita, A., J.Biochem (Tokyo) 82, 1722-
    1730 (1977).
4   Djangmah, J.S., Gabbott, P. A., and Wood, E. J., Comp. Biochem.
    Physiol. 60B, 245-250 (1978).
5. Imai, K. and Yonetani, T., J.Biol.Chem. 250, 7093-7098.
6. Ikeda-Saito, M., Yonetani, T., Chiancone, E., Verzili, D., Asco-
    li, F., and Antonini, E., J.Mol.Biol. 170, 1009-1018 (1983).
7. Antonini, E., Chiancone, E., and Ascoli, F., in "Structure and
    Function Relatioships in Biochemical System" (Bossa, F. et al.
    eds.), pp. 67-73 (1982).
8. Antonini, E., Ascoli, F., Brunori, M., Chiancone, E., Verzili,
    D., Morris, R. J., and Gibson, Q. H., J.Biol.Chem. 259, 6730-6738
    (1984).
9. Gattoni, M., Verzili, D., Chiancone, E., and Antonini, E., Bio-
    chim.Biophys.Acta 743, 180-185 (1983).
10. Petruzzelli, R., Goffredo, B. M., Barra, D., Bossa, F., Boffi,
    A., Verzili, D., Ascoli, F., and Chiancone, E., FEBS Lett. 184,
    338-332 (1985).
11. Boffi, A., Ascoli, F., Hermans, J., and Chiancone, E., 8th Int.
    Biophys.Congr., Bristol, Abstr. 46 (1984).
12. Eisenberg, D., Annu.Rev.Biochem. 53, 595-623 (1984).
13. Verzili, D., Citro, G., Ascoli, F., and Chiancone, E., FEBS Lett.
    181, 347-352 (1985).
14. Royer, W. E., Love, W. E., and Fenderson, F. F., Nature in press.
15. Furuta, H., and Kajita, A., Biochemistry 22, 917-922 (1983).
16. Como, P. F., and Thompson, E. O. P., Aust. J.Biol.Sci. 33, 643-
    652 (1980).
17. Fisher, W. K., Gilbert, A. T., and Thompson, E. O. P., Aust.J.
    Biol.Sci. 37, 191-203 (1984).
18. Gilbert, A. T. and Thompson, E. O. P., Aust.J.Biol.Sci. in press.
19. Chou, P. Y., and Fasman, G. D., Biochemistry 13, 222-245 (1974).
20. Blagdon, D. E., and Goodman, M., Biopolymers 14, 241-245 (1975).
21. Verzili, D., Santucci, R., Ikeda-Saito, M., Chiancone, E., Asco-
    li, F., Yonetani, T., and Antonini, E., Biochim.Biophys.Acta 704,
    215-220 (1982).
22. Bolognesi, M., Coda, A., Gatti, G., Ascenzi, P., and Brunori, M.,
    J.Mol.Biol. 183, 113-115 (1985).

THE LOW RESOLUTION STRUCTURES OF THE COOPERATIVE HEMOGLOBINS FROM THE BLOOD CLAM SCAPHARCA INAEQUIVALVIS

William E. Royer, Jr.* and Warner E. Love
The Thomas C. Jenkins Dept. of Biophysics
The Johns Hopkins University, 3400 N. Charles St.
Baltimore, MD 21218, USA

*Present address:
Columbia University, Dept. of Biochemistry & Molecular Biophysics
New York, NY 10032 USA

INTRODUCTION

The clams of the Arcid family (the blood clams), unlike most molluscs, possess intracellular hemoglobins (1-6). In Scapharca inaequivalvis these hemoglobins are formed from three distinct polypeptide chains, two of which associate to form heterotetramers (A_2B_2), while a third associates to form homodimers (5). Both dimers and tetramers show cooperative oxygen binding with Hill coefficients of 1.5 and 2.1, respectively (6). Amino acid sequences for dimer chains from this and related clams (7-9) and for tetramer chains from a related clam (10,11) align functionally with sequences from other hemoglobins that possess the myoglobin fold and they show additional residues at the amino termini.

The presence of these cooperative hemoglobins raises interesting questions about their origin and relationship to vertebrate hemoglobins. Comparison of sequences of hemoglobins and myoglobins from many species has led to the conclusion that the split of the genes for the α and β chains occurred after the emergence of the vertebrates (12). Since both the α and β chains are essential for cooperativity in vertebrate hemoglobin (13), it would appear that its cooperativity arose after the emergence of the vertebrates. Therefore, the development of cooperativity in an invertebrate hemoglobin would be evolutionarily independent of the development of the cooperative mechanism in vertebrate hemoglobin.

We describe here our low resolution crystal structures for both the dimeric and tetrameric CO-liganded hemoglobins from S. inaequivalvis. The structure determination has been presented previously (14).

Invertebrate Oxygen Carriers
Ed. by Bernt Linzen
© Springer–Verlag Berlin Heidelberg 1986

STRUCTURE DESCRIPTION

Our final 5.5A clam tetramer electron density map clearly shows the course of each polypeptide chain. The subunits have the "myoglobin fold" with what appears to be an additional helix at each amino terminus. We call this helix "pre-A". The pre-A helix lies between the H-helix and the E-F corner and is oriented approximately parallel to, and in the same plane as, the G and H helices. In each subunit, this brings the amino terminus within 10A of the carboxy terminus.

Despite the similarity of the clam tetramer subunits to those of vertebrate hemoglobins, their arrangement to form a cooperative tetramer is radically different. In the vertebrate hemoglobin tetramer, the E and F helices are external, exposed to solvent, while the G and H helices are largely internal, involved in subunit interactions. In the clam hemoglobin tetramer the E and F helices are largely internal, involved in extensive subunit interactions while the G and H helices are largely external. In this way, the clam hemoglobin tetramer can be considered "back to front" relative to the vertebrate hemoglobin tetramer.

The clam hemoglobin tetramer is formed from two dimers with the subunits in each dimer arranged similarly to the subunits in the clam homodimeric hemoglobin. An extensive contact holding the dimer together is formed from the E-F surfaces of two subunits facing each other. The tetramer is then formed from two dimers with less extensive contacts involving the A helix, the A-B corner and the G-H corner. The two dimers are related by a molecular two-fold axis. Within each dimer a local two-fold relates the two subunits; this local dyad is not perpendicular to the molecular dyad but forms an angle of about 75°. Thus, unlike vertebrate hemoglobins, the clam tetramer does not have approximate 222 point group symmetry.

DISCUSSION

Cooperativity in hemoglobin is believed to be triggered by the changes that occur within an individual subunit upon oxygenation. In vertebrate hemoglobin there is particular interest in the FG corner (15). The greatest intrasubunit displacements found in the comparison of liganded and unliganded hemoglobin and myoglobin structures have been in the F helix; in human hemoglobin subunits, the F helix moves by about 1A upon CO binding (15) while in sperm whale myoglobin the F helix moves

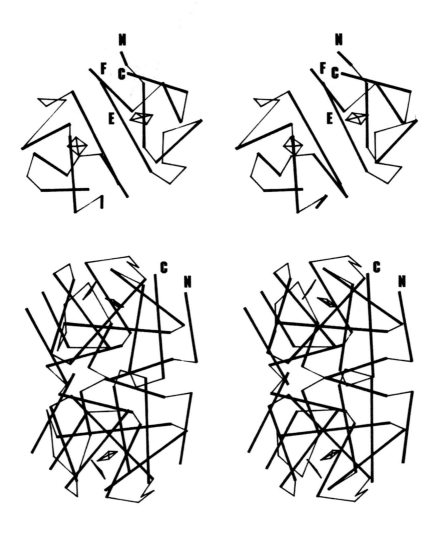

Figure 1. Stereoscopic view showing arrangement of subunits in the clam hemoglobin dimer (top) and tetramer (bottom). The pre-A, A, B, E, F, G and H helices are drawn with heavy lines. One subunit in each molecule is labelled N at the amino terminus and C at the carboxy terminus. The view of the dimer is looking down the dyad axis relating its two subunits. In the tetramer, the molecular dyad is oriented horizontally in the plane of the paper. (Reprinted from Nature 316, 278 (1985), with permission).

in the same direction upon oxygen binding but only by about 0.3A (16).
In both the clam tetrameric and dimeric hemoglobins, heme-heme communication could easily be modulated through the F helix as it is involved
in an extensive contact between subunits. We think that a deoxy
contact, similar to the contact in these liganded structures, would
"clamp" the subunits in a low affinity configuration. Upon binding of
oxygen to one subunit, movement of the F helix would disrupt the deoxy
contact releasing the clamp on the second subunit and allowing it to
attain a higher affinity configuration.

    With this reasoning, we expect that any disruption of the E-F con-
tact would result in a higher affinity species. Indeed, Furuta and
colleagues have found that binding of p-chloromercuri benzoate to cys 92
in the dimer of Anadara broughtonii leads to an 85-fold increase in
oxygen affinity (10,17). Given the position of this residue near the
beginning of the F helix, binding of this mercury compound could easily
disrupt the E-F dimer contact.

    Recently, several amino acid sequences have been determined from
subunits of both dimeric and tetrameric clam hemoglobins (10-14). The
three dimer sequences show from 90-98% homology. Comparison of the
tetramer chains with the dimer chains shows overall homologies of 42 to
53% but in the E and F regions 83 to 100% of the residues are identical.
A comparison of the two distinct chains in the tetramer from the clam
Anadara trapezia shows overall homology of 53% with 84% of the residues
in the E and F regions identical. Clearly, the residues in the E and F
helices have been conserved during the evolution of these dimeric and
tetrameric hemoglobins. From these structures of S. inaequivalvis
hemoglobins we conclude that this sequence conservation is necessary to
preserve a functional contact that is present in both dimers and tetramers and that is probably responsible for the cooperative oxygen binding
in these molecules.

    The similarity between the homodimer and half of the heterotetramer
suggests that the clam tetramer is formed from two dimers that would be
cooperative on their own. This is far different from the vertebrate
tetramer in which a cooperative tetramer is formed from two non-cooperative dimers. This difference is reflected in the extent of different
contact regions. In the vertebrate tetramer the cooperative, or "sliding", $\alpha_1\beta_2$ contacts are much less extensive than the non-cooperative, or
"packing", $\alpha_1\beta_1$ contacts. In the clam tetramer the likely cooperative
E-F contacts are much more extensive than the likely non-cooperative
contacts involving the AB and GH regions. The use of two non-cooperative dimers associated in such a way to form a cooperative tetramer

appears to be a more effective mechanism to obtain heme-heme interaction given the higher cooperativity of the vertebrate hemoglobin (n=2.8).

The radically different assemblage of these clam hemoglobins compared to vertebrate hemoglobins is consistent with the notion that the development of cooperativity occurred on evolutionarily distinct tracks and inconsistent with postulates that clam hemoglobins represent some primitive precursor to the vertebrate hemoglobins. The parallel evolution of clam and vertebrate hemoglobins offers an important illustration of how tetrameric cooperativity can be obtained in two radically different ways starting with similar myoglobin like monomers. Both probably utilize the same ligand induced changes in subunit structure, but in different ways, to trigger cooperativity.

We thank Dr. E. Chiancone for providing the hemoglobins used in this study, F. F. Fenderson for growing the first crystals of clam hemoglobins and Drs. J. L. Smith and W. A. Hendrickson for providing programs and helpful discussions. This work was supported by a grant, AM02528, from the National Institutes of Health. WER was supported in part by NIH training grant GM07231.

REFERENCES

1. Ohnoki, S., Mitomi, Y., Hata, R. and Satake, K., J. Biochem. (Tokyo) 73, 717-725 (1973).
2. Djangmah, J.S., Gabbott, P.A. and Wood, E. J., Comp. Biochem. Physiol. 60B, 245-250 (1978).
3. Furuta, H., Ohe, M. and Kajita, A., J. Biochem. (Tokyo) 82, 1723-1730 (1977).
4. Como, P. F. and Thompson, E.O.P. Aust. J. Biol. Sci. 33, 643-652 (1980).
5. Chiancone, E., Vecchini, P., Verzili, D., Ascoli, F. and Antonini, E., J. Molec. Biol. 152, 577-592 (1981).
6. Ikeda-Saito, M., Yonetari, T., Chiancone, E., Ascoli, F., Verzili, D. and Antonini, A., J. Molec. Biol. 170, 1009-1018 (1983).
7. Furuta, H. And Kajita, A., Biochemistry 22, 917-922 (1983).
8. Fisher, W. K., Gilbert, A. T., and Thompson, E.O.P., Aust. J. Biol. Sci. 37, 191-203 (1984).
9. Petruzzelli, R., Goffredo, B. M., Barra, D., Bossa, F., Boffi, A., Verzili, D. Ascoli, F. and Chiancone, E., FEBS Lett. 184, 328-332 (1985).
10. Como, P. F., and Thompson, E.O.P., Aust. J. Biol. Sci. 33, 653-664 (1980).
11. Gilbert, A. T. and Thompson, E.O.P., Aust. J. Biol. Sci., submitted.
12. Goodman, M., Moore, G. W., and Matsuda, G., Nature 253, 603-608 (1975).
13. Benesch, R. and Benesch, R. E., Science 185, 905-908 (1974).
14. Royer, W. E., Love, W. E. and Fenderson, F. F., Nature 316, 277-280 (1985).
15. Baldwin, J. and Chothia, C., J. Molec. Biol. 129, 175-220 (1979).
16. Phillips, S. E. V., J. Molec. Biol. 142, 531-554 (1980).
17. Furuta, H., Ohe, M., and Kajita, A., Biochem. Biophys. Acta 625, 318-327 (1980).

STRUCTURE AND FUNCTION OF THE DIMERIC AND TETRAMERIC HEMOGLOBINS FROM THE BIVALVE MOLLUSC, ANADARA BROUGHTONII

H. Furuta and A. Kajita
Department of Biochemistry, Dokkyo University School of Medicine,
Mibu, Tochigi 321-02, JAPAN

I. Introduction

Clams of the primitive family Arcidae (the so-called blood clam) have dimeric and tetrameric hemoglobin components in the nucleated corpuscles (1-5). A few species of the Arcidae contain homodimeric (Hb I) and $\alpha_2\beta_2$ tetrameric forms (Hb II) [Anadara broughtonii (2), Anadara trapezia (4), Scapharca inaequivalvis (5)]. Both hemoglobin components bind oxygen cooperatively, without the alkaline Bohr effect. The tetrameric hemoglobin polymerizes in the deoxygenated state (5,6). Very recently, Royer et al. (7) have determined the crystal structures of dimeric (Hb I) and tetrameric hemoglobins (Hb II) from S. inaequivalvis to a 5.5°A resolution. The subunits of both molecules assemble 'back to front', relative to mammalian hemoglobins. Thus, the studies show that certain Arcidae pigments share distinct structural and functional characteristics with each other. We previously reported the amino acid sequence of the homodimeric Hb I from A. broughtonii (8). The sequences of A. trapezia globins (9,10) have also been determined.

In this report, we focus on the blocked N-terminal structures of A. broughtonii Hb II, the sequence of α chain from A. broughtonii Hb II, and the structual differences between Hb I and Hb II.

II. Materials & Methods

Removal of the protoheme group from Hb II was carried out using acid-acetone. Each α and β globin chain was separated with a CM-cellulose column in 2-mercaptoethanol-8M urea as previously described (2). Two N-terminal blocked amino acids were extracted from exhaustive proteolytic digests (pronase, trypsin, CPase) and were purified with Dowex 50W-X2 and Dowex 1-X2 resins.

Invertebrate Oxygen Carriers
Ed. by Bernt Linzen
© Springer–Verlag Berlin Heidelberg 1986

Mass spectra on samples were recorded with a Shimadzu-LKB 9000 GC/MS system. Both samples were also analyzed by using paper-chromatography in n-butanol/acetic acid/water (4:1:1 V/V). N-Acetyl-valine was synthesized from L-valine and anhydrous acetic acid and N-acetyl-L-serine was obtained from Calbiochem.

The α globin chain separated from Hb II was modified with iodo-acetic acid or 4-vinylpyridine. The modified chain was specifically cleaved using enzymatic and chemical methods; peptides were produced through cleavage at methionine, tryptophan, lysine residues and at an aspartylproline bond. The peptides were separated by gel-filtration. The selected fragments were sequenced manually and in a JEOL JAS-47K spinningcup sequencer, according to the method of Edman & Begg (11).

The amino acid composition was determined as follows: S-modified globin and selected fragments were hydrolyzed in vacuum-sealed Pyrex tubes containing constant boiling HCl, for 24h, at 110°C. All analyses were performed on a Hitachi KLA-5 amino acid analyzer.

III. Results & Discussion

A. N-terminal blocked structures of α and β chains.

The Cm-α chain was digested with pronase and an acid peptide was extracted by using a Dowex 50W-X2 resin column. An X-valine was isolated by the subsequent carboxypeptidase digestion and purified with Dowex resin.

A tryptic peptide map of the Cm-β chain showed an unusual spot which was ninhydrin negative and Sakaguchi positive (2). The amino acid composition of the peptide extracted from the paper was $Ser_{1.1}$ and $Arg_{1.0}$. Y-Serine was isolated following CPase B digestion.

Both blocked structures (designated as X- and Y-) were determined as acetyl groups through mass-spectroscopy and paper chromatography.

Acetyl serine termini are seen in a relatively high number of acetyl blocked hemoglobins of other species. One reason can be attributed to fish hemoglobins which generally have acetyl serine in their α chains. Most of the acetyl blocking reported so far in $\alpha_2\beta_2$ tetramers has been seen in either the α or β chains but never in both. The acetylated $\alpha_2\beta_2$ structure seen in A. broughtonii Hb II is a very rare case among invertebrate and vertebrate hemoglobins.

The results of oxygen equilibrium studies of Hb II with chlor-

ide anion in concentrations ranging from 0.006 to 0.3 M showed that
HbII oxygen affinity was not strongly affected by the chloride anion.
HbII has no Bohr effect and is not affected by the organic phosphates. This suggests that there is little probability of the occurrence
of heterotropic regulation in Hb II, unlike that of mammalian hemoglobins.

B. The primary structure of Hb II α chain.

The primary structure of the Hb II α chain was determined by sequencing the peptide fragments produced by enzymatic and chemical cleavage. The Hb II α chain has 149 amino acid residues with acetyl-
valine at the NH_2 terminus and leucine at the COOH terminus.

The comparison of the Hb II α chain and the Hb I dimer (8) in
A. broughtonii is shown in Figure 1. Each chain contains only two
histidine residues, which correspond to the distal and proximal heme-
linked positions. Verzili et al. reported from their EPR studies
that two histidine residues in S. inaequivalvis Hb I are the apical
heme ligands (12). The maximum homology of the α chain with A. broughtonii Hb I is 45%. In contrast, the homology of the A. broughtonii α chain with A. trapezia α chain (9) is 82%. Comparison of the

```
                        10                20                30                40
A.b. α    AcV D A A V A K V C G S E A I K A N L R R S W G V L T A D I E A S G L M L M S N L
A.b. I        P S V Q G · A A Q L T A D V · K D · · D · · K · I G S · K K G N · V A · · T T ·


                            50                60                70                80
A.b. α    F T L R P D T K T Y F T R L G D V Q K G K A N S L R G H A I T L T Y(A L B W]F
A.b. I    · A D N Q E · I G · · K · · · N · S Q · M · · D · · · · · S · · · M · · · Q N ·


                          90               100               110               120
A.b. α    V B S L B)D P S R L K C V V E K F A V N H I N R K I S G D A F G A I I E P M K E
A.b. I    I D Q · D N T D D · V · · · · · · · · · · · · · T · · · · A A E · · K · N G · I · K


                        130             140
A.b. α    T L K A R M G N Y Y S D D V A G A W A A L V G V V Q A A L
A.b. I    V · A S K N F G D K Y A N - - - - - - · K · · A · · · · · ·
```

Figure 1. Comparison of Hb I and Hb II α chain from A. broughtonii.

two sequences from A. broughtonii and A. trapezia α chain shows that the homology in the E and F helix region is particularly high. Crystallographic studies of S. inaequivalis Hbs' by Royer et al.(7) indicated that those E and F helices are involved in subunit interfaces. This implies that the subunit interaction sites in both dimer and tetramer molecules are well-conserved.

References

1. Ohnoki, S., Mitomi, Y., Hata, R., & Satake, K. (1973) J. Biochem. 73, 717-725.

2. Furuta, H., Ohe, M., & Kajita, A. (1977) J. Biochem. 82, 1723-1730.

3. Djangmah, J.S., Gabbott, P.A., & Wood, E.J. (1978) Comp. Biochem. Physiol. B 60, 245-250.

4. Thompson, E.O.P. (1980) in The Evolution of Protein Structure and Function (Sigman, D.S., & Brazier, M.A.B., eds.), UCLA Forum No. 21, pp. 267-298, Academic Press, New York.

5. Chiancone, E., Vecchini, P., Verzili, D., Ascoli, F., & Antonini, E. (1981) J. Mol. Biol. 152, 577-592.

6. Furuta, H., Ohe, M., & Kajita, A. (1981) Biochim. Biophys. Acta 668, 448-455.

7. Royer, W.E., Love, W.E., & Fenderson, F.F., submitted.

8. Furuta, H., & Kajita, A. (1983) Biochemistry 22, 917-922.

9. Como. P.F., & Thompson, E.O.P. (1980) Aust. J. Biol. Sci. 33, 653-664.

10. Fisher, W.K., Gilbert, A.T., & Thompson, E.O.P. (1984) Aust. J. Biol. Sci. 37, 191-203.

11. Edman, P., & Begg, G. (1967) Eur. J. Biochem. 1, 80-91.

12. Verzili, D., Santucci, R., Ikeda-Saito, M., Chiancone, E., Ascoli, F., Yonetani, T., & Antonini, E. (1982) Biochim. Biophys. Acta 704, 215-220.

COSOLVENT EFFECTS ON HEMOPROTEINS' OXYGEN UPTAKE AND EQUILIBRIA

Cordone Lorenzo*, Di Stefano Lucio** and Russo Gian Carlo**
* Istituto di Fisica,via Archirafi 36 - 90123 Palermo (ITALY)
**Istituto di Istologia ed Embriologia,via Archirafi 20 - Palermo (ITALY)

INTRODUCTION

The effects of monohydric alcohols and amides on the reaction of haemoglobin (Hb) with oxygen have been reported in a series of papers appeared in the last few years (see e.g. 1,2). The data analyzed within the framework of the Monod Wyman Changeux (MWC) model (3) indicated that cosolvents affect the allosteric equilibrium of Hb through two main contributions: 1) Contributions lineary related to the inverse bulk dielectric constant of the solvent (bulk-electrostatic contributions). These are dominant at low cosolvent concentration and were thought to arise from the varied strenght of electrostatic interaction among charged groups on the protein surface (e.g. Perutz's salt bridges) that stabilize the T conformation of Hb. Indeed these contributions are positive for cosolvents that decrease the bulk dielectric constant (alcohols) and negative for those that increase it (amides). 2) Other contributions (non bulk electrostatic) become relevant by increasing cosolvent concentration; these always stabilize the Hb R conformation. The dependence of these last contributions on cosolvent concentration and alkyl group size allowed to conclude that they are related to the smaller free energy needed to expose hydrophobic surfaces to the solvent (following T→R transition) when cosolvents are present in the solution medium. This interpretation was confirmed by the fact that entropy and enthalpy values related to non bulk electrostatic contributions exhibit features typical of hydrophobic interactions (7).

In this paper we present preliminary results on the effects of the above cosolvents on the oxygen uptake of Spirographis spallanzanii chlorocruorin (Chl;ca. 3×10^6 DA).

MATERIAL AND METHODS

Chl was prepared as described by Di Stefano et al.(4). Oxygen uptake measurements were performed as already described (5). The percentage saturation of Chl with oxygen was determined by reading absorbance values at 605,577 and 490 nm (6). Samples contained K-Phosphate buffer 0.1 M, pH 7.4 (in water at 20°C). Oxygen uptake experiments were performed at 20°C using a Perkin-Elmer 550 S spectrophotometer. Methanol,ethanol,n-propanol and formamide (Carlo Erba RP-grade) were used without further purification. As already reported (2) the alcohols do not affect pH values within concentrationsused in our experiments; on the contrary formamide has a quite sizeable effect on the pH of our buffered solutions. In this paper P50 values obtained in the presence of formamide have not been corrected for Bohr effect due to pH variation.

RESULTS AND DISCUSSION

Fig.1 shows Log P50 as a function of cosolvent concentration. P50 in the absence of cosolvent was 39.5 mm Hg. As it can be seen from fig.1 the presence of alcohols causes a P50 increase while the presence of formamide makes P50 to decrease. This is in full analogy with findings already reported for human Hb (2,7).

Invertebrate Oxygen Carriers
Ed. by Bernt Linzen
© Springer-Verlag Berlin Heidelberg 1986

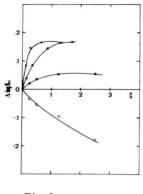

Fig.1

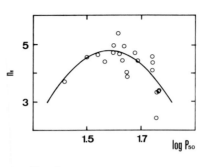

Fig.2

Fig.1 – Log P50 vs. cosolvent concentration. ● methanol; ▲ ethanol; ★ n-propanol; ⬤ formamide.

Fig.2 – Hill's constant at 50% saturation (n_H) vs. Log P50

Fig.2 shows n_H (Hill constant at 50% saturation) as a function of Log P50. As it can be seen points in this fig. fit a bell shaped curve. This behaviour is predicted (8) by the MWC model when K_R and K_T are not affected by the agents that make P50 to vary. Analogous behaviour has been reported for the effects of these cosolvents on human Hb (2,1). Data in fig.2 therefore indicate that MWC model (considering K_R not affected) can be safely applied to the presently reported data.

According to MWC model one has for human Hb:

$$P50 = L^{1/4} K_R \qquad (1)$$

Since for Chl the number of subunits is not known, we shall write for this haemoprotein, according to equation (1):

$$P50 = L^{1/n} K_R \qquad (2)$$

where n indicates the unknown number of subunits.

From equation (2) considering K_R not affected by cosolvents one has:

$$1/n \, RT \, \ln \frac{P50(C)}{P50(O)} = RT \, \ln \frac{L(C)}{L(O)} = \Delta G(C) - \Delta G(O) = \Delta\Delta G \qquad (3)$$

Here $\Delta G(C)$ is the free energy difference between R and T states in the presence of a concentration C of cosolvent. Since n value is not known, the reported $\Delta\Delta G$ are expressed in arbitrary units.

In fig.3 we report $\Delta\Delta G$ values vs. the variation of the inverse of the bulk dielectric constant of the solvent medium. Data in fig.3 are very similar to the analogous ones reported for human Hb when no correction for pH effect due to the presence of amides is performed (1) i.e. a single straight line fits data points to all the alcohols,at low cosolvent concentration,while a different straight line fits data points relative to formamide. For human Hb correction for pH effect makes a single straight line to fit points at low cosolvent concentration relative both to alcohols and formamide (2). The features of data in fig.3 indicate that contributions linearly related to the bulk dielectric constant of the medium are present; these can be expressed as

$$\Delta\Delta G_{es} = A \ (1/\varepsilon) \qquad\qquad (4)$$

In analogy with what is found for human Hb this fact suggests that following the T→R transition salt bridges on the protein surfaces are broken.

Data in fig.3 show that by increasing alcohol concentration other contributions become relevant. These can be expressed as

$$\Delta\Delta G_{nbes} = \Delta\Delta G - \Delta\Delta G_{bes} \qquad\qquad (5)$$

These last contributions are reported in fig.4 and are in extremely good agreement with analogous ones reported for human Hb; indeed they increase both by increasing alcohol concentration and alkyl group size. The presence of these contributions that for human Hb act as to stabilize R conformation indicates that also for Chl, hydrophobic surfaces are exposed to the solvent following the transition to the conformation of higher oxygen affinity.

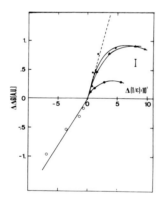

Fig.3

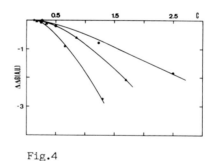

Fig.4

Fig.3 – $\Delta\Delta G$(arbitrary units) vs. the inverse bulk dielectric constant at the sol-medium. Symbols as in Fig.1

Fig.4 – Non bulk electrostatic contributions to $\Delta\Delta G$ (arbitrary units) vs. cosolvent concentration. Symbols as in Fig.1

The autors wish to thank prof. A.Cupane for useful discussion.

This work was supported by contribution of Ministero della Pubblica Istruzione.

REFERENCES
1. Cordone L., Cupane A., San Biagio P.L., Vitrano E.
   Biopolymers 20, 53-63 (1981a).
2. Vitrano E., Cupane A., Cordone L.
   J.Mol.Biol. 180, 1157-1171 (1984).
3. Monod J., Wyman J., Changeux J.P.
   J.Mol.Biol. 12, 88-118 (1965).
4. Di Stefano L., Mezzasalma V., Piazzese S., Russo G.C., Salvato B.
   FEBS LETTERS 79, 337-339 (1977).
5. Cordone L., Cupane A., San Biagio P.L., Vitrano E.
   Biopolymers 18, 1975-1988 (1979).
6. Benesch R., McDuff G., Benesch R.E.
   Anal.Biochem. 11, 81-87 (1965).
7. Cordone L., Cupane A., San Biagio P.L., Vitrano E.
   Biopolymers 20, 39-51 (1981b).
8. Baldwin J.M.
   Prog.Biophys.Mol.Biol. 29, 225-320 (1975).

HEMOGLOBIN FROM THE PARASITIC BARNACLE, BRIAROSACCUS CALLOSUS

Robert C. Terwilliger, Nora B. Terwilliger and Eric Schabtach
Department of Biology, University of Oregon
Eugene, OR 97403, USA
Oregon Institute of Marine Biology
Charleston, OR 97420, USA

Hemoglobin occurs widely amongst the invertebrates. Its presence in the largest of the animal phyla, the Arthropoda, is curiously limited; it is found in only a few insects and in four subgroups of crustaceans, the Branchiopoda, Ostracoda, Copepoda and Cirripedia (2). Branchiopod hemoglobin, a large extracellular protein, has been extensively studied, both structurally and functionally (3,4,5). No comparable studies describing hemoglobins from the other crustacean groups are available. We are, therefore, investigating an extracellular hemoglobin from the rhizocephalan cirriped, Briarosaccus callosus, a barnacle which is parasitic on the red king crab, and we report some preliminary observations of this unusual pigment.

Blood samples (2-5 ml) were obtained from Briarosaccus callosus (Boschma) which is parasitic on the red king crab Paralithodes camtschatica (Tilesius). The blood was removed from the stalk of the parasite's externa by syringe and immediately centrifuged at 10,000g for 10 min ($5^{o}C$). The supernatant which contained about 5 mg hemoglobin/ml (6) was analysed by column chromatography, gel electrophoresis, and electron microscopy (7).

A freshly obtained, unpurified sample of Briarosaccus blood shows a typical oxyhemoglobin spectrum (Table I).

Table I. Absorption maxima of Briarosaccus callosus hemoglobin

ligand stage	λ^{γ} max (nm)	λ^{β} max (nm)		λ^{α} max (nm)
oxy	417	541		578
deoxy	432		555	
carboxy	423	541		572

The ratio of absorbance at 578 to 541 was about 1.04, indicating that the sample contains little methemoglobin. The hemoglobin is readily deoxygenated to a typical deoxyhemoglobin spectrum with a broad absorbance maximum near 555 nm and a sharp peak at 432 nm, and it binds oxygen reversibly. The carboxy-spectrum is also typical of other hemoglobins.

Chromatography of a sample of blood, clarified by centrifugation, on a column of BioGel A-5m in equilibrium with 0.05 I Tris-HCl buffer (pH 7.5), 0.1 M NaCl, 0.01 M $CaCl_2$, 0.05 M $MgCl_2$, showed a broad asymmetric peak of heme-containing pro-

Invertebrate Oxygen Carriers
Ed. by Bernt Linzen
© Springer–Verlag Berlin Heidelberg 1986

tein as well as some other 280 nm absorbing material (Fig. 1). In this and exper-

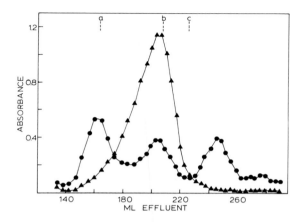

Fig. 1. Chromatography of <u>Briarosaccus callosus</u> hemoglobin on a 1.8 x 109 cm
column of BioGel A-5m (5°C). Absorbance at 280 (●) and 417 (▲) nm.

iments on blood samples from other individual barnacles, the hemoglobin, as identified by absorbance at 417 nm, ranges in apparent molecular weight from about 250,000
to greater than 4×10^6. Earthworm hemoglobin (3.8×10^6 molecular weight) and
<u>Cancer magister</u> hemocyanin (900,000 molecular weight) each give a sharp peak when
chromatographed on the same column. The broad, asymmetric peak of barnacle hemoglobin suggests that the hemoglobin exists as molecules of different sizes. Absorption spectrum measurements of aliquots at different elution positions indicate
that there is no measurable difference in methemoglobin content in the different
molecular weight fractions. Thus the size heterogeneity is not a result of methemoglobin formation.

Gel electrophoresis of <u>Briarosaccus</u> whole blood, which had been clarified by
centrifugation at 10,000g (5°C) to remove debris, was carried out in the absence

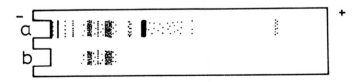

Fig. 2. Regular gel electrophoresis of <u>B. callosus</u> whole blood (a) and purified
hemoglobin (b) with hemoglobin in the oxy state, on a 7.5% polyacrylamide
gel, pH 8.9.

of denaturants and is shown in Fig. 2. The electrophoresis pattern, like the chromatography, shows hemoglobin heterogeneity as well as the presence of other proteins. Electrophoretic heterogeneity has also been demonstrated by Shirley et al. (8).

Negatively stained samples of fresh blood and purified hemoglobin were examined in the electron microscope. The results are coincident with the apparent heterogeneity seen by column chromatography and gel electrophoresis. Molecules which have been identified as hemoglobin appear in various sizes suggesting different polymerization states. Other protein aggregates are also present.

SDS gel electrophoresis of the purified hemoglobin results in two polypeptide chains with molecular weights of 17,000 and 19,000. Similar hemoglobin subunit molecular weights are obtained when fresh blood is added directly to boiling SDS solutions in the presence of reducing agent and proteolytic inhibitors.

The hemoglobin of this parasitic barnacle shows a quaternary structure that is not seen in any other animal group. Furthermore the hemoglobin is strikingly unlike the other arthropod extracellular hemoglobins which have been studied, not only with respect to its quaternary structure, but also its small subunit molecular weight.

Acknowledgements

We are grateful to Dr. Tom and Susan Shirley, School of Fisheries and Science, University of Alaska, Juneau, for their suggestion that we work on this project and for supplying us with animals. The study was supported in part by NSF grant PCM 82-07548 and Grant # RC-84-04C from the Fisheries Research Center of the University of Alaska.

References

1. Terwilliger, R.C., Amer. Zool. 20, 53-67 (1980).
2. Fox, H.M., Nature 179, 148 (1957).
3. Ilan, E. and Daniel, E., Comp. Biochem. Physiol. 633, 303-308 (1979).
4. Dangott, L.J. and Terwilliger, R.C., Comp. Biochem. Physiol. 70B, 549-557 (1981).
5. Wolf, G., Van Pachtenbeke, M., Moens, L. and van Hauwaert, M., Comp. Biochem. Physiol. 76B, 731-736 (1983).
6. Shirley, S. and Shirley, T., American Zool. 24, 121A (1984).
7. Terwilliger, N.B. and Terwilliger, R.C., J. exp. Zool. 221, 181-191 (1982).
8. Shirley, S.M., Shirley, T.C. and Meyers, T.R., Proc. Internat'l. King Crab Symposium, Lowell Wakefield Symposia Series (1985) in press.

SULFIDE-BINDING BY AN EXTRACELLULAR HEMOGLOBIN

Alissa J. Arp
University of California, San Diego
Scripps Institution of Oceanography, A-002
La Jolla, California, USA, 92093

The blood of the deep-sea, hydrothermal vent tube worm Riftia pachyptila contains a large, extracellular hemoglobin (Hb) that binds sulfide with a high affinity and capacity both in vitro and in vivo (Arp and Childress, 1981; Arp and Childress, 1983). The animal appears to utilize this binding ability to transport sulfide to internal bacterial symbionts that use it as an energy source. My recent work indicates that this sulfide-binding protein is the Hb molecule itself and that sulfide-binding is a nonheme phenomenon that may occur at disulfide bridges that are the forces holding the submultiples of the giant molecule together. Sulfide-binding capacity may be dependent on the assembly state of this molecule, and assembly state appears to differ between the two Hb-containing fluids in this animal.

The deep-sea, hydrothermal vent tube worm Riftia pachyptila has tentatively been placed in the phylum Pogonophora (Jones; personal communication), a phylum of worm-like animals that have many characteristics in common with the annelids. The blood of R. pachyptila has an extracellular, high molecular weight hemoglobin (Hb) that has several characteristics in common with annelid extracellular, high molecular weight Hbs. These include: intact molecular weights of 1,700,000 to 3,000,000; a two tiered, hexagonal array of submultiples; subunit molecular weights of 15,000 and 30,000; a heme content of one heme per 23,000 g protein; and a similar amino acid composition of purified fractions.

Riftia pachyptila has two discrete blood pools that both contain extracellular Hb. The circulating vascular blood has an average heme concentration of 3.5 mM and constitutes approximately 4% of the weight of the animal, whereas the coelomic fluid has an average heme concentration of 1.9 mM and constitutes approximately 26% of the weight of the animal (Childress et al., 1984). Both of these Hb-containing bloods possess proteins that bind sulfide with a high affinity, and have sulfide-binding capacities that vary from 0.4 to 2.0 mmol sulfide per mmol heme (Table 1). All of the protein in the vascular blood and coelomic fluid detectable by polyacrylamide gel electrophoresis (PAGE) appears to be Hb with comparative heme and

Invertebrate Oxygen Carriers
Ed. by Bernt Linzen
© Springer–Verlag Berlin Heidelberg 1986

protein staining of nondenaturing PAGE, and comparative spectrophotometric analysis of gel filtration products show parallel elution profiles when examined at a heme absorbance wavelength or at a protein
absorbance wavelength (Figure 1). Further, Hb purified on gel filtration columns shows a sulfide-binding capacity comparable to the whole
blood (Table 2). These data suggest that the sulfide-binding protein
in the blood and coelomic fluid of this animal is the extracellular
Hb molecule itself.

Table 1. Comparative characteristics of fresh coelomic fluid and
vascular blood (C = coelomic fluid; V = vascular blood).

Animal number	% FA (1,500,000)	% FB (400,000)	mM heme	mmol sulfide / mmol heme
16-1 C	55	45	1.20	0.97
16-1 V	77	23	4.07	1.93
16-2 C	29	71	1.48	1.04
16-2 V	80	20	4.00	2.00
19-1 C	0	100	0.47	(no data)
19-1 V	52	48	2.98	1.69 FA only
30-16 C	17	83	1.64	0.43 FB only
30-16 V	69	31	3.99	1.02 FA only

Table 2. Gel filtration products of coelomic fluid (MW = molecular
weight; C-1 and C-2 are coelomic fluid samples from separate animals).

Gel filtration fraction	MW	mmol sulfide / mmol heme	disulfide bridge / mmol heme
A. Riftia C-1			
FA	1,500,000	1.58 + .14 (3)	(no data)
FB	400,000	.78 + .12 (3)	(no data)
FC	50,000	0	(no data)
B. Riftia C-2			
FA	1,500,000	1.83	1.0
FB	400,000	.90	.5

Experimentation with metHb and HbCO formation indicates that the
heme, or the oxygen binding site on the heme, is not involved with
sulfide-binding. OxyHb, deoxyHb, MetHb and COHb all show similar
sulfide-binding capacities (Table 3). Sulfide-binding by the blood
does not affect oxygen-binding either, and the visible spectrum of the
Hb is not altered in the presence of sulfide (Childress et al., 1984).

Table 3. Effect of HbCO and MetHb formation on sulfide-binding (CO = carbon monoxide; PF = potassium ferricyanide).

Sample	mM CO	mM PF	mmol sulfide/ mmol heme
Riftia C	0.14/1.07	0	1.10/1.15
Riftia V	0.10/1.04	0	1.09/1.04
Riftia C-1	0	0/1.5	1.21/1.23
Riftia C-2	0	0/10.0	1.04/1.04

The extracellular Hb of Riftia pachyptila blood shows a considerable amount of heterogeneity in both the vascular blood as well as the coelomic fluid. The vascular Hb of R. pachyptila contains predominantly one peak on a Sepharose 4B gel filtration column that corresponds to an apparent molecular weight of 1,500,000 (FA), but contains variable amounts of a lower molecular weight fraction of approximately 400,000 (FB) also. The coelomic fluid, however, contains predominantly the lower molecular weight fraction (FB) but has varying quantities of the larger molecular weight fraction (FA) also. Heterogeneity varies from 0 to 50% and occurs between animals and between samples taken from the same animal (Table 1; Figure 1). The question of whether the heterogeneity in the two blood types is a naturally occurring phenomenon, or if it is present due to intermingling of blood samples during collection remains.

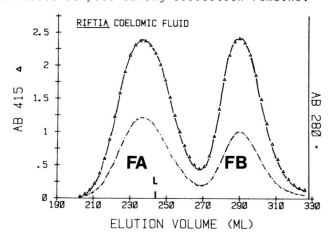

Figure 1. Elution profile of Riftia pachyptila coelomic fluid on a gel filtration column (Sepharose 4B). FA has an approximate molecular weight of 1,500,000 and FB has an approximate molecular weight of 400,000. Triangles are absorbance at 415 nm and dots are absorbance at 280 nm. Calibrant L = Lumbricus sp. Hb.

The two fractions resulting from gel filtration of R. pachyptila coelomic fluid have different sulfide-binding capacities. The lower molecular weight fraction (FB: 400,000) consistently has half the sulfide-binding capacity of the higher molecular weight fraction (FA: 1,500,000) and dissociation into a 50,000 molecular weight species causes a loss of binding ability altogether (Table 2A). Data on disulfide bond location correlate with the sulfide-binding data. The 1,500,000 molecular weight aggregates have twice as many easily reduceable disulfides as the 400,000 molecular weight fraction - the exact proportion of sulfide-binding between the two fractions (Vetter and Arp, unpublished; Table 2B).

Sulfide-binding by the whole blood is variable in both the vascular blood and coelomic fluid (sulfide-binding varies from 1.0 to 2.0 mmols sulfide per mmol heme for the vascular blood and from 0.4 to 1.0 mmols sulfide per mmol heme in the coelomic fluid; Table 1). This variability in binding capacity may be due to different amounts of these two Hb species being present in the individual blood samples. As my data indicate, the vascular blood contains predominantly the higher molecular weight Hb species (FA) and has a higher sulfide-binding capacity, whereas the coelomic blood contains predominantly the lower molecular weight Hb species (FB) and shows a lower sulfide-binding capacity. Thus it appears that sulfide-binding capacity may vary with the assembly state of the Hb molecule present.

In conclusion, the data to date indicate that the sulfide-binding protein present in the blood of Riftia pachyptila is the Hb molecule, and that sulfide-binding does not occur at the oxygen-binding site of the heme portion of this molecule, but may occur at disulfide bridges that hold the submultiples of the molecule together. There appear to be two distinct Hb species in this worm, a larger 1,500,000 molecular weight Hb that occurs in the vascular blood, and a smaller 400,000 molecular weight Hb that occurs in the coelomic fluid. This conclusion is supported by the facts that: a large amount of variability in distribution of the two types of Hb is witnessed in the coelomic fluid and the vascular blood, and that there is considerable variability in heme content and sulfide-binding capacity for both blood types as well.

Literature

1. Arp, A.J., and J.J. Childress, Science 213, 342-344 (1981).
2. Arp, A.J., and J.J. Childress, Science 219, 295-297 (1983).
3. Childress, J.J., A.J. Arp, and C.R. Fisher Jr., Mar. Biol. 83, 109-124 (1984).

Structure of arthropodan hemocyanins

THREE-DIMENSIONAL STRUCTURE OF HAEMOCYANIN FROM THE SPINY LOBSTER,
PANULIRUS INTERRUPTUS, AT 3.2 Å RESOLUTION

Anne Volbeda and Wim G.J. Hol
Laboratory of Chemical Physics
Nijenborgh 16
9747 AG Groningen
The Netherlands

1. Introduction

Haemocyanins are the non-haem, copper-containing oxygen
transporting molecules occurring freely dissolved in the haemolymph
of a large number of invertebrate species. The molecular architectures
of the two known classes of haemocyanins are entirely different.
Molluscan haemocyanins have the form of cylinders, with 10-20 subunits forming the complete molecules with molecular weights up to
about ten million daltons. The subunits are made up by maximally 8
"repeated" domains, each of which has one dinuclear copper site.
Arthropodan haemocyanins are composed of hexamers, or multi-hexamers,
with individual subunits having molecular weights in the order of
75.000 daltons, each subunit containing one pair of copper ions.
Complete molecules range from single hexamers of ~ 460.000 daltons to
octa-hexamers with molecular weights of something like 3.7 million.
All haemocyanins are thus large molecules, but some are larger than
others (1-3).
In spite of the considerable difference in overall architecture,
significant similarities exist between the spectroscopic features of
the dinuclear copper sites occurring in molluscan and arthropodan
haemocyanins (see reviews 2 and 3 for further references). It may
therefore well be that the dinuclear copper-containing domains of
the molluscan haemocyanins resemble those of their arthropodan
counterparts to a considerable extent, inspite of the fact that these
domains have a molecular weight of ~ 50.000 which is considerably
less than the 75.000 for the arthropodan subunits (2,3). The

Invertebrate Oxygen Carriers
Ed. by Bernt Linzen
© Springer–Verlag Berlin Heidelberg 1986

relationship, if any, between the polypeptide foldings of molluscan and arthropodan haemocyanins remains a most interesting question which may perhaps be resolved to some degree in the near future when amino acid sequence information of molluscan haemocyanins becomes available.

In our laboratory, the three-dimensional structure of the haemocyanin from the spiny lobster, *Panulirus interruptus*, has been determined by means of X-ray diffraction techniques to a resolution of 3.2 Å (4-7). *Panulirus interruptus* haemocyanin is a single-hexameric molecule and therefore one of the smallest haemocyanins known with, still, a respectable molecular weight of ~ 470.000 daltons. Some of its characteristics are given in Table 1, which shows that this

Table 1   Characteristics of *Panulirus interruptus* haemocyanin

Molecular weight	6 x 77.000
Residues per subunit	657
Copper ions per subunit	2
Carbohydrate moieties per subunit	1
No. of disulfide bridges per subunit	3
Subunit types native Hc	*a*, *b* & *c*
Amino acid sequence difference *a* vs *b*	~ 3%

haemocyanin is a metallo-glycoprotein. The function of its carbohydrate chain is unknown and several haemocyanins from arthropodan origin have been reported which contain no carbohydrate. It is therefore likely that the sugar chain plays no role in the oxygen transport process.

Like many other arthropodan haemocyanins, *Panulirus* haemocyanin displays subunit heterogeneity (Table 1). In several multi-hexameric haemocyanins this subunit variation has been linked with the formation of precise connections between hexamers (see refs. 2 and 3 for references). As *Panulirus* haemocyanin is, however, single-hexameric, this functional explanation is not valid for this case. One possibility is that we are dealing with an evolutionary relic i.e. *Panulirus* has lost its capacity to form multi-hexamers, but still displays a non-functional diversity in subunits. Another possibility is that the different subunits are used in connection with fine-tuning the oxygen-transport properties of the haemolymph, but in absence of further information it is probably best to refrain here from speculations.

Electron-microscopy provided evidence for a "standard" hexameric building block for all arthropodan haemocyanins (1-3). This suggestion has received recently magnificent support from amino acid sequence comparison studies (6,8). Consequently, the general features of single-hexameric *Panulirus* haemocyanin are to be encountered in all arthropodan haemocyanins. Let us therefore see how this molecule actually looks like.

2. X-ray structure determination

For the X-ray diffraction studies, monoclinic crystals were grown from native *Panulirus interruptus* haemocyanin solutions, containing three subunit types (4,5,9). They diffracted reasonably well to ~ 3.2 Å resolution, once they were mounted in a capillary entirely filled with buffer and 4% gelatin, and kept at 4°C during data collection. Upon dissolving some of the crystals, SDS gel electrophoresis revealed that they contained only subunits a and b, in roughly equal amounts.

The central "phase problem" in crystallography was solved by collecting data of two heavy atom derivatives out to 4 Å resolution. They gave good phase information only to about 4.5 Å but the isomorphous replacement phases could be greatly improved by using the fact that the entire hexamer occurs in the asymmetric unit. Sixfold averaging of the electron density, employing a suite of computer programs kindly made available by Dr. Gérard Bricogne, proved to increase the quality of the 4 Å electron density distribution to such an extent that the site of both coppers could be established unequivocally (5). Encouraged by this success, the density averaging procedure was also used to obtain phases beyond 4 Å - where no isomorphous phase information was available. After numerous cycles of averaging, a 3.2 Å map was obtained which was of a surprisingly good quality. It enabled us to build a model of a complete subunit using amino acid sequence information which became available for subunit a from the group of Dr. Beintema (Biochemistry Department, University of Groningen). Some crystallographic information is given in Table 2. Detailed accounts of the structure determination will be published elsewhere (7,10).

Table 2 Structure determination of *Panulirus interruptus* haemocyanin

pH	4.5
Buffer	0.01 M acetate
Temperature	4°C
Space group	P2$_1$
Cell dimensions	119.8 x 193.1 x 122.2 Å; β=118.1°
Daltons per asymmetric unit	~ 470.000
Subunit types in crystal	*a* & *b*
No. of sites of Pt-derivative	36
No. of sites of Hg-derivative	70
No. of refls with MIR phases	32721
No. of refls with only MR phases	31121
Total no. of phased reflections	63842
No. of atoms per hexamer	~ 32.000

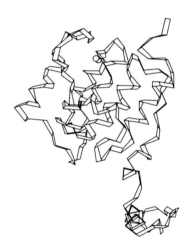 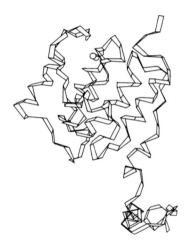

HEMOCYANIN DOMAIN 1   HEMOCYANIN DOMAIN 1

Figure 1 Stereo diagram of domain 1 of *Panulirus interruptus*
haemocyanin. The open spheres indicate the disulphide
bridge. This picture as well as the other figures in this
article were produced by using a computer program written
by Lesk and Hardman (11).

3. The three-dimensional structure of a single subunit

The ~ 657 residues of a single subunit are folded into three
distinct domains (Figs. 1-4). The first domain comprises ~ 175
residues of which the first three are invisible in the electron
density distribution, probably because they are rather mobile. This
domain can be subdivided into two parts: (i) a larger globular part
containing six α-helices and one disulphide bridge; and, (ii) a small
"appendix" formed by one α-helix and one β-strand. The single carbohydrate moiety is attached to asparagine 167 which is part of the C-
terminal β-strand. At least three sugar residues are visible in the
electron density map indicating a relative immobility of the carbohydrate chain.

Domain 2 contains ~ 220 residues and, like domain 1, is mainly
helical. Its shape is quite globular (Fig. 2). The dinuclear copper

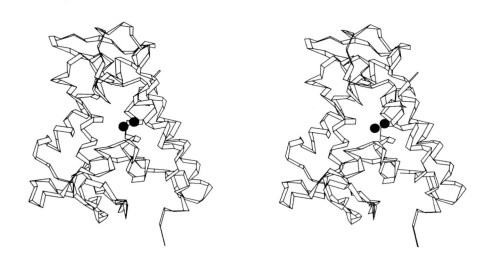

HEMOCYANIN DOMAIN 2   HEMOCYANIN DOMAIN 2

Figure 2   Stereo diagram of domain 2, viewed from the same direction
           as domain 1 in Fig. 1. The coppers are depicted by two
           black spheres.

site resides, surrounded by four helices, close to the centre of
this domain. Three other helices occur in addition to these four
"copper binding helices" whereas there are also two two-stranded
twisted anti-parallel β-sheets. Domain 2 makes extensive contacts
with domains 1 and 3, much more so than domains 1 and 3 do with each
other. The oxygen-binding domain is thus truly the central domain of
each subunit.

The third domain is the largest of the three- it contains ~ 260
amino acid residues. Its shape is not globular at all, as can be seen
in Fig. 3. The core of this domain is formed by a seven-stranded
anti-parallel β-barrel, comprising only ~ 60 residues. The remaining
200 residues occur in several long loops which contain a total of
six α-helices. The first and most extended loop contains two anti-
parallel β-strands which form a three-stranded β-sheet with the β-
strand located in the "appendix" of domain 1. Two disulphide bridges
can be found in domain 3, the first one linking together residues
483 and 502, the second one residues 562 and 609.

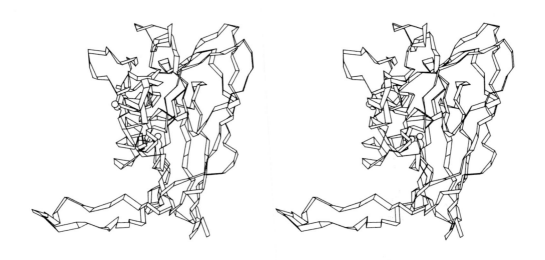

HEMOCYANIN DOMAIN 3   HEMOCYANIN DOMAIN 3

Figure 3   Stereo diagram of domain 3. The direction of view is the
           same as in Figs. 1, 2 and 4. Disulphide bridges are indica-
           ted by open spheres.

Three segments are less well defined in the electron density map than the remaining parts: (i) the three C-terminal residues; (ii) residues 548-560 in the second long loop extending from the β-barrel; and (iii) residues 597-607. Probably these segments are quite flexible. Together with the 3 N-terminal residues, there are thus four segments with great mobility in the hexamer. All of them are fully accessible to solvent. In view of the emerging evidence that high mobility may be related to good immunogenicity (12,13), one would expect that antibodies raised against peptides corresponding with these flexible regions in the *Panulirus interruptus a* and *b* subunits would bind excellently with complete subunits as well as complete hexamers. These four segments might be major antigenic determinants.

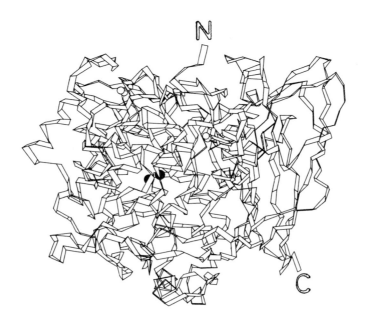

Figure 4   Ribbon diagram of one subunit of *Panulirus interruptus* haemocyanin

4. Structure of the hexamer

The point group symmetry of the hexamer has been established by rotation function studies at 10 Å and 5 Å resolution (4,10). The

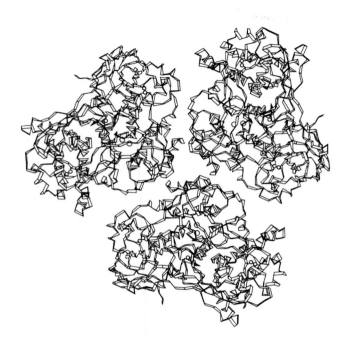

Figure 5 A drawing of a trimer of subunits, viewed along the molecular three-fold axis. Open spheres represent copper atoms

point group appeared to be "32" with the molecular three-fold axis
intersecting three molecular twofolds in the centre of the molecule.
The 3.2 Å electron density distribution showed that the six **subunits**
form "two layers", each layer formed by a trimer as depicted in Fig.
5. It appeared that two subunits, one from the "upper" trimer and one
from the "lower" trimer, related by a molecular twofold, have quite
extensive interactions. In fact the inter-subunit interactions within
these dimers are more numerous than between two subunits of the
trimers - as can be deduced to some extent from comparing Figures 5
and 6, and which has been quantified by Gaykema et al. (7). It is
therefore more appropriate to consider the arthropodan haemocyanin
hexamer as a trimer of dimers than as a dimer of trimers.

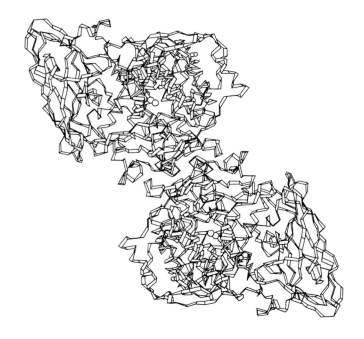

Figure 6   Drawing of a dimer of subunits viewed along a molecular
          twofold. Open spheres represent copper atoms.

5. The oxygen binding site

In our current haemocyanin model, the copper to copper distance
is 3.7 ± 0.3 Å. This is in quite good agreement with recent EXAFS results (14) which gave distances of ~ 3.45 Å for deoxy haemocyanin,
~ 3.6 Å for oxy-haemocyanin and ~ 3.4 Å for metaquo haemocyanin.
Similar distances between the coppers have been reported by Brown
et al. (15).

The 3.2 Å electron density map clearly shows that six histidines
are involved in copper coordination: Cu(A) is interacting with His-
194, His-198 and His-224 and Cu(B) with His-344, His-348 and His-384
(Fig. 7). Comparison of seven arthropodan haemocyanin sequences (8)
shows that these six residues are strictly conserved, as expected
for amino acids crucial for the proper functioning of a protein.
Spectroscopic studies have been ambiguous as to the number of residues coordinating each copper ion (2,3,14,15). The X-ray results

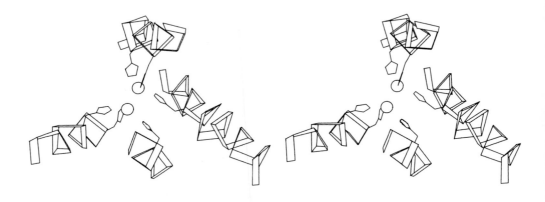

HEMOCYANIN COPPER SITE HEMOCYANIN COPPER SITE

Figure 7   The dinuclear copper site of *Panulirus* haemocyanin. The
            spheres represent copper ions. The six coordinating histi-
            dines are provided by four helices.

agree well with studies on cobalt-substituted haemocyanins (16,17)
as well as with the EXAFS results obtained for deoxy-haemocyanin by
Brown et al. (15).

     The crystals used for the X-ray studies were colourless and re-
vealed no absorption maximum at 340 nm when investigated by a single-
crystal spectrophotometer (M. Vincent, Biozentrum, Basle & WGJH, unpublished results). Consequently, the crystals contain the deoxy or
met form of haemocyanin.

     The crystallographic studies at 3.2 Å do, sofar, not allow
statements regarding the presence or absence of a small ligand, such
as a μ-hydroxo group, bridging the copper ions. The electron density
distribution does not show any evidence for an amino acid side chain
functioning as a bridging ligand. As tyrosines have frequently been
proposed to perform such a function (2,3) it may be of interest to
report that the tyrosine hydroxyl group nearest to the dinuclear
copper site is still ~ 10 Å removed from the copper centre. The
shortest distances of a conserved tyrosine (7) to the copper centre
is ~ 17 Å. Consequently, it seems unlikely that tyrosine will assume

a role in bridging the copper ions in the oxy as well as in the deoxy state of arthropodan haemocyanins.

6. Conclusions

The "standard" hexameric building block of arthropodan haemocyanins contains six kidney shape subunits with three distinct domains per subunit. The second domain contains near its centre the dinuclear copper site where oxygen is reversible and cooperatively bound. Each copper is surrounded by three histidine residues whereas no evidence for a protein side chain functioning as a bridging ligand can be obtained from the X-ray map. Due to the considerable distances of tyrosine hydroxyls to the oxygen binding centre, the chances that a tyrosine will be a bridging ligand in either deoxy or oxy haemocyanins from arthropods seem remote.

Panulirus haemocyanin contains three disulfide bridges and one well-defined carbohydrate chain attached to an asparagine from a β-strand in the first domain. The third domain has three rather mobile regions which, together with the flexible three N-terminal residues, may be major antigenic determinants of this large molecule.

Numerous questions concerning the structure and function of haemocyanins remain still to be settled, but it is intriguing to see how different the three oxygen-transporting molecules with known three-dimensional structure are with regards to their architecture, while their functions are so very similar.

Acknowledgements

We like to thank Lies van Schaick, Wilma Schutter and Wil Gaykema for their numerous contributions to this X-ray investigation. The close cooperation and exchange of information with the amino acid sequence group of Prof. Beintema has been very stimulating indeed. We are also indebted to Profs. J. Drenth and E.F.J. van Bruggen for their continuous interest and encouragement. This research was supported by the Dutch Foundation for Chemical Research (SON) with financial aid from the Dutch Organisation for the Advancement of

Pure Research (ZWO).

References

1. Van Bruggen, E.F.J., Schutter, W.G., Van Breemen, J.F.L., Bijlholt, M.M.C. and Wichertjes, T. in "Electron Microscopy of Proteins", M. Harris (ed.), Academic Press, New York, pp. 1-37 (1981).

2. Van Holde, K.E. and Miller, K.I., Q. Rev. Biophys. 15, 1-129 (1982).

3. Ellerton, H.D., Ellerton, N.F. and Robinson, H.A., Progr. Biophys. Mol. Biol. 41, 143-248 (1983).

4. Van Schaick, E.J.M., Schutter, W.G., Gaykema, W.P.J., Schepman, A.M.H. and Hol, W.G.J.,J. Mol. Biol. 158, 457-485 (1982).

5. Gaykema, W.P.J., Van Schaick, E.J.M., Schutter, W.G. and Hol, W.G.J., Chemica Scripta 21, 19-23 (1983).

6. Gaykema, W.P.J., Hol, W.G.J., Vereijken, J.M., Soeter, N.M., Bak, H.J. and Beintema, J.J., Nature 309, 23-29 (1984).

7. Gaykema, W.P.J., Volbeda, A. and Hol, W.G.J., submitted for publication.

8. Linzen, B., Soeter, N.M., Riggs, A.F., Schneider, H.-J., Schartau, W., Moore, M.D., Yokota, E., Behrens, P.Q., Nakashima, H., Takagi, T., Nemoto, T., Vereijken, J.M., Bak, H.J., Beintema, J.J., Volbeda, A., Gaykema, W.P.J. and Hol, W.G.J., Science 229, 519-524 (1985).

9. Kuiper, H.A., Gaastra, W,, Beintema, J.J., Van Bruggen, E.F.J., Schepman, A.M.H. and Drenth, J., J. Mol. Biol. 99, 619-629 (1975).

10. Hol, W.G.J., Volbeda, A. and Gaykema, W.P.J., Proceedings of the Daresbury Meeting on Molecular Replacement (1985) in press.

11. Lesk, A.M. and Hardman, K.D., Science 216, 539-540 (1982).

12. Westhof, E., Altschuh, D., Moras, D., Bloomer, A.C., Mondragon, A., Klug, A. and Van Regenmortel, M.H.V., Nature 311, 123-126 (1984).

13. Tainer, J.A., Getzoff, E.D., Alexander, H.A., Houghton, R.A., Olson, A.J., Lerner, R.A. and Hendrickson, W.A., Nature 312, 127-133 (1984).

14. Woolery, G.L., Powers, L., Winkler, M., Solomon, E.I. and Spiro, T.G., J. Am. Chem. Soc. 106, 86-92 (1984).

15. Brown, J.M., Powers, L., Kincaid, B., Larrabee, J.A. and Spiro, T.G., J. Am. Chem. Soc. <u>102</u>, 4210-4216 (1980).

16. Suzuki, S., Kino, J., Kimura, M., Mori, W. and Nakahara, A., Inorg. Chim. Acta <u>66</u>, 41-47 (1982).

17. Suzuki, S., Kino, J. and Nakahara, A., Bull. Chem. Soc. Jpn. <u>55</u>, 212-217 (1982).

PRIMARY STRUCTURE OF THE a CHAIN OF *PANULIRUS INTERRUPTUS* HEMOCYANIN

H.J. BAK, N.M. SOETER, J.M. VEREIJKEN, P.A. JEKEL, B. NEUTEBOOM, and J.J. BEINTEMA
Biochemisch Laboratorium, Nijenborgh 16, 9747 AG Groningen, The Netherlands

ABSTRACT

The primary structure of the a chain of *Panulirus interruptus* hemocyanin (657
residues) has been completed except for one overlap that remains to be proven
chemically, and two amide positions, Glx-372 and Asx-405, that are uncertain at
this moment. By fitting the sequence to the electron-density map (1, 8), all six
copper ligands have been unequivocally established to be histidines.

INTRODUCTION

Hemocyanin of the spiny lobster *Panulirus interruptus* is a hexamer containing at
least three different types of subunits, designated a, b, and c (1). In order to
elucidate its structure at the atomic level, both X-ray diffraction studies and
amino acid sequence studies are being performed (1, 8).

Limited proteolysis of the a chain with trypsin produced an 18 kDa fragment
derived from the N-terminus, a 71 kDa fragment derived from the C-terminus, and a
glycopeptide located in between (2). In contrast to the 18 kDa fragment, the

molecular weight of the 71 kDa fragment is overestimated by sodium dodecyl sulfate
polyacrylamide gel electrophoresis (1). The sequence of the first 230 residues,
comprising the 18 kDa fragment, the glycopeptide and 55 N-terminal residues of the
71 kDa fragment, has already been reported (3, 4).

In this paper, the elucidation of the sequence of the 71 kDa fragment will be
described.

MATERIALS AND METHODS

The a chain and the products of limited proteolysis were isolated as described in
previous papers (2, 5). The reduced and carboxymethylated 71 kDa fragment was both
cleaved with CNBr and digested with trypsin after citraconylation. Additional
evidence was obtained from a digest with protease from *S. aureus* V8 and one
fragment isolated from a hydroxylamine cleavage (6) of complete a chain. Initial
fractionation of peptides was performed on Sephadex columns: G-100 for CNBr and
hydroxylamine digests, and G-50F for tryptic and *S. aureus* V8 protease digests.
The peptides were purified on appropiate Bio-gel columns and/or by reversed-phase
HPLC (RP-HPLC) using a nucleosil 10C18 column (300 x 4.6 mm), from which the
peptides usually were eluted with a linear 0-67% gradient of acetonitrile in 0.1%
trifluoroacetic acid or 0.1% ammonium acetate pH 6.0, over 60 minutes at a flow
rate of 1 ml/min. Other gradient conditions used were: 0-80% organic solvent

Invertebrate Oxygen Carriers
Ed. by Bernt Linzen
© Springer–Verlag Berlin Heidelberg 1986

(acetonitrile : 2-propanol = 3 : 1) in 0.1% trifluoroacetic acid, in 0.1% ammonium acetate pH 6.0, or in 10% formic acid, over 80 minutes at a flow rate of 1 ml/min. Amino acid analysis, enzymatic degradations and CNBr cleavage were performed as described in a previous paper (3). Large fragments were sub-digested with chymo-trypsin, thermolysin, trypsin or protease from *S. aureus* V8 as described (3). Isolation of sub-fragments was done with RP-HPLC as described above. Sequences were determined either automatically or manually using the DABITC method of Chang (7) and/or the dansyl-Edman procedure.

RESULTS AND DISCUSSION
Fig. 1 schematically shows the elucidation of the amino acid sequence of the 71 kDa fragment. The elucidation was mainly based on two digests (CNBr and trypsin) from both of which a complete set of peptides was obtained. Additional sequence information was obtained from a digest with *S. aureus* V8 protease, only the relevant peptides of which are shown in Fig. 1, and one fragment obtained by cleaving the intact a chain with hydroxylamine. The latter fragment was necessary because of the Asn-Gly sequence at position 548-549 (Fig. 2) which was found to be very resistant to Edman degradation.

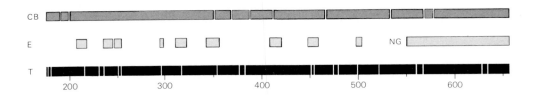

Fig. 1. Graphical representation of the fragments isolated from a CNBr (CB), a tryptic (T) and a *S. aureus* V8 protease (E) digest of the 71 kDa fragment, and one fragment obtained by a hydroxylamine cleavage of the a chain (NG). The numbering is the same as in Fig. 2. The fragments shown were used to elucidate the complete amino acid sequence of the 71 kDa fragment.

From the first tryptic digest, one large peptide (383-451) was lost due to its very low solubility in the eluent used for the first fractionation (30% acetic acid). Therefore, a second tryptic digest was made and the initial fractionation on Sephadex G-50F was carried out with 8 M urea in 0.25% formic acid as eluent.

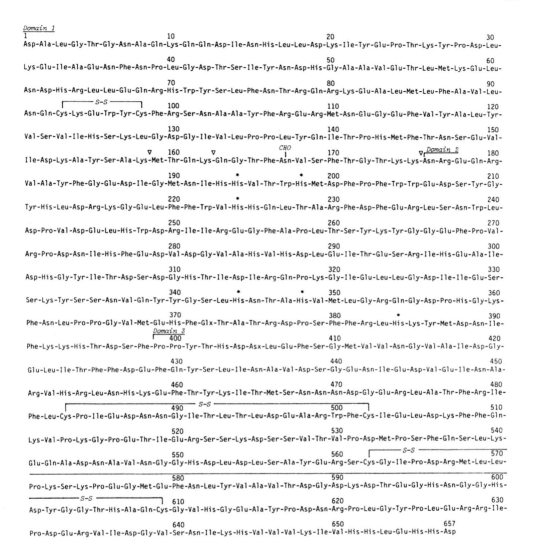

Domain 1
1   10   20   30
Asp-Ala-Leu-Gly-Thr-Gly-Asn-Ala-Gln-Lys-Gln-Gln-Asp-Ile-Asn-His-Leu-Leu-Asp-Lys-Ile-Tyr-Glu-Pro-Thr-Lys-Tyr-Pro-Asp-Leu-

                                           40   50   60
Lys-Glu-Ile-Ala-Glu-Asn-Phe-Asn-Pro-Leu-Gly-Asp-Thr-Ser-Ile-Tyr-Asn-Asp-His-Gly-Ala-Ala-Val-Glu-Thr-Leu-Met-Lys-Glu-Leu-

                                           70   80   90
Asn-Asp-His-Arg-Leu-Leu-Glu-Gln-Arg-His-Trp-Tyr-Ser-Leu-Phe-Asn-Thr-Arg-Gln-Arg-Lys-Glu-Ala-Leu-Met-Leu-Phe-Ala-Val-Leu-

                ┌——— S-S ————┐   100   110   120
Asn-Gln-Cys-Lys-Glu-Trp-Tyr-Cys-Phe-Arg-Ser-Asn-Ala-Ala-Tyr-Phe-Arg-Glu-Arg-Met-Asn-Glu-Gly-Glu-Phe-Val-Tyr-Ala-Leu-Tyr-

                                          130   140   150
Val-Ser-Val-Ile-His-Ser-Lys-Leu-Gly-Asp-Gly-Ile-Val-Leu-Pro-Pro-Leu-Tyr-Gln-Ile-Thr-Pro-His-Met-Phe-Thr-Asn-Ser-Glu-Val-

                                 ▽   160   ▽   CHO   170   ▽┌Domain 2   180
                                                              |
Ile-Asp-Lys-Ala-Tyr-Ser-Ala-Lys-Met-Thr-Gln-Lys-Gln-Tyr-Phe-Asn-Ser-Phe-Thr-Gly-Thr-Lys-Lys-Asn-Arg-Glu-Gln-Arg-

                                          190   *   *   200   210
Val-Ala-Tyr-Phe-Gly-Glu-Asp-Ile-Gly-Met-Asn-Ile-His-His-Val-Thr-Trp-His-Met-Asp-Phe-Pro-Phe-Trp-Trp-Glu-Asp-Ser-Tyr-Gly-

                                          220   *   230   240
Tyr-His-Leu-Asp-Arg-Lys-Gly-Glu-Leu-Phe-Phe-Trp-Val-His-His-Gln-Leu-Thr-Ala-Arg-Phe-Asp-Phe-Glu-Arg-Leu-Ser-Asn-Trp-Leu-

                                          250   260   270
Asp-Pro-Val-Asp-Glu-Leu-His-Trp-Asp-Arg-Ile-Ile-Arg-Glu-Gly-Phe-Ala-Pro-Leu-Thr-Ser-Tyr-Lys-Tyr-Gly-Gly-Glu-Phe-Pro-Val-

                                          280   290   300
Arg-Pro-Asp-Asn-Ile-His-Phe-Glu-Asp-Val-Asp-Gly-Val-Ala-His-Val-His-Asp-Leu-Glu-Ile-Thr-Glu-Ser-Arg-Ile-His-Glu-Ala-Ile-

                                          310   320   330
Asp-His-Gly-Tyr-Ile-Thr-Asp-Ser-Asp-Gly-His-Thr-Ile-Asp-Ile-Arg-Gln-Pro-Lys-Gly-Ile-Glu-Leu-Leu-Gly-Asp-Ile-Ile-Glu-Ser-

                                          340   *   *   350   360
Ser-Lys-Tyr-Ser-Ser-Asn-Val-Gln-Tyr-Tyr-Gly-Ser-Leu-His-Asn-Thr-Ala-His-Val-Met-Leu-Gly-Arg-Gln-Gly-Asp-Pro-His-Gly-Lys-

                                          370   380   *   390
Phe-Asn-Leu-Pro-Pro-Gly-Val-Met-Glu-His-Phe-Glx-Thr-Ala-Thr-Arg-Asp-Pro-Ser-Phe-Phe-Arg-Leu-His-Lys-Tyr-Met-Asp-Asn-Ile-
                                    ┌Domain 3
                                    ┌——— 400   410   420
Phe-Lys-Lys-His-Thr-Asp-Ser-Phe-Pro-Pro-Tyr-Thr-His-Asp-Asx-Leu-Glu-Phe-Ser-Gly-Met-Val-Val-Asn-Gly-Val-Ala-Ile-Asp-Gly-

                                          430   440   450
Glu-Leu-Ile-Thr-Phe-Phe-Asp-Glu-Phe-Gln-Tyr-Ser-Leu-Ile-Asn-Ala-Val-Asp-Ser-Gly-Glu-Asn-Ile-Glu-Asp-Val-Glu-Ile-Asn-Ala-

                                          460   470   480
Arg-Val-His-Arg-Leu-Asn-His-Lys-Glu-Phe-Thr-Tyr-Lys-Ile-Thr-Met-Ser-Asn-Asn-Asn-Asp-Gly-Glu-Arg-Leu-Ala-Thr-Phe-Arg-Ile-

                ┌——————— S-S ———————┐   490   500   510
Phe-Leu-Cys-Pro-Ile-Glu-Asp-Asn-Asn-Gly-Ile-Thr-Leu-Thr-Leu-Asp-Glu-Ala-Arg-Trp-Phe-Cys-Ile-Glu-Leu-Asp-Lys-Phe-Phe-Gln-

                                          520   530   540
Lys-Val-Pro-Lys-Gly-Pro-Glu-Thr-Ile-Glu-Arg-Ser-Ser-Lys-Asp-Ser-Ser-Val-Thr-Val-Pro-Asp-Met-Pro-Ser-Phe-Gln-Ser-Leu-Lys-

                                          550   560   ┌——————— S-S ———————┐   570
Glu-Gln-Ala-Asp-Asn-Ala-Val-Asn-Gly-Gly-His-Asp-Leu-Asp-Leu-Ser-Ala-Tyr-Glu-Arg-Ser-Cys-Gly-Ile-Pro-Asp-Arg-Met-Leu-Leu-

———580   590   600
Pro-Lys-Ser-Lys-Pro-Glu-Gly-Met-Glu-Phe-Asn-Leu-Tyr-Val-Ala-Val-Thr-Asp-Gly-Asp-Lys-Asp-Thr-Glu-Gly-His-Asn-Gly-Gly-His-

————————— S-S ————————┐   610   620   630
Asp-Tyr-Gly-Gly-Thr-His-Ala-Gln-Cys-Gly-Val-His-Gly-Glu-Ala-Tyr-Pro-Asp-Asn-Arg-Pro-Leu-Gly-Tyr-Pro-Leu-Glu-Arg-Arg-Ile-

                                          640   650   657
Pro-Asp-Glu-Arg-Val-Ile-Asp-Gly-Val-Ser-Asn-Ile-Lys-His-Val-Val-Val-Lys-Ile-Val-His-His-Leu-Glu-His-His-Asp

Fig. 2. Complete amino acid sequence of the a chain of *P. interruptus* hemocyanin. The disulfide bridges (*S-S*), the carbohydrate moiety (*CHO*), the copper ligands (*) and the cleavage sites of limited trypsinolysis (▽) are indicated. The disulfide bridges and the copper ligands have been deduced by fitting the amino acid sequence to the electron-density map (1, 4). The starting points of the domains (1) are indicated by (┌—).

The isolated large fragment was found to be soluble only in watery solutions containing at least 4 M urea.

In Fig. 2, the complete amino acid sequence of the a chain of *Panulirus interruptus* hemocyanin is shown with the sites of limited trypsinolytic cleavage, the disulfide bridges, and the copper ligands. Except for an Arg-Phe sequence (pos. 230-231) all overlaps have been chemically proven by Edman degradation. Two amide positions are not yet certain. Glx-372 was twice identified as Glu (in a tryptic peptide and a peptide from *S. aureus* V8 protease digest). However, during the DABITC degradation of the CNBr fragment containing Glx-372, a very weak spot at the position of Gln was also seen. It is unknown whether this is an artefact or deamidation plays a role. The same kind of contradictory evidence was obtained for position Asx-405.

The abnormal behaviour of the 71 kDa fragment on SDS polyacrylamide gels is not yet completely understood. Its real molecular weight is 55,414 Da. The C-terminal part of the peptide seems to be responsible for the abnormal behaviour, since both the C-terminal CNBr fragment (residue 579-657) as well as a large tryptic peptide (residue 568-628) also behave abnormally on SDS polyacrylamide gels.

REFERENCES

1) Gaykema, W.P.J., Hol, W.G.J., Vereijken, J.M., Soeter, N.M., Bak, H.J., and Beintema, J.J., Nature, 309, 23-29 (1984).

2) Vereijken, J.M., Schwander, E.H., Soeter, N.M., and Beintema, J.J., Eur. J. Biochem., 123, 283-289 (1982).

3) Vereijken, J.M., Vlieg, J. de, and Beintema, J.J., Biochim. Biophys. Acta, 788, 298-305 (1984).

4) Linzen, B., Soeter, N.M., Riggs, A.F., Schneider, H.J., Schartau, W., Moore, M.D., Yokato, E., Behrens, P.Q., Nakashima, H., Takagi, T., Nemoto, T., Vereijken, J.M., Bak. H.J., Beintema, J.J., Volbeda, A., Gaykema, W.P.J., and Hol, W.G.J., Science, 229, 519-524 (1985).

5) Schwander, E.H., and Vereijken, J.M., Life Chem. Rep. 1, suppl. 1, 103-106. (1983).

6) Bornstein, P., and Balian, G., Methods Enzymol., 47, 132-145 (1977).

7) Chang, J.Y., Brauer, D., and Wittmann-Liebold, B., FEBS Lett. 93, 205-214 (1978)

8) Volbeda, A., and Hol, W.G.J., this volume, pp. 135-147.

SUBUNITS a, b AND c OF <u>PANULIRUS INTERRUPTUS</u> HEMOCYANIN AND EVOLUTION OF ARTHROPOD
HEMOCYANINS

NELL M. SOETER, JAAP J. BEINTEMA, PETER A. JEKEL, HENK J. BAK, JOHAN J. VEREIJKEN
and BEN NEUTEBOOM
Biochemisch Laboratorium, University of Groningen, Nijenborgh 16, 9747 AG
Groningen, The Netherlands

INTRODUCTION

Hemocyanins are large, non-haem, oxygen-carrying proteins occurring freely
dissolved in the hemolymph of molluscs and arthropods.

In the arthropods, hemocyanins from two groups, Crustacea and Chelicerata, have
been investigated extensively (1,2). Recently, structural information on the
hemocyanin of a uniramous arthropod, the centipede, <u>Scutigera coleoptrata</u>, has
been published (3).

Crustacean hemocyanins consist of hexamers or dodecamers. Cheliceratan hemocyanins
may be 24-mers and 48-mers as well (4,5), while hemocyanin of the centipede was
found to possess a 6x6 structure (3). Arthropodan hemocyanin subunits have a
molecular weight of about 75,000, and each contains a pair of copper ions that bind
one oxygen molecule. Sometimes, the subunits are connected covalently by disulfide
bridges (4,5).

Hemocyanin of the spiny lobster, <u>Panulirus interruptus,</u> occurs as hexamers, which
consist of different subunits (6). Three different subunits, a, b and c (7),
formerly called 94K, 90K and 80K (8) after their apparent molecular weight on
SDS-polyacrylamide gels, have been found.

The three-dimensional structure of <u>Panulirus</u> hemocyanin has been determined at a
resolution of 3.2 Å. Each subunit of the hexamer contains 660 amino acids and is
folded into three domains of roughly equal size. The first and the second domains
are mainly helical and globular, whereas the third domain contains a beta-barrel
structure and two long loops. The active site is located in the middle of the
second domain. The two copper ions are each ligated by three histidines, two of
which are provided by a -His-X-X-X-His- sequence located in a helix (7).

In February 1984, several research groups sequencing arthropodan hemocyanin
subunits had a meeting in Groningen. Complete and partial sequences from two
crustacean and five cheliceratan chains were aligned and compared with the
three-dimensional structure of Panulirus hemocyanin. The polypeptide folding
established for <u>Panulirus</u> hemocyanin appeared to be common to all arthropodan
hemocyanins, although some differences between crustacean and cheliceratan
hemocyanins were found. These findings were published recently (9).

Since at that time only part of the sequence of <u>Panulirus</u> subunit a was available,
a comparison of complete chain a with the other arthropodan hemocyanin chains is

Invertebrate Oxygen Carriers
Ed. by Bernt Linzen
© Springer–Verlag Berlin Heidelberg 1986

```
Panulirus interruptus a   DOMAIN 1                      α α α α α α 1.1 α α α α α                              α α
                          -6                1                  10              20                30
Panulirus interruptus a               D-A-L-G-T-G-N-A-Q-K-Q-Q-D-I-N-H-L-L-D-K-I-Y-E-P-T-K-Y-P-D-L-
Panulirus interruptus b               D-A-L-G-T-G-N-A-N-K-Q-Q-D-I-N-H-L-L-D-K-I-Y-E-P-T-K-Y-P-D-L-
Panulirus interruptus c   A-D-C-Q-A-G-D-S-A-D-K-L-L-A-Q-K-Q-H-D-V-N-Y-L-V-Y-K-L-Y-G-D-I-R-D-D-H-L-
Astacus leptodactylus b               D-A-S-G-A-T-L-A-K-R-Q-Q-V-V-N-H-L-L-E-H-I-Y-D-H-T-H-F-T-D-L-


Pi a       α 1.2 α α                        α α α α α α α 1.3 α α α α α
                 40                     50             60                     70
Pi a   K-E-I-A-E-N-F-N-P-L-G-D-T-S-I-Y-N-D-H-G-A-A-V-E-T-L-M-K-E-L-N-D-H-R-L-L-E-Q-R-H-W-Y-S-L-F-
Pi b   K-D-I-A-E-N-F-D-P-                       -K-E-L-N-D-H-R-L-L-E-Q-R-H-W-F-S-L-F-
Pi c   K-E-L-G-E-T-F-N-P-Q-G-D-L-L-L-Y-H-D-N-G-A-S-V-N-T-L-M-               -H-W-F-S-L-F-
Al b   K-N-I-A-G-T-F-S-P-E-A-D-T-S-I-                       -L-L-E-Q-H-H-W-F-S-L-F-


Pi a   α α α α α α α 1.4 α α α α α α α         α α α α α 1.5 α α α α α             α α α 1.6 α
               80                     90        ┌───────────┐ 100           110                120
Pi a   N-T-R-Q-R-K-E-A-L-M-L-F-A-V-L-N-Q-C-K-E-W-Y-C-F-R-S-N-A-A-Y-F-R-E-R-M-N-E-G-E-F-V-Y-A-L-Y-
Pi b   N-T-R-Q-R-                       -E-W-Y-C-F-R-S-N-A-A-Y-F-R-E-R-
Pi c   N-T-R-             -V-L-N,M-C-K-                       -M-N-E-G-E-Y-L-Y-A-L-Y-
Al b   N-T-R-                   -S-W-E-C-F-L-D-N-A-A-Y-F-R-       -M-N-E-G-E-F-V-Y-A-I-


Pi a   α α α α α              β β β β 1 A β β β             α α α α α α 1.7 α α α α α         β β β
               130                   140                 150               160
Pi a   V-S-V-I-H-S-K-L-G-D-G-I-V-L-P-P-L-Y-Q-I-T-P-H-M-F-T-N-S-E-V-I-D-K-A-Y-S-A-K-M-T-Q-K-Q-G-T-
Pi b                               -M-F-T-N-S-E-V-I-D-K-A-Y-S-A-K-M-T-Q-K-P-G-T-
Pi c   V-S-L-I-H-S-G-L-G-E-G-V-V-L-P-P-L-Y-E-V-T-P-H-M-F-T-N-S-E-V-I-H-E-A-Y-K-A-Q-M-T-N-T-P,S,K-
Al b   V-A-V-I-H-S-G-I-G-H-G-I-V-I-P-P-I-Y-E-V-T-P-H-K-F-T-N-S-E-V-I-N-K-A-Y-S-G-K-M-T-Q-T-P-G-R-


Pi a   β β 1 B β          ┌DOMAIN 2              α α α α α α α 2.1 α α α α α α
           CHO    170      |            180           190   *     *  200                210
Pi a   F-N-V-S-F-T-G-T-K-K-N-R-E-Q-R-V-A-Y-F-G-E-D-I-G-M-N-I-H-H-V-T-W-H-M-D-F-P-F-W-W-E-D-S-Y-G-
Pi b   F-N-V-S-F-T-G-T-K-K-N-R-E-Q-R-V-A-Y-
Pi c   F-E-S-H-F-T-G,S,K-K-N-P-E-Q-H-V-A-Y-F-G-E-D-V-G-M-
Al b   F-N-M-D-F-T-G-T-K-K-N-K-Z-Q-R-V-A-Y-F-G-E-D-I-G-M-N-I-H-H-V-T-W-H-M-D-F-P-F-      -I-Y-G-


Pi a       α α α α α α α α α α α α 2.2 α α α α α α α α α α        β β 2 A β β        β β β β
                   220     *           230             240               250
Pi a   Y-H-L-D-R-K-G-E-L-F-F-W-V-H-H-Q-L-T-A-R-F-D-F-E-R-L-S-N-W-L-D-P-V-D-E-L-H-W-D-R-I-I-R-E-G-
Pi b           -G-E-L-F-F-W-V-H-H-Q-L-T-A-R-F-D-F-E-R-L-S-N-W-L-D-P-V-D-E-L-H-W-D-R-I-I-R-E-G-
Pi c                                                               -A-I-D-E-G-
Al b   Y-G-I-   -K-G-E-L-F-F-W-V-H-H-Q-L-T-A-R-F-D-S-E-R-I-S-N-W-I-D-V-V-D-E-G-H-W-S-   -I-E-G-


Pi a   β β 2 B β β β     β β β β β 2 C β β β β β β                α α α α α α α 2.3 α α α α α
               260             270               280               290               300
Pi a   F-A-P-L-T-S-Y-K-Y-G-G-E-F-P-V-R-P-D-N-I-H-F-E-D-V-D-G-V-A-H-V-H-D-L-E-I-T-E-S-R-I-H-E-A-I-
Pi b   F-A-P-L-T-S-Y-K-                                           -I-H-D-A-I-
Pi c   F-A-P-H-T-A-Y-K-
Al b   F-A-P-H-T-S-Y-K-Y-G-G-E-F-P-A-R-P-D-N-V-H-F-E-D-V-D-G-V-A-R-V-R-D-       -S-R-I-R-D-A-L-


Pi a   α α β 2 D β β       β β 2 E β        α α α 2.4 α α α                               α α α α
               310             320             330                   340     *
Pi a   D-H-G-Y-I-T-D-S-D-G-H-T-I-D-I-R-Q-P-K-G-I-E-L-L-G-D-I-I-E-S-S-K-Y-S-S-N-V-Q-Y-Y-G-S-L-H-N-
Pi b   D-H-G-Y-I-T-D-S-D-G-H-T-I-D-I-R-Q-P-K-
Pi c                   -M-N-S-H-G-I-E-F-L-G-D-I-I-E-S-S-G-Y-S-A-N-P-G-F-Y-G-S-L-H-N-
Al b   A-H-G-Y-L-L-D-N-S-G-N-K-                       -I-F-N-V-Q-Y-Y-G-A-I-H-N-


Pi a   α α α 2.5 α α α α                                       α α α 2.6 α α α α        α
           *   350                 360                 370                 380   *             390
Pi a   T-A-H-V-M-L-G-R-Q-G-D-P-H-G-K-F-N-L-P-P-G-V-M-E-H-F-E-T-A-T-R-D-P-S-F-F-R-L-H-K-Y-M-D-N-I-
Pi b               -Q-G-D-P-H-G-K-F-N-L-P-P-G-V-M-E-H-F-E-T-A-T-R-D-P-S-F-F-R-L-H-K-Y-M-D-N-I-
Pi c   T-A-H-I-M-                                           -L-H-K-Y-M-D-N-I-
Al b   T-A-H-I-M-I-G-R-Q-G-D-H-----K-F-D-M-P-P-G-V-M-E-H-F-E-T-A-T-R-D-P-S-F-F-R-L-H-K-Y-M-D-N-I-


Pi a   α α α 2.7 α α        α α 3.1 α α      β β β β β β β β 3 A β β β β β β      β 3 β β
           DOMAIN 3┐ 400              410               420               430
Pi a   F-K-K-H-T-D-S-F-P-P-Y-T-H-D-D-L-E-F-S-G-M-V-V-N-G-V-A-I-D-G-E-L-I-T-F-F-D-E-F-Q-Y-S-L-I-N-
Pi b   F-K-
Pi c   F-R-
Al b   F-K-E-H-K-D-S-I-P-P-Y-T-K-N-D-I-A-V-P-G-V-V-I-D-S-V-A-V-F-I-E-S-P-
```

```
Pi a                    β β β 3 C β β β        β β β β β β β 3 D β β β β β β            β β β β 3 E β
                        440              450              460              470                    480
Pi a    A-V-D-S-G-E-N-I-E-D-V-E-I-N-A-R-V-H-R-L-N-H-K-E-F-T-Y-K-I-T-M-S-N-N-N-D-G-E-R-L-A-T-F-R-I-
Pi b                                        -L-N-H-N-E-F-T-Y-K-I-T-M-S-N-N-N-D-G-E-R-L-A-T-F-R-
Pi c                                                    -I-T-M-S,B,B,B,B,G,E,R-
Al b                    -I-N-H-E-E-F-S-Y-N-I-D-I-S-N-T-D-K-                              -L-


Pi a    β β β β β β β        β 3 F    α α α 3.2 α α      β β β β 3 G β β β β      β β β β 3 H β β β β
                      490              500┐              510                    520
Pi a    F-L-C-P-I-E-D-N-N-G-I-T-L-T-L-D-E-A-R-W-F-C-I-E-L-D-K-F-F-Q-K-V-P-K-G-P-E-T-I-E-R-S-S-K-D-
Pi b                      -W-F-C-I-E-L-D-K-F-F-Q-K-V-P-S-G-P-E-T-I-E-R-S-S-K-
Pi c                      -W,F,C,I,E,L,D,K,F,F,Q-K-
Al b    F-L-C-P-V-                    -M-D-K-F-Y-K-S-I-A-P-G-T-N-H-I-V-R-K-S-T-D-


Pi a            β β β 3 I β β β α α α α α 3.3 α α α α α                        α α α 3.4 α
                      530              540              550              560 ┌─────── 570
Pi a    S-S-V-T-V-P-D-M-P-S-F-Q-S-L-K-E-Q-A-D-N-A-V-N-G-G-H-D-L-D-L-S-A-Y-E-R-S-C-G-I-P-D-R-M-L-L-
Pi b                      -E-Q-A-D-N-A-V-N-G-G-H-D-L-D-L-S-A-Y-E-R-S-C-G-I-P-D-R-M-L-L-
Pi c
Al b    S-S-V-T-V-P-D-R-                        -L-B-L-H-M-F-Q-R-S-C-G-I-P-D-R-M-I-I-


Pi a    α           β β β β β 3 J β β β β β      α α 3.5 α                              β β β
                      580              590              600              610
Pi a    P-K-S-K-P-E-G-M-E-F-N-L-Y-V-A-V-T-D-G-D-K-D-T-E-G-H-N-G-G-H-D-Y-G-G-T-H-A-Q-C-G-V-H-G-E-A-
Pi b    P-K-S-K-P-E-G-M-K-
Pi c            -M-E-F-N-L-V-V-A-V-T-D-G-R-T-D-A-A-L-D-------D-L-H-E-N-T-K-F-I-H-Y-G-Y-D-R-Q-
Al b    I-E-S-R-P-D-G-M-D-F-A-I-                                              -E-K-


Pi a    β 3 K β β                              α α α 3.6 α α α      β β β β β β 3 L β β β β β β
                620              630              640                    650              657
Pi a    Y-P-D-N-R-P-L-G-Y-P-L-E-R-R-I-P-D-E-R-V-I-D-G-V-S-N-I-K-H-V-V-V-K-I-V-H-H-L-E-H-H-D
Pi b                      -I-P-D-E-R-V-I-D-G-V-S-N-I-K-H-V-V-V-K-I-V-H-H-L-E-H-H-D
Pi c    Y-P-D-K-R-P-H-G-Y-P-L-D-R-R-V-D-D-E-R-I-F-E-A-L-P-N-F-K-Q-R-T-V-K-L-Y-S-H-E-G-V-D-G-G
Al b    Y-P-D-K-K-P-M-G-Y-P-V-D-R-S-I-P-D-N-R-V-F-L-E-S-P-N-I-K-R-T-Y-V-K-V-F-H-D-E-H-G-G-E-Q-H
```

Fig. 1   Aligned amino acid sequences of crustacean hemocyanins. Complete sequence of subunit a of <u>Panulirus</u>, Pi a (13); partial sequences of subunits b, Pi b (14) and c, Pi c (15,16) of <u>Panulirus</u>; partial sequence of subunit b of <u>Astacus leptodactylus</u>, Al b (9). The one-letter code for amino acid residues (17) has been used. Dashes represent gaps introduced for alignment, blank spaces are parts not yet sequenced and the sequence of residues separated by a comma is presumptive. The secondary-structure elements (alpha helices 1.1-3.6 and beta strands 1A-3L), disulfide bridges, carbohydrate attachment site (CHO), copper ligands (*) and domain borders of subunit a of <u>Panulirus</u> <u>interruptus</u> hemocyanin are indicated. Amino acid compositions and other experimental evidence on the tryptic peptides of subunits b and c will be published elsewhere.

made in this paper.

Recently, amino acid sequence studies of subunits b and c have been started, and parts of the primary structures of the three subunits are compared among each other, with the sequence from Astacus, and with cheliceratan hemocyanins.

COMPARISON OF CRUSTACEAN HEMOCYANIN SUBUNITS

Before presenting structural data about subunits a, b and c of Panulirus interruptus hemocyanin, some general functional characteristics will be summarized. The three subunits a, b and c had already been isolated by van den Berg et al. (6), who called their three components respectively Fraction I, Fraction IIa and Fraction IIb. Later, van Eerd (8) found on alkaline gels two 94K components (94K I and II) and two 90K components (90K I and II) migrating with a slight difference in charge. Our structural studies described below, have been performed with unfractionated 94K (a) and 90K (b) subunits. So far, no indications of heterogeneity within the subunits a, b and c have been observed. Possibly, the additional heterogeneity observed by van Eerd originates from deamidation.

The major subunits, a and b, each account for 40-45 percent of the monomers. The hexamers of Panulirus hemocyanin are heterogeneous, consisting of different proportions of the subunits per molecule, but the distribution of the subunits among the hexamers in vivo is not known. After dissociation of Panulirus

subunit	a	b	c
no. of residues	657		
M.W. polypeptide chain	75.7K		
apparent M.W. on SDS-PAGE	94K	90K	80K
nomenclature according to ref. 6	Fraction I	Fraction IIa	Fraction IIb
carbohydrate moieties	1 (Asn 167)	1 (Asn 167)	unknown, not at Asn 167
S-S bridges, SH groups	3 S-S, no SH	3 S-S, no SH	unknown, SH possible
percent of the monomers	40-45%	40-45%	10-20%
sequence difference compared with a	-	about 5%	about 45%
N-terminal extension compared with a	-	no	6 residues
C-terminal extension compared with a	-	no	1 residue
internal deletion compared with a	-		3 residues (597-599)
ability to form homohexamers	yes	yes	yes
ability of the homohexamers to bind O_2	yes	n.d.	yes
ability of the homohexamers to bind O_2 cooperatively	yes	n.d.	no

Table 1   Comparison of structural and functional characteristics of subunits a, b and c of Panulirus interruptus hemocyanin. n.d.: not determined.

		DOMAIN 1	DOMAIN 2	DOMAIN 3	SUBUNIT
No. of residues	in Pi a	175	223	259	657
No. of residues sequenced (percent)					
	in Pi a	175 (100)	223 (100)	259 (100)	657 (100)
	in Pi b	109 (62)	118 (53)	117 (45)	344 (52)
	in Pi c	144 (80)	73 (33)	104 (41)	321 (49)
	in Al b	137 (78)	188 (84)	153 (59)	478 (73)
No. of residues and percentual difference between	Pi a and Pi b	5 (5)	1 (1)	3 (3)	9 (3)
	Pi a and Pi c	59 (41)	19 (26)	48 (46)	126 (39)
	Pi a and Al b	40 (29)	35 (19)	75 (49)	150 (31)
	Pi c and Al b	53 (42)	12 (23)	36 (55)	101 (41)

Table 2   Differences between crustacean hemocyanin subunits: <u>Panulirus</u> interruptus chain a, Pi a, chain b, Pi b and chain c, Pi c; <u>Astacus</u> <u>leptodactylus</u> chain b, Al b. The alignments presented in Fig. 1 were used.

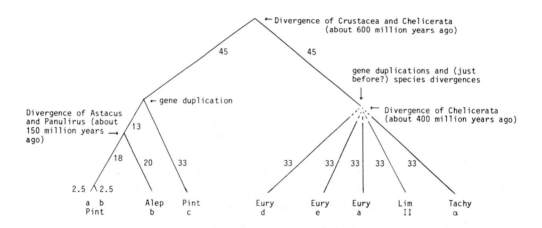

Fig. 2   Tentative evolutionary tree of hemocyanin chains of arthropods. The alignment presented in Fig. 1 of ref. 9 was used for calculating the distances in PAM's for the cheliceratan chains. The alignment of Fig. 1 of this paper was used for the four crustacean chains (see also Table 2).

hemocyanin and subsequent reassociation, three or four different hexamers are observed, indicating that molecules with certain subunit compositions are preferentially formed (6).

In vitro, homohexamers of subunits a, b and c can be formed, according to electron microscopy (10,11) and analytical ultracentrifuge studies. Homohexamers of subunit a were found to bind oxygen cooperatively, like homohexamers of Limulus subunit IV

	SUBUNIT a		SUBUNIT b	
	position	amino acid residue	amino acid residue	position according to the numbering scheme of Fig. 1 in ref. 9
DOMAIN 1	9	Gln	Asn	9
	32	Glu	Asp	32
	38	Asn	Asp	38
	72	Tyr	Phe	72
	163	Gln	Pro	163
DOMAIN 2	298	Glu	Asp	305
DOMAIN 3	458	Lys	Asn	473
	514	Lys	Ser	529
	579	Glu	Lys	594

Table 3   Amino acid replacements established in subunit b (14) as compared with subunit a of _Panulirus interruptus_ hemocyanin. About ten other substitutions are expected in the larger tryptic peptides starting at positions 39, 110, 181, 264, 320, 480 and 525 (14; not shown in Fig. 1) from amino acid compositions.

hemocyanin (12). Homohexamers of subunit c do bind oxygen as well, but not cooperatively. The role of subunit c in native hexamers remains unclear since homohexamers of subunit a only, bind oxygen at least as cooperatively as native hemocyanin.

Some characteristics of the subunits are summarized in Table 1.

Of the crustacean hemocyanins, only chain a of _Panulirus interruptus_ has been sequenced completely (13). Partial sequences of chain b (14) and chain c (15,16) of _Panulirus_, and of chain b of the crayfish _Astacus leptodactylus_ (9) are available. Fig. 1 shows these cructacean sequences aligned. In Table 2, for each domain the number of residues, the percentage sequenced, the number of residues and the percentual difference between the chains are given. These data have been used in the construction of a tentative evolutionary tree (Fig. 2).

Subunits a and b are very similar, with only about five percent of the amino acids different. Table 3 shows substitutions established in subunit b. Substitutions occur in all three domains. The structural resemblance of subunits a and b explains why there were no problems in determining an electron-density map using crystals of a mixture of subunits a and b.

Subunit c is strongly different from a and b, with at least 40% of the residues substituted. The percentage may become higher upon completion of the sequence determination of this subunit since many shorter tryptic peptides (16), together containing about 100 residues, cannot yet be aligned. With the data available, subunit c is the most deviating one of the four crustacean chains.

The six Cys residues occurring in subunit a form three disulfide bridges as deduced from the electron-density map (7). In the accompanying paper (14) evidence has been presented for identical positions of the disulfide bonds in subunit b.

In Astacus hemocyanin, the same three disulfide bridges may be present as well since there are cysteine residues at three of the six half-cystine positions in subunit a. The other three positions have not yet been sequenced. In subunit c the situation is different. Two of the disulfide bonds may be present since cysteines

have been detected at positions 93 and 502 (16), but the third disulfide bridge of subunit a, between residues 562 and 609, cannot be present at the same position in subunit c since Cys 609 is substituted (15,16). One extra Cys was found at position -4 (Fig. 1). We do not yet know whether this Cys has a free sulfhydryl group or is connected to another half-cystine. Possibly, the presence of a free sulfhydryl group in subunit c has not been detected in native hemocyanin (18) because subunit c constitutes only 10-20% of the monomers.

Both subunits a (19) and b (14) contain carbohydrate attached at asparagine at position 167. At this position, a carbohydrate moiety can be attached neither to chain b of Astacus (9) nor to subunit c of Panulirus (15) since these do not possess a sequence -Asn-X-Thr/Ser- around this position. We cannot exclude that carbohydrate is attached at another position in subunit c, but so far, no sequence capable of Asn-linked carbohydrate attachment has been encountered.

As shown in Fig. 1, the N-terminus and C-terminus are the same for subunits a and b. Astacus chain b starts at the same position, whereas subunit c has an N-terminal extension of six residues.

The molecular weight as estimated with SDS-PAA gel electrophoresis differs for the three Panulirus hemocyanin chains, even for the very similar subunits a and b. No sufficient explanation is available for the difference in apparent molecular weight between these two chains. As pointed out in the accompanying paper (14), no deletions in subunit b have been found in the 80% of the chain for which sequence information is available.

The deviance in molecular weight on SDS-PAA gels is most extreme for subunits a and b of Panulirus hemocyanin. We know regions which are responsible for the overestimation of the molecular weight and which are not found in other hemocyanins (9). Part of the overestimation is due to the carbohydrate moiety (about 4K according to the difference in molecular weight of N-terminal fragments 23K and 18K (14,19)). Another part is due to abnormal behaviour of the C-terminal peptide (13) starting at position 579 (594 in the numbering scheme of ref. 9). The difference in binding of SDS by this peptide of Panulirus hemocyanin as compared with other arthropodan hemocyanins (9) may be located in two very variable regions starting at positions 610 and 647, respectively, in the numbering scheme of ref. 9, and possibly in the region starting at position 631 (same numbering scheme), where Panulirus chain a is the most deviating chain.

In the next paragraph the substitutions in subunits b and c will be compared with those in other arthropodan hemocyanins.

CRUSTACEAN VERSUS CHELICERATAN HEMOCYANINS

Alignment of six hemocyanin sequences of one crustacean (Astacus) and several cheliceratan species (Eurypelma, Limulus, Tachypleus) with chain a of Panulirus, and correlation of these primary structures with the three-dimensional structure of Panulirus (9), revealed that the polypeptide folding is the same for all arthropods. This structure is, therefore, at least as old as the estimated time of divergence of crustaceans and chelicerates, 540 to 600 million years ago.

In Table 1 of ref. 9, a comparison of the percentual differences between amino acid sequences of domains and complete subunits of hemocyanins was given. At that time, only an incomplete sequence of the third domain of Panulirus chain a was available. In Table 4, percentages calculated from the complete sequence of chain a are shown. The percentual difference between Panulirus chain a and the cheliceratan chains is about 70%. The largest percentual difference, about 80%, is found in the first domain; in the third domain, the difference is about 70% and in the second domain, only 60%. To complete Table 2 of ref. 9: conversion of these percentual differences to PAM's and PAM's per 10^8 years are, 172 and 14.3, respectively, for Panulirus chain a versus the chelicerates in domain 3, and 159 and 13.3, respectively, for the whole subunit.

With partial sequences of subunits b and c of Panulirus available now, we can compare these with the sequences used in Fig. 1 of ref. 9.

Of the nine substitutions in subunit b as compared with subunit a (Fig. 1, Table 3), Phe 72, Pro 163, Asp 298, and Ser 514 in chain b are more in agreement with residues occurring at the same position in hemocyanin chains from other species than the respective residues of chain a. For Asn 9, Asp 32, Asp 38, Asn 458 and Lys 579 this difference thus does not exist.

Although subunit c seems to be the most deviating of all crustacean chains, the general pattern of its conserved and replaced residues is similar to that for other arthropodan chains (9). Of the residues conserved in other chains, only those at positions 346 and 646 (numbering scheme of Fig. 1, ref. 9) are replaced in subunit c, in both cases by an isofunctional residue. In addition, the isofunctional residues at positions 43, 172, 330, 598 and 638 in other chains are replaced by a residue from another group in subunit c.

The generalizations derived from comparison of the seven chains (9) also remain the same if subunits b and c of Panulirus are included.

So far, about 17% of the positions are identical for all nine chains, and about 29% possess identical or isofunctional residues. Many identical residues are clustered. The similarity is greatest in the second domain of the subunit, which contains the six histidines ligating the oxygen-binding copper ions.

Like the other crustacean chains, chain c of Panulirus has the sequence

-His-Asn-Thr-Ala-His- at the second copper-binding site, where the chelicerates have the sequence -His-Asn-Trp-Gly-His-.

More than half of the glycines occur at identical positions. Many of the completely conserved glycines occur at, or close to, the beginning of a beta strand or an alpha helix.

Most differences between crustacean and chelicerate hemocyanins occur in the first domain. The occurrence of a 21-residue loop from Tyr 22 to Asp 42, which contains helix 1.2, in the four crustacean chains is striking. A second shorter crustacean loop occurs just before the third disulfide bridge in _Panulirus_ hemocyanin (Fig.

DOMAIN 1

	Pint a	Eury d	Eury e	Eury a	Lim II
Eury d	81%				
Eury e	78%	56%			
Eury a	84%	61%	60%		
Lim II	80%	58%	54%	50%	
Tachy α	79%	46%	46%	60%	52%
NPC*	166-180	141-154			

DOMAIN 2

	Pint a	Eury d	Eury e	Eury a	Lim II
Eury d	59%				
Eury e	59%	35%			
Eury a	59%	37%	33%		
Lim II	63%	37%	32%	31%	
Tachy α	58%	38%	37%	37%	36%
NPC	226-227	224-227			

DOMAIN 3

	Pint a	Eury d	Eury e	Lim II
Eury d	73%			
Eury e	74%	49%		
Lim II	68%	51%	51%	
Tachy α	71%	53%	46%	45%
NPC	240-264	238-257		

COMPLETE SUBUNIT

	Pint a	Eury d	Eury e	Lim II
Eury d	70%			
Eury e	70%	45%		
Lim II	69%	47%	45%	
Tachy α	69%	46%	43%	43%
NPC	647-666	618-632		

* NPC, the numbers of positions compared.

Table 4 Comparison of the percentual differences between amino acid sequences of domains and complete subunits of hemocyanins. The percentual differences between amino acid sequences of the first and the second domain are from Table 1 of ref. 9. For the third domain and the complete subunit the alignment of Fig. 1 in ref. 9 was used but with the complete sequence of _Panulirus_ subunit a. The percentual differences between sequences were derived according to Dayhoff et al. (20). The numbers of differences between the amino acid sequences of each domain were counted for those hemocyanin chains for which complete or nearly complete sequences of that domain are known. Positions where in one sequence there is an amino acid and in the other a deletion were counted as differences, but positions where both sequences contain deletions were ignored. Residues that may be either aspartic acid or asparagine, or glutamic acid or glutamine, were counted as identical residues.

1, ref 9), in subunits a and b of <u>Panulirus</u> and chain b of <u>Astacus</u>. No sequence information on this region of subunit c of <u>Panulirus</u> is available.

Additional crustacean features are an N-terminal extension (eleven residues in subunit c of <u>Panulirus</u> and five residues in the other three crustacean chains) and the characteristic replacements near the active site already mentioned.

At the end of the first domain, near the carbohydrate attachment site of <u>Panulirus</u> subunits a and b, which is near the inter-subunit contact area, <u>Eurypelma</u> chain a and <u>Limulus</u> chain II, subunits occupying identical positions in the multihexamers, have similar additions (9), supporting the orthologous nature of these identically positioned subunits.

The sequence data from ref. 9 and from this paper (Fig. 1) have been used for the construction of a tentative evolutionary tree (Fig. 2).

Subunits a and b of <u>Panulirus</u> <u>interruptus</u> hemocyanin exhibit the characteristics of extracellular proteins synthesized on membrane-bound polyribosomes (RER). They are glycosylated, and have only disulfide bonds and no free sulfhydryl groups. These structural characteristics indicating secretory proteins are less clear in the cheliceratan hemocyanin chains. Here, sulfhydryl groups are frequent, and in only one cheliceratan hemocyanin, <u>Androctonus</u> <u>australis</u>, carbohydrate has been found hitherto (J.P.Kamerling, personal communication).

It would be interesting to have in the future amino acid sequence information of the third group of arthropods, the Uniramia.

ACKNOWLEDGEMENTS

We thank Drs. J.M. van der Laan and R. Torensma for performing analytical ultracentrifuge measurements and oxygen-binding experiments. We thank Dr. R.N. Campagne for carefully reading the manuscript. This work was supported in part by the Netherlands Foundation for Chemical Research (SON) with financial aid from the Netherlands Organization for the Advancement of Pure Research (ZWO).

REFERENCES

1. Van Holde, K.E. and Miller, K.I., Q. Rev. Biophys. 15, 1-129 (1982).
2. Ellerton, H.D., Ellerton, N.F. and Robinson, H.A., Prog. Biophys. molec. Biol. 41, 143-248 (1983).
3. Mangum, C.P., Scott, J.L., Black, R.E.L., Miller, K.I. and van Holde, K.E., Proc. Natl. Acad. Sci. USA 82, 3721-3725 (1985).
4. Markl, J., Hofer, A., Bauer, G., Markl, A., Kempter, B., Brenzinger, M. and Linzen, B., J. Comp. Physiol. 133, 167-175 (1979).
5. Markl, J., Markl, A., Schartau, W. and Linzen, B., J. Comp. Physiol. 130, 283-292 (1979).
6. van den Berg, A.A, Gaastra, W. and Kuiper, H.A., in Structure and function of hemocyanin (ed. Bannister, J.V.) 6-12, Springer, Berlin (1977).
7. Gaykema, W.P.J., Hol, W.G.J., Vereijken, J.M., Soeter, N.M., Bak, H.J. and

Beintema, J.J., Nature 309, 23-29 (1984).

8. van Eerd, J.P. and Folkerts, A., in Invertebrate Oxygen-binding proteins (eds. Lamy, J. and Lamy, J.) 139-149, Marcel Dekker, New York (1981).

9. Linzen, B., Soeter, N.M., Riggs, A.F. et al., Science 229, 519-524 (1985).

10. Keegstra, W. and van Bruggen, E.F.J., in Electron microscopy at molecular dimension (eds. Baumeister, W. and Vogell, W.) 318-327, Springer, Berlin (1980).

11. van Breemen, J.F.L., Schutter, W.G., Keegstra, W., Bijlholt, M.M.C. and van Bruggen, E.F.J., Electron microscopy, Vol. 2, 580-581 (1980).

12. Brenowitz, M., Bonaventura, C. and Bonaventura J., Biochemistry 22, 4707-4713 (1983).

13. Bak, H.J., Soeter, N.M., Vereijken, J.M., Jekel, P.A., Neuteboom, B. and Beintema, J.J., in Invertebrate Oxygen Carriers (ed. Linzen, B.), Springer, Berlin, pp. 149-152.

14. Soeter, N.M., Jekel, P.A. and Beintema, J.J., in Invertebrate Oxygen Carriers (ed. Linzen, B.) Springer, Berlin, pp. 165-168.

15. Neuteboom, B., Beukeveld, G.J.J. and Beintema, J.J., in Invertebrate Oxygen Carriers (ed. Linzen, B.), Springer, Berlin, pp. 169-172.

16. Soeter, N.M., unpublished tryptic peptides.

17. Dayhoff, M.O., Atlas of protein sequence and structure, Vol. 5, Suppl. 1, S10, National Biomedical Research Foundation, Washington, D.C. (1973).

18. Kuiper, H.A., Gaastra, W., Beintema, J.J., van Bruggen, E.F.J., Schepman, A.M.H. and Drenth, J., J. Mol. Biol. 99, 619-629 (1975).

19. Vereijken, J.M., Schwander, E.H., Soeter, N.M. and Beintema, J.J., Eur. J. Biochem. 123, 283-289 (1982).

20. Dayhoff, M.O., Atlas of protein sequence and structure, Vol 5, D6, National Biomedical Research Foundation, Washington, D.C., (1972).

PRIMARY STRUCTURE OF SUBUNIT b OF PANULIRUS INTERRUPTUS HEMOCYANIN

NELL M. SOETER, PETER A. JEKEL and JAAP J. BEINTEMA
Biochemisch Laboratorium, Rijksuniversiteit, Nijenborgh 16, 9747 AG Groningen, The Netherlands

INTRODUCTION

Hemocyanin of the spiny lobster, Panulirus interruptus, consists of hexamers of three different subunits, named a, b and c (1). Subunits a and b are the major ones (2). The molecular weights of the three subunits as estimated from SDS-PAA gels are different (2).

Differences in amino acid sequence of 18% between subunit a, and subunit b and of more than 30% between subunit a and subunit c, have been predicted (3) from rocket-line immunoelectrophoresis of the subunits and the relationship between immunological distance and sequence difference as derived by Champion et al. (4). However, a 3.2 Å electron-density map obtained using crystals containing subunits a and b in roughly equal amounts, could be interpreted without serious problems (1).

While the amino acid sequence determination of subunit a was proceeding, a pilot study to estimate the degrees of difference in sequence between subunits a, b and c was undertaken using the strategy described below.

Since then, the primary structure of chain a has been elucidated completely (5). Subunits b and c are being sequenced entirely now using the results of the pilot studies.

The present state of the elucidation of the amino acid sequence of subunit c can be found in the accompanying communications (6,7); that of subunit b will be described in this paper.

STRATEGY

Subunits b and c were isolated as described earlier (2), reduced and carboxymethylated, and cleaved with trypsin. Each digest was fractionated on a Sephadex G-50 Fine column (1.5x300 cm) in 30% acetic acid. Pools of fractions with peptides of an expected length of seven to fifteen residues were further fractionated by reversed-phase HPLC using a Nucleosil 10C18 column (300x4.6 mm) with a 0-67% linear gradient of acetonitril in 0.1% trifluoroacetic acid over 60 minutes, or a 0-80% linear gradient of a mixture of acetonitril and 2-propanol (3:1) in 0.1% trifluoroacetic acid over 80 minutes, at a flow rate of 1 ml/min. Impure peptides were rechromatographed using the first system, but with 0.1% ammonium acetate pH 6.0 instead of trifluoroacetic acid.

Invertebrate Oxygen Carriers
Ed. by Bernt Linzen
© Springer–Verlag Berlin Heidelberg 1986

Pure peptides were submitted to amino acid analysis and determination of their N-terminal residues by dansylation, and aligned with homologous sequences of subunit a. At first, only those peptides in which substituted residues were expected from the amino acid compositions, were sequenced by the Dansyl-Edman procedure or the DABITC-method of Chang (8), later on the other peptides were sequenced.

RESULTS AND DISCUSSION

Peptides with about half of the residues of subunit b have been sequenced now (Fig. 1). Alignment of these peptides with the sequence of subunit a was easy since only few differences have been found so far. In addition, six large peptides: viz. 39-58, 110-143, 181-216 (including copper ligands 194 and 198), 264-295, 320-353 (including copper ligands 344 and 348), 480-499 and 525-540 were isolated. These were aligned easily with the sequence of subunit a by amino acid composition and N-terminal residue analysis. These peptides account for another 30% of the length of chain a. They are not shown in Fig. 1, but from their amino acid compositions, about ten substitutions are expected to occur in these 190 residues. Substituted amino acids are dispersed along the entire length of the chain.

The differences found are far fewer than predicted from an immunological comparison (3), and it explains why a 3.2 Å electron-density map could be determined using crystals containing subunits a and b in roughly equal amounts (1).

Both the N-terminus and the C-terminus are identical in subunits a and b, so the difference in molecular weight found in PAA-SDS gel electrophoresis (2) is not due to terminal extensions or deletions. So far, no internal deletions in subunit b have been found either.

N-Acetylglucosamine has been found upon amino acid analysis of peptide 159-174, so carbohydrate is present in subunit b. It must be attached to Asn 167, like it is in subunit a, since no other Asn-X-Ser/Thr sequence occurs in this peptide and the DABTH derivative of Asn 167 could not be identified during sequence analysis of the peptide.

Since Fraction I (subunit a) and Fraction II (containing mainly subunit b) contained similar amounts of carbohydrate (9), we do not expect other glycosylation sites in subunit b.

Subunit a contains three disulfide bridges. Half-cystines of each of these three bridges at positions 98, 502 and 562 have been identified in subunit b. Since the three disulfide bridges are well defined in the electron density map (1), the same

```
  DOMAIN 1                        10                              20                            30
a Asp-Ala-Leu-Gly-Thr-Gly-Asn-Ala-Gln-Lys-Gln-Gln-Asp-Ile-Asn-His-Leu-Leu-Asp-Lys-Ile-Tyr-Glu-Pro-Thr-Lys-Tyr-Pro-Asp-Leu-
b                                          -Asn-
  |----------------------------------------------| |-------------------------------------------| |----------------------------
                                                                                                         |-----------------|
                                 40                              50                            60
a Lys-Glu-Ile-Ala-Glu-Asn-Phe-Asn-Pro-Leu-Gly-Asp-Thr-Ser-Ile-Tyr-Asn-Asp-His-Gly-Ala-Ala-Val-Glu-Thr-Leu-Met-Lys-Glu-Leu-
b    -Asp-                       -Asp-
  --| |------------------------|                                                                              |--------
                                 70                              80                            90
a Asn-Asp-His-Arg-Leu-Leu-Glu-Gln-Arg-His-Trp-Tyr-Ser-Leu-Phe-Asn-Thr-Arg-Gln-Arg-Lys-Glu-Ala-Leu-Met-Leu-Phe-Ala-Val-Leu-
b                                              -Phe-
  |--------------| |------------------------| |-------------------------------------| |-----|
                   |--------------------|       100                           110                          120
a Asn-Gln-Cys-Lys-Glu-Trp-Tyr-Cys-Phe-Arg-Ser-Asn-Ala-Ala-Tyr-Phe-Arg-Glu-Arg-Met-Asn-Glu-Gly-Glu-Phe-Val-Tyr-Ala-Leu-Tyr-
b
      |----------------------| |---------------------| |-----|
                                 130                            140                           150
a Val-Ser-Val-Ile-His-Ser-Lys-Leu-Gly-Asp-Gly-Ile-Val-Leu-Pro-Pro-Leu-Tyr-Gln-Ile-Thr-Pro-His-Met-Phe-Thr-Asn-Ser-Glu-Val-
b                                                                                  |------------------------DOMAIN 2
                                 160          CHO               170          rDOMAIN 2     180
a Ile-Asp-Lys-Ala-Tyr-Ser-Ala-Lys-Met-Thr-Gln-Lys-Gln-Gly-Thr-Phe-Asn-Val-Ser-Phe-Thr-Gly-Thr-Lys-Lys-Asn-Arg-Glu-Gln-Arg-
b                                                              -Pro-
  ----------| |------------------| |--------------------------------------------| |-| |-----| |------------
                                 190       *                   *       200                    210
a Val-Ala-Tyr-Phe-Gly-Glu-Asp-Ile-Gly-Met-Asn-Ile-His-His-Val-Thr-Trp-His-Met-Asp-Phe-Pro-Phe-Trp-Trp-Glu-Asp-Ser-Tyr-Gly-
b
  |----------|
                                 220          *                 230                           240
a Tyr-His-Leu-Asp-Arg-Lys-Gly-Glu-Leu-Phe-Phe-Trp-Val-His-His-Gln-Leu-Thr-Ala-Arg-Phe-Asp-Phe-Glu-Arg-Leu-Ser-Asn-Trp-Leu-
b                     |---------------------------------------------| |------------------| |-------------
                                 250                            260                           270
a Asp-Pro-Val-Asp-Glu-Leu-His-Trp-Asp-Arg-Ile-Ile-Arg-Glu-Gly-Phe-Ala-Pro-Leu-Thr-Ser-Tyr-Lys-Tyr-Gly-Gly-Glu-Phe-Pro-Val-
b
  ----------------------------------| |----------| |-------------------------------------|
                                 280                            290                           300
a Arg-Pro-Asp-Asn-Ile-His-Phe-Glu-Asp-Val-Asp-Gly-Val-Ala-His-Val-His-Asp-Leu-Glu-Ile-Thr-Glu-Ser-Arg-Ile-His-Glu-Ala-Ile-
b                                                                                                           -Asp-
                                                                                          |--------------------
                                 310                            320                           330
a Asp-His-Gly-Tyr-Ile-Thr-Asp-Ser-Asp-Gly-His-Thr-Ile-Asp-Ile-Arg-Gln-Pro-Lys-Gly-Ile-Glu-Leu-Leu-Gly-Asp-Ile-Ile-Glu-Ser-
b
  --------------------------------------------------| |----------|
                                 340          *                 350                           360
a Ser-Lys-Tyr-Ser-Ser-Asn-Val-Gln-Tyr-Tyr-Gly-Ser-Leu-His-Asn-Thr-Ala-His-Val-Met-Leu-Gly-Arg-Gln-Gly-Asp-Pro-His-Gly-Lys-
b                                                                                          |-------------------------
                                 370                            380       *                   390
a Phe-Asn-Leu-Pro-Pro-Gly-Val-Met-Glu-His-Phe-Glu-Thr-Ala-Thr-Arg-Asp-Pro-Ser-Phe-Phe-Arg-Leu-His-Lys-Tyr-Met Asp-Asn-Ile-
b
  |--------------------------------------------------| |------------------| |---------| |--------------------
                                 rDOMAIN 3          410                     420
a Phe-Lys-Lys-His-Thr-Asp-Ser-Phe-Pro-Pro-Tyr-Thr-His-Asp-Asn-Leu-Glu-Phe-Ser-Gly-Met-Val-Val-Asn-Gly-Val-Ala-Ile-Asp-Gly-
b
  ------|
                                 430                            440                           450
a Glu-Leu-Ile-Thr-Phe-Phe-Asp-Glu-Phe-Gln-Tyr-Ser-Leu-Ile-Asn-Ala-Val-Asp-Ser-Gly-Glu-Asn-Ile-Glu-Asp-Val-Glu-Ile-Asn-Ala-
b
                                 460                            470                           480
a Arg-Val-His-Arg-Leu-Asn-His-Lys-Glu-Phe-Thr-Tyr-Lys-Ile-Thr-Met Ser-Asn-Asn-Asn-Asp-Gly-Glu-Arg-Leu-Ala-Thr-Phe-Arg-Ile-
b                                 -Asn-
          |--------------------------------------------| |---------------------------------| |-----------------|
                      |-----------490-----------|              |-------500-------|           510
a Phe-Leu-Cys-Pro-Ile-Glu-Asp-Asn-Asn-Gly-Ile-Thr-Leu-Thr-Leu-Asp-Glu-Ala-Arg-Trp-Phe-Cys-Ile-Glu-Leu-Asp-Lys-Phe-Phe-Gln-
b                                                                                 |---------------------------
                                                                                  |------------------------------| |---------
                                 520                            530                           540
a Lys-Val-Pro-Lys-Gly-Pro-Glu-Thr-Ile-Glu-Arg-Ser-Ser-Lys-Asp-Ser-Ser-Val-Thr-Val-Pro-Asp-Met-Pro-Ser-Phe-Gln-Ser-Leu-Lys-
b --| |       -Ser-
  ---| |--------------------------------------| |---------|
                                 550                            560                           r--------570-
a Glu-Gln-Ala-Asp-Asn-Ala-Val-Asn-Gly-Gly-His-Asp-Leu-Asp-Leu-Ser-Ala-Tyr-Glu-Arg-Ser-Cys-Gly-Ile-Pro-Asp-Arg-Met-Leu-Leu-
b
  |--------------------------------------------| |-------------------------| |-----------
                      |-------580-------|              |-------590-------|           600-
a Pro-Lys-Ser-Lys-Pro-Glu-Gly-Met-Gly-Phe-Asn-Leu-Tyr-Val-Ala-Val-Thr-Asp-Gly-Asp-Lys-Asp-Thr-Glu-Gly-His-Asn-Gly-Gly-His-
b                                        -Lys-
  ------| |------------------------|
                                 610                            620                           630
a Asp-Tyr-Gly-Gly-Thr-His-Ala-Gln-Cys-Gly-Val-His-Gly-Glu-Ala-Tyr-Pro-Asp-Asn-Arg-Pro-Leu-Gly-Tyr-Pro-Leu-Glu-Arg-Arg-Ile-
b                                                                                                           |---
                                 640                            650                  657
a Pro-Asp-Glu-Arg-Val-Ile-Asp-Gly-Val-Ser-Asn-Ile-Lys-His-Val-Val-Val-Lys-Ile-Val-His-His-Leu-Glu-His-His-Asp
b
  --------------| |----------------------------------| |----------------| |-----------------------------|
```

Fig. 1   Amino acid sequence of subunit a of <u>Panulirus</u> <u>interruptus</u> hemocyanin and
partial sequence of subunit b. Only residues of subunit b different from the
homologous ones in subunit a are given. Peptides of subunit b which were isolated
and sequenced are indicated beneath the sequence by dashed lines.
Disulfide bridges, carbohydrate attachment site (CHO), copper ligands (*) and
boundaries between the domains of subunit a are indicated.
Additional evidence for the sequence of subunit b will be published elsewhere.

three disulfide bridges are expected in subunit b.

Glutamine at position 163 in subunit a is replaced by proline in subunit b. This
substitution is located near an inter-subunit contact region. Perhaps this
substitution (partly) explains why, at pH 8.5, subunit a is in the monomeric form
and subunit b largely in the hexameric form (10).

In the same region are the sites, behind positions 158, 162 and 175, susceptible
to limited proteolysis (11). Native hemocyanin is not cleaved, the sites are only
susceptible in the monomers. Under mild conditions, plasmin cleaves chain a behind
Lys 175, generating a 23K fragment. Cleavage behind Lys 162 then results in an 18K
fragment. This second cleavage behind Lys 162 does not occur in subunit b, where
glutamine 163 is substituted by proline. This explains why after limited cleavage
of subunit b with plasmin only a 23K fragment was observed and no 18K fragment.

In the accompanying paper (7), the substitutions in subunit b are discussed in
relation to a general comparison of arthropod hemocyanin chains (12).

REFERENCES

1. Gaykema, W.P.J., Hol, W.G.J., Vereijken, J.M., Soeter, N.M., Bak, H.J. and
   Beintema, J.J., Nature 309, 23-29 (1984).
2. van Eerd, J.P. and Folkerts, A., in Invertebrate Oxygen-binding proteins (eds.
   Lamy, J. and Lamy, J.) 139-149, Marcel Dekker, New York (1981).
3. Folkerts, A. and van Eerd, J.P., in Invertebrate Oxygen-binding proteins (eds.
   Lamy, J. and Lamy, J.) 215-225, Marcel Dekker, New York (1981).
4. Champion, A.B., Soderberg, K.L. Wilson, A.C. and Ambler R.P., J. Mol. Evol. 5,
   291-305 (1975).
5. Bak, H.J., Soeter, N.M., Vereijken, J.M., Jekel, P.A., Neuteboom, B. and
   Beintema, J.J., in Invertebrate Oxygen Carriers (ed. Linzen, B), Springer,
   Berlin, pp. 149-152.
6. Neuteboom, B., Beukeveld, G.J.J. and Beintema J.J., in Invertebrate Oxygen
   Carriers (ed. Linzen, B.), Springer, Berlin, pp. 169-172.
7. Soeter, N.M. and Beintema, J.J., in Invertebrate Oxygen Carriers (ed. Linzen,
   B.), Springer, Berlin, pp. 153-163.
8. Chang, J.Y., Brauer, D. and Wittmann-Liebold, B., FEBS Lett. 93, 205-214
   (1978).
9. van den Berg, A.A., Gaastra, W. and Kuiper, H.A., in Structure and function of
   hemocyanin (ed. Bannister, J.V.) 6-12, Springer, Berlin (1977).
10. Vereijken, J.M. et al., unpublished results.
11. Vereijken, J.M., Schwander, E.H., Soeter, N.M. and Beintema, J.J., Eur. J.
    Biochem. 123, 283-289 (1982).
12. Linzen, B., Soeter, N.M., Riggs, A.F. et al., Science 229, 519-524 (1985).

AMINO ACID SEQUENCE STUDIES ON THE c CHAIN OF PANULIRUS INTERRUPTUS HEMOCYANIN

B. NEUTEBOOM, G.J.J. BEUKEVELD, J.J. BEINTEMA. Biochemisch Laboratorium, University of Groningen, Nijenborgh 16, 9747 AG, The Netherlands.

ABSTRACT. A method is described to isolate the c chain in amounts sufficient for sequencing. The present state of the elucidation of the amino acid sequence is described. A comparison with the a chain is made.

INTRODUCTION

Native Panulirus interruptus hemocyanin consists of 45% a chain, 45% b chain and 10% c chain. The isolation procedure as described by van Eerd and Folkerts (1) was modified to obtain c chain in a sufficiently high yield. Isolated c chain was cleaved with CNBr. The amino acid sequence of five CNBr fragments is presented in this paper. The high percentage of substituted amino acids is in agreement with the absence of immunological cross-reactivity between a chain and c chain (2).

METHODS AND RESULTS

Procedure for the isolation of c chain

Native hemocyanin was dialysed for two days against 50 mM Tris-HCl pH 8.7 containing 5 mM EDTA. At this pH the hexamers are largely dissociated into monomers. For separation of a chain from b and c chains, ion-exchange chromatography with a KCl gradient was used (Fig. 1-I).
The b + c mixture was dialysed for two days against 50 mM Tris-HCl pH 8.7 containing 5 mM EDTA and 2 M urea. Ion-exchange chromatography was performed with a NaCl gradient (Fig. 1-II). The purity of the isolated chains was tested on alkaline polyacrylamide gels (3).

Sequence determination

3.2 micromoles (240 mg) of c chain was reduced and carboxymethylated (4) and cleaved with cyanogen bromide (5). The mixture of CNBr fragments was applied to a Sephadex G-100 column (Fig. 2). Fragments 1, 2, 3 and 4 were purified by reversed-phase HPLC. Fragment 5 was purified on a Bio-gel P-10 column (200 x 1.2 cm, eluent 0.25% formic acid). The fragments were sequenced by Edman degradation. To elucidate the C-terminal sequences the fragments were sub-digested with Staphylococcus aureus protease V8 (5), and sub-fragments were sequenced manually with the DABITC method (6). Homology with the a chain was used to align the sub-fragments.

The amino acid sequence of the five fragments, together comprising 254 residues, is presented in Fig. 3.

Invertebrate Oxygen Carriers
Ed. by Bernt Linzen
© Springer–Verlag Berlin Heidelberg 1986

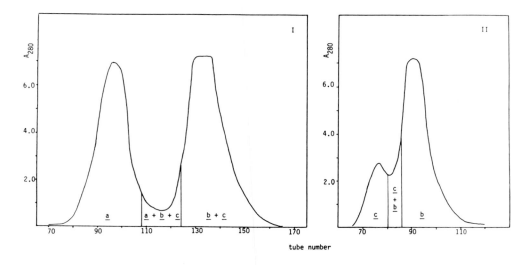

Fig. 1. Isolation of the a, b and c chains of the hemocyanin from Panulirus in-
terruptus by ion-exchange chromatography on DEAE-Sepharose CL-6B (35 x 4.0 cm).
I : Elution of the a + b + c mixture with 50 mM Tris-HCl pH 8.7 + 5 mM EDTA
    with a KCl gradient (0.1-0.5 M, 1.5 + 1.5 l). Fraction volume 15 ml.
II: Elution of the b + c mixture with 50 mM Tris-HCl pH 8.7 + 5 mM EDTA + 2 M
    urea, with a NaCl gradient (0.1-0.4 M, 1 + 1 l). Fraction volume 10 ml.

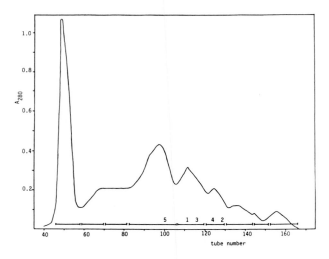

Fig. 2. Fractionation of a CNBr digest of carboxymethylated c chain. Two columns of Sephadex G-100 (200 x 2.0 cm) in series were used. Eluent, 30% acetic
acid; fraction volume 6.7 ml; flow rate, 15 ml/h. Pools are indicated by hori-
zontal bars. The positions of the five fragments are indicated.

1: Ala-Asp-Cys-Gln-Ala-Gly-Asp-Ser-Ala-Asp-Lys-Leu-Leu-Ala-Gln-Lys-Gln-His-Asp-Val-Asn-Tyr-Leu-Val-Tyr-Lys-
   Leu-Tyr-Gly-Asp-Ile-Arg-Asp-Asp-His-Leu-Lys-Glu-Leu-Gly-Glu-Thr-Phe-Asn-Pro-Gln-Gly-Asp-Leu-Leu-Leu-Tyr-
   His-Asp-Asn-Gly-Ala-Ser-Val-Asn-Thr-Leu-Met.[57]

2: Asn-Glu-Gly-Glu-Tyr-Leu-Tyr-Ala-Leu-Tyr-Val-Ser-Leu-Ile-His-Ser-Gly-Leu-Gly-Glu-Gly-Val-Val-Leu-Pro-Pro-
   Leu-Tyr-Glu-Val-Thr-Pro-His-Met.[144]

3: Phe-Thr-Asn-Ser-Glu-Val-Ile-His-Glu-Ala-Tyr-Lys-Ala-Gln-Met-Thr-Asn-Thr-Pro,Ser,Lys-Phe-Glu-Ser-His-Phe-
   Thr-Gly,Ser,Lys-Lys-Asn-Pro-Glu-Gln-His-Val-Ala-Tyr-Phe-Gly-Glu-Asp-Val-Gly-Met.[190]

4: Asn-Ser-His-Gly-Ile-Glu-Phe-Leu-Gly-Asp-Ile-Ile-Glu-Ser-Ser-Gly-Tyr-Ser-Ala-Asn-Pro-Gly-Phe-Tyr-Gly-Ser-
   Leu-His-Asn-Thr-Ala-His-Ile-Met.[350]

5: Glu-Phe-Asn-Leu-Val-Val-Ala-Val-Thr-Asp-Gly-Arg-Thr-Asp-Ala-Ala-Leu-Asp-Asp-Leu-His-Glu-Asn-Thr-Lys-Phe-
   Ile-His-Tyr-Gly-Tyr-Asp-Arg-Gln-Tyr-Pro-Asp-Lys-Arg-Pro-His-Gly-Tyr-Pro-Leu-Asp-Arg-Arg-Val-Asp-Asp-Glu-
   Arg-Ile-Phe-Glu-Ala-Leu-Pro-Asn-Phe-Lys-Gln-Arg-Thr-Val-Lys-Leu-Tyr-Ser-His-Glu-Gly-Val-Asp-Gly-Gly.[657]

Fig. 3. The amino acid sequence of five CNBr fragments of the c chain. The numbering refers to the homologous sequences in the a chain (7). Underlined residues are different in the a chain.

DISCUSSION

The amino acid sequences of the five fragments are homologous with sequences of the a chain (7). The mean percentage of substitutions is 45%, which is in agreement with the predicted percentage of at least 30% based on immunological experiments (2). There is an N-terminal extension of six residues. The carbohydrate attachment site of the a chain (positions 167-169) has been substituted completely. The C-terminal CNBr fragment is completely different at positions 593-615. A deletion of three amino acid residues had to be introduced in the c chain for alignment of the C-terminal positions 616-657, starting with the constant Tyr-Pro-Asp sequence. The most striking feature is that cysteine residue 609 in the a chain, which is in a disulfide linkage with cysteine residue 562, is not present in the c chain.

REFERENCES

1) J.P. van Eerd and A. Folkerts, in Invertebrate Oxygen-binding Proteins (eds. U. Lamy and J. Lamy), 139-149 (Marcel Decker, New York, 1981).

2) A. Folkerts and J.P. van Eerd, in Invertebrate Oxygen-binding Proteins (eds. J. Lamy and J. Lamy), 215-225 (Marcel Decker, New York, 1981).

3) H.R. Maurer, in Disc Electrophoresis and Related Techniques of Polyacrylamide Gel Electroforesis, 2nd ed. p. 44, De Gruyter, Berlin, New York.

4) A.M. Crestfield, S. Moore and W.H. Stein (1963).J. Biol. Chem. 238, 622-627.

5) J.M. Vereijken, J. de Vlieg and J.J. Beintema (1984). Bioch. Bioph. Acta 788, 298-305.

6) J.Y. Chang, D. Brauer, and B. Wittmann-Liebold (1978) FEBS Lett. 93, 205-214.

7) H.J. Bak, N.M. Soeter, J.M. Vereijken, P.A. Jekel, B. Neuteboom and J.J. Beintema. This volume, pp. 149-152.

PARTIAL AMINO ACID SEQUENCE OF CRAYFISH (ASTACUS LEPTODACTYLUS) HEMOCYANIN

H.-J. Schneider, W. Voll, L. Lehmann, R. Grißhammer, A. Goettgens and B. Linzen

Zoologisches Institut, Universität München, Luisenstr. 14, 8000 München, F.R.G.

Introduction

There is a marked heterogeneity of arthropodan hemocyanin (Hc) subunits (1) which raises the question of the functional and evolutionary significance of such diversity. In particular it has often been contended, mainly on the basis of SDS electrophoresis, that the crustacean Hc subunits have a much higher molecular weight than the chelicerate Hc chains. Following the elucidation of two chelicerate Hc amino acid sequences (2,3) we have therefore attempted to sequence a Hc subunit from a crayfish, Astacus leptodactylus. This Hc partially dissociates at alkaline pH into a heterodimer (7S) and four different monomers (a, b_1, b_2, c; 5S; Mr about 75,000) (4,5). b_1 and b_2 are closely related; their mixture, b, served as starting material for the present investigation.

Experimental

Astacus blood was sampled by tapping the abdominal blood sinus. After clotting, the blood was cleared by centrifugation. For dissociation the hemolymph proteins (5 mg/ml) were dialysed against glycine/NaOH buffer, 10 mM EDTA, I=0,05, pH 9.6 at 4 °C for 48 h. The Hc was then concentrated to 10-20 mg/ml and monomers, dimer and undissociated Hc separated on Ultrogel AcA 44, 5.3x130 cm).

The monomers were fractionated on a DEAE Sepharose CL 6B column (2x32 cm) equilibrated with 50 mM Tris/HCl, pH 8.8, containing 1 mM EDTA and 1 M urea. Elution of the protein was carried out with a linear gradient of 0.1 M to 0.4 M KCl in starting buffer at a flow rate of 10-20 ml/h. Chain b is the first to be eluted at a salt concentration of about 0.23 M KCl. The protein was transferred to 6 M guanidine hydrochloride, 1 M Tris/HCl, pH 8.2, reduced and carboxymethylated.

Enzymatic digestions were performed with trypsin, Staphylococcus aureus V8 protease or chymotrypsin. Chemical cleavage at the methionine residues was performed in 70% formic acid with cyanogen bromide. The peptide mixtures were fractionated over Sephacryl S 200 sf and Biogel P 10 in 5% acetic acid. Final purification was usually achieved by HPLC. The purity of the peptides was checked by end group analysis of a small aliquot, using the method of Chang et al. (6). The amino acid analyses were done on 1-2 nmol samples (kindly run by Dr. W. Schartau). Manual sequence analysis was by the method of Chang et al. (6). For automatic sequencing (kindly performed by Dr. F. Lottspeich, MPI for Biochemistry, Martinsried), the Edman degradation method (7) in the sequenator version was employed. The PTH derivatives were identified by isocratic HPLC (8).

Invertebrate Oxygen Carriers
Ed. by Bernt Linzen
© Springer–Verlag Berlin Heidelberg 1986

174

Results and Discussion

Figs. 1 and 2 illustrate the isolation of subunit b from the dissociation mixture. The purity of the fractions was checked by SDS-PAGE. The yield of subunit
b was low: at best ca. 7% of the starting Hc, normally only ca. 5%.

In contrast to Eurypelma Hc limited proteolysis could not be applied to
Astacus Hc. Therefore the whole chain was reduced, carboxymethylated and cleaved
enzymatically or by cyanogen bromide. The resulting peptides were purified by
gel filtration and HPLC.

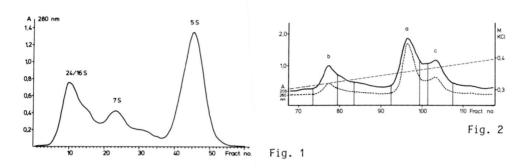

Fig. 1

Fig. 2

Fig. 1. Gel filtration of the dissociated hemolymph on Ultrogel AcA 44, with gly-
cine/NaOH, pH 9.6, I=0.05 as eluent. Sedimentation coefficients are indicated.

Fig. 2. Fractionation of the monomer (5 S) peak of Fig. 1 (see Experimental).

The peptides were arranged by means of overlapping sequences or by alignment
with homologous sequences of chelicerate Hc's or Panulirus interruptus Hc chain
a (9,10). 562 position have been determined.

The partial sequence is shown in Fig. 3. By comparing it to the sequence of
Panulirus interruptus Hc, subunit a (10) one finds 366 (65%) identical residues
and 64 (11%) residues with isofunctional side chains so that overall homology is
high. This agrees with the finding that P. interruptus a and A. leptodactylus b
both belong to the alpha type of crustacean Hc subunits (11). The highest degree
of homology was found in domain 2 which comprises the active site with the two
groups of histidine residues. The active site sequences reveal features, which
in comparison with chelicerate Hc might be termed typically crustacean: HH**V**TWH
in place of HH**W**HWH at the Cu(A) site, and HNT**A**H in place of HN**W**GH at the Cu(B)
site. The removal of a Trp from both sites and inclusion of a Thr residue is notable. Other features common to the crustacean chains include a N-terminal extension of 5 residues, an insertion of 21 residues (starting at pos. 20) and another

```
1                     10                    20                    30                    40
D A S G A T L A K R Q Q V V N H L L E H I Y D H T H F T D L K N I A G T F S P E
         A                           50                    60                    70                    80
A   T S(I)Y T D D . . . . . . .(M)E E L R D G R L L E Q H H W F S L F N T R Q R
    D
                           90                   100                   110                   120
. . . .(M)L F E V L I H C K S W E C F L D N A A Y F R E R M N E G E F V Y A(I)Y
                      130                   140                   150                   160
V A V(I)H S G(I)G H G(I)V(I)P P(I)Y E V T P H(I)F T N S E V(I)N K A Y S   K M T
                                                                             A
                                                                             G
                      170                   180                   190                   200
Q T P G R F N M D F T G T K K N K E Q R V A Y F G E D I G M N I H H V T W H M D
                      210                   220                   230                   240
F P F W W K D S Y G Y H L D R K G E L F F W V H H Q L T A R F D S E R(I)S N W(I)
         (I)          250                   260                   270                   280
D V V D E   H W S C I - - E G F A P H T S Y K Y G G E F P A R P D N V H F E D V
         G
                      290                   300                   310                   320
D G V A R V R D . . . . . S R I R D A L A H G Y L L D N S G N K . . . . . . . .
                      330                   340                   350                   360
. . . . . . . . . S S V Y S P N V Q Y Y G A(I)H N T A H(I)M(I)G R Q G D - H - K
                      370                   380                   390                   400
F D M P P G V M E H F E T A T R D P S F F R L H K Y M D N I F K E H K D S L P P
                      410                   420                   430                   440
Y T K N D(I)A V P G V V(I)D S V A . . . Q L K T F F D T F E V N L G N A K . . .
                      450                   460                   470                   480
. . V A D V A I S A D V H R(I)N H E E F S Y N(I)D(I)S N T D K . . . . . . . .(I)
                      490                   500                   510                   520
F(I)C P V K D D N G I . . . . . . . . . . . . . . M D K F Y K S(I)  P G T N H(I)V
                                                                   A
                      530                   540                   550                   560
R K S V D S S V T V P D R Q Y A . . . . . . . . . . . . . . . . . L D L H M F E R
                      570                   580                   590                   600
S C G I P N R D M(I I I)E S R P D G M D F A L F V . . . . . . . . T V - - D D P
                      610                   620                   630                   640
E E(I)G A T H S Q(H)G(I)K . . K Y P D K K P M G Y P V D R S I P D N R V F L E S
                      650                   660
P N I K R T Y V K V F H D E H G G E Q H -COOH
```

Fig. 3. Partial amino acid sequence of <u>Astacus leptodactylus</u> hemocyanin, subunit <u>b</u>. (I), Ile ore Leu (not distinguished); ( ) identification tentative; two symbols at any given position show true microheterogeneity.

one of 8 residues (starting at pos. 563). All these segments are at the surface
of the subunit, are exposed to the solvent and do not make contacts to other subunits. Their deletion in the chelicerate Hc's might have occurred without major
consequences. On the other hand, in both crustacean Hc's there is a deletion of
5 residues starting at pos. 464. The clustering of His near the C-terminus is
more pronounced in the crustacean Hc's than in the chelicerate Hc's. It appears
that the chain length of Astacus b will be very similar to that of P. interruptus
Hc a which would result in a Mr of about 75,000. Thus, the previous estimate of
ca. 78,000 was too high, although the error has not been as great as in the case
of the Panulirus Hc (4).

One notable difference between the Astacus b and the Panulirus a sequence is
the carbohydrate attachment site -N-V-S-F- at position 167 in P. interruptus. In
Astacus b this has mutated to -N-M-D-F, and carbohydrate is no longer present.
Up to now, carbohydrate has not been detected at any other position in Astacus b.

Acknowledgement. This work was supported by grants from the Deutsche Forschungsgemeinschaft (Li 107/22-24, Schn 226/2, Scha 317/3) and from the Fonds der Chemi-
schen Industrie.

References

1. Linzen, B., Life Chemistry Rep. suppl. 1 (Wood, E.J., ed.). Harwood Acad.
   Publ., Chur-London-New York, pp. 27-38 (1983).
2. Schneider, H.-J., Drexel, R., Feldmaier, G., Linzen, B., Lottspeich, F., and
   Henschen, A., Hoppe-Seyler's Z. Physiol. Chem. 364, 1357-1381 (1983).
3. Schartau, W., Eyerle, F., Reisinger, P., Geisert, H., Storz, H., and Linzen,
   B., Hoppe-Seyler's Z. Physiol. Chem. 364, 1383-1409 (1983).
4. Markl, J., Hofer, A., Bauer, G., Markl, A., Kempter, B., Brenzinger, M., and
   Linzen, B., J. Comp. Physiol. 133, 167-175 (1979).
5. Stöcker, W., Doctoral dissertation, Univ. Munich (1984).
6. Chang, J.Y., Brauer, D., and Wittmann-Liebold, B., FEBS Lett. 93, 205-214
   (1978).
7. Edman, P., and Henschen, A., in Protein Sequence Determination, 2nd edn.
   (Needleman, S.B., ed.). Springer-Verlag, Berlin, pp. 232-279 (1975).
8. Lottspeich, F., Hoppe-Seyler's Z. Physiol. Chem. 361, 1829-1834 (1980).
9. Linzen, B., Soeter, N.M., Riggs, A.F., et al., Science 229, 519-524 (1985).
10. Bak, H.J., Soeter, N.M., Vereijken, J.M. Jekel, P.A., Neuteboom, B., and
    Beintema, J.J., this volume, pp. 149-152 (1986).
11. Markl, J., Stöcker, W., Runzler, R., and Precht, E., this volume, pp. 281-292
    (1986).

HEMOCYANIN OF THE SPIDER EURYPELMA CALIFORNICUM: AMINO ACID SEQUENCE OF SUBUNIT a AND OF THE SMALLER CB-PEPTIDES OF SUBUNITS b AND c

W. Schartau, W. Metzger, P. Sonner and W. Pysny
Zoologisches Institut, Universität München, Luisenstr. 14, 8000 München 2, F.R.G.

Introduction

Hemocyanins (Hcs) of arthropods are high molecular weight oxygen carrying proteins (M_r up to 3.5 million) which are composed of different subunits (for reviews see 1,2). The smallest functional unit is the hexamer and depending on the species, 1-, 2-, 4-, and 8-hexamers occur. Heterogeneity of the subunits varies greatly; up to 8 different subunit types per species have been reported (2,3). From models of the quarternary structure (4,5) it becomes clear that each subunit type occupies a specific place within the oligomer. Recently the first amino acid sequences of different Hc-subunits (**Eury d** (6), **Eury e** (7), **Tachy α** (8), **Lim II** (9), **Pint a** (10)) became available, and X-ray crystallography of **Panulirus** subunit a revealed the first Hc tertiary structure (11). Comparison of the amino acid sequences with the tertiary structure of **Panulirus** Hc strongly supports a common tertiary structure for all arthropod Hcs (12). Nevertheless, for the understanding of the complex structure-function relations in Hcs, elucidation of the primary structure of each individual subunit type of one species becomes essential.

In the spider **Eurypelma californicum**, 7 different subunits form a 4-hexamer structure. The amino acid sequences of subunit d (6), and e (7) have recently been established, and the complete covalent structure of subunit d has been determined (13). In this paper we present the nearly complete amino acid sequence of subunit a and sequences of the smaller CB-peptides of subunits b and c.

Materials and Methods

Subunit a was isolated from the monomer fraction by a two-step ion exchange chromatography. The heterodimer bc was separated by ion exchange chromatography from the homodimer ff. Dialysis against TRIS-HCl-buffer, 4 M urea, 1 mM EDTA for 48 h and subsequent ion exchange chromatography in the same buffer yielded the pure subunits b and c. Cyanogen bromide (CB) cleavage was performed according to Gross and Witkop (14). Enzymatic digestion with trypsin, chymotrypsin, S. aureus proteinase and Astacus proteinase were performed at 37 °C as previously described (6). The peptide mixtures were purified by reversed phase HPLC with an acidic and/or neutral system. Manual sequence determination was performed with the "DABITC"-reagent according to Chang et al. (15).

Invertebrate Oxygen Carriers
Ed. by Bernt Linzen
© Springer–Verlag Berlin Heidelberg 1986

Results and Discussion

Sequence strategy: The polypeptide chains were split either by limited proteolysis or with cyanogen bromide. The resulting fragments were subcleaved with the appropriate enzymes, yielding peptides suitable for manual sequence determination with the "DABITC"-reagent.

I. Subunit a

By limited chymotrypsinolysis subunit a is split mainly into two fragments with molecular weights of about 25 (a-CH_n 25) and 40 (a-CH_n 40) kDa, respectively. After separation by cation exchange chromatography (Fig. 1) they were subcleaved with trypsin, chymotrypsin and the S. aureus proteinase, respectively. The peptide mixtures were purified on reversed phase HPLC.

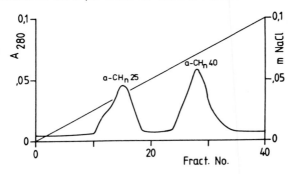

Fig. 1. Separation of a-CH_n 25 and a-CH_n 40 by cation exchange chromatography. 0.05 M Na-acetate buffer, pH 4.5, 8 M urea. Gradient: 0-0.15 M NaCl (200 + 200 ml).

After CB-cleavage, the peptide mixture was prepurified by gel filtration on Bio Gel P 10 (Fig. 2) and the smaller CB-peptides separated by HPLC (fractions III-VIII in Fig. 2).

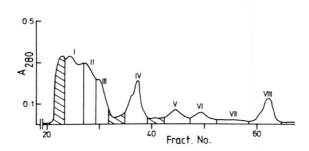

Fig. 2. Gel filtration of the CB-peptides of subunit a on Bio Gel P 10 in 5 % acetic acid.

Due to the high degree of homology in amino acid sequences between subunit a and subunit d (6) and e (7), the larger CB-peptides of subunit a (fractions I and II in Fig. 2) were not further purified but the mixture directly subcleaved with the common enzymes. The resulting peptide mixtures were also purified by HPLC.

Fig. 3 shows the amino acid sequence of **Eurypelma** subunit a. Domain 1, 2, and 3 and the copper A and copper B binding sites, which show the typical sequence of two histidine residues separated by three amino acid residues and the third histidine necessary for copper liganding some 20 amino acid residues apart (11) are indicated. There are only two small gaps at pos. 104 and 115 respectively, comprising about 7 amino acid residues. The position of the peptide starting at pos. 108 is not clear yet: after cleavage with S. aureus protease the expected peptide starting with Gly-Leu-... has not yet been found. A section which presented particular difficulties to sequence analysis is the one starting at pos. 156 with a threonine. Two steps later there are three threonine residues in sequence, followed by a glutamine which, in the DABITC determination, is at the same position as the third threonine spot (threonine appears as three spots on the DABITC-chromatogram).

If the sequence of subunit a is compared to **Eury** d, **Eury** e, **Tachy** α and **Lim** II, there are about 55 % identical residues; if the isofunctional exchanges are taken into account, homology increases by further 11-12 %. If we compare domain per domain, a significantly higher degree of homology occurs between domain 1 of **Eury** a and **Lim** II (Table 1). Out of 175 compared amino acid residues there are 81 identities which makes a plus of 14 % if compared to the other subunits. In the quaternary structure (4,5), **Eury** a and **Lim** II are homologous , i.e. they occupy the same place within the oligomer. Therefore we conclude that the sequence information for the localization of a subunit within the aggregate should be contained in the first domain. This conclusion is supported by comparison of **Eury** e to the other Hcs: it is again domain 1 which shows remarkable differences in the degree of homology. Identities compared to **Tachy** α are about 10 % higher than those to other Hcs (Table 2). **Eury** e and **Tachy** α are immunologically related (18) and therefore might be placed at homologous positions within the molecule. Furthermore the 3-D-structure of **Panulirus** Hc shows that domain 1 makes contacts to other subunits (11).

Table 1. Comparison of domain 1 of **Eury** a to the other cheliceratan Hcs

	ident. AA.	%
Eury d	59	41
Eury e	63	44
Lim II	81	56
Tachy α	59	41

Table 2. Comparison of domain 1 of **Eury** e to the other cheliceratan Hcs.

	ident. AA.	%
Eury a	65	43
Eury d	66	44
Lim II	69	45
Tachy α	82	55

Domain 1   67
T L I H D K Q V Q A L S L F E K I S V A A T G E P V P I D Q D R L R N L T T L G
a-CH$_n$ 25
                                                                              107
P N E F F S C F Y P D H L E Q A K R V Y E V F C H A A N F D D F V S I A

                                                                              147
(K G F N E G L)   S A E V A L L H R E D C R G V T V P P V Q E V F A D R F I P

                                                            |Domain 2   187
A D S I N I A F T I A T T T Q P G D E S D I I V D V K D T G N L L|D P E Y K L A

                                                                              227
Y F R E D I G V N A H H̊ W H W H̊ V V Y P S T Y D P A F F G K V K D R K G E L F Y

                                                                              267
Y M H̊ Q Q M C A R Y D C E R L S N G L N R M I P F H N F N E P L G G Y|A A H L T
                                                             |a-CH$_n$ 40
                                                                              308
H V A S G R H Y A Q R P D G L A M H D V R E V D V Q D M E R W T E R I M E A I D

                                                                              348
L R R V I S P T G E Y L P I D E E H G A D I L G A L I E S S Y E S K N R G Y Y G

                                                                              390
S L H̊ N W G H̊ V M M A Y I H D P D G R F R E N P G V M T D T A T S L R D P L F Y

                                                    |Domain 3   431
R Y H̊ R F I D N V F Q E Y K K T L|P V Y S K D N L D F P Q V T I T D V K V K A K

                                                                              475
I P N V V H T F L R E D E L E L S H C L H F K L Y I Q V D S F Y H H L D H E S F

                                                                              515
S Y I L S A Q N N S N A D K Q A T V R I F L A P T Y D E L G N D L S L D E Q R R

                                                                              555
L Y I E M D K F Y H T I R P G K N T L V R S S T D S S V T L S S V T V K E L L R

                                                                              603
G E D L V E G Q T E F C S C G K P Q H L T V P R G N E K G M Q F E L F V M L T D

                                                                              643
A S V D R V Q S G D G T P V C A D A L S Y C G V L D Q K Y P D K R A M G Y P F D

                                                    669
R K I T A D T H E E F L T G N M N I S H V T V R F Q

Fig. 3. Amino acid sequence of **Eurypelma** subunit a̲. Numbering of amino acids is according to (12). a-CH$_n$ 25 and a-CH$_n$ 40 are the̲ fragments generated by limited chymotrypsinolysis. Histidine residues involved in copper binding (copper A and copper B site) are marked with asterisks.

II. Heterodimer bc

In the 4-hexamer structure of **Eurypelma** Hc, the subunits b and c are present as the heterodimer bc. It was shown by reassembly experiments with isolated subunits that it functions as a linker between each two hexamers (16). For this specific role in the architecture of **Eurypelma** Hc, elucidation of the primary structure of both subunits is of specific interest.

The heterodimer bc was isolated from the dimer fraction (17) by a two step ion exchange chromatography (Fig. 4). 48 h dialysis against TRIS-HCl buffer pH 8.8, 4 M urea, 1 mM EDTA, and subsequent ion exchange chromatography in the same buffer yielded the pure subunits b and c (Fig. 5).

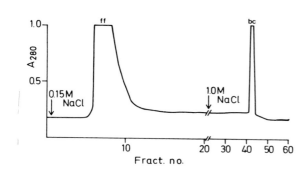

Fig. 4. Isolation of the heterodimer bc from the dimer fraction. Two step ion exchange chromatography in 0.1 M carbonate buffer, pH 9.6. 0.15 M NaCl: homodimer ff; 1.0 M NaCl: heterodimer bc. Note that the abszissa is changed at fraction 20.

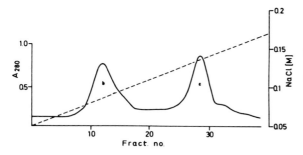

Fig. 5. Isolation of subunits b and c by ion exchange chromatography. Dissociation buffer (see text). Gradient: 0.05 - 0.2 M NaCl (200 + 200 ml).

Since dissociation of the heterodimer bc can be achieved simply by dialysis against alkaline buffer containing 4 M urea and 1 mM EDTA, both subunits are not linked covalently by SS-bridges but only by amino acid side chain interactions. Subunits b and c were cleaved with CB and the cleavage products submitted to gel-filtration on Bio Gel P10 in 5 % acetic acid (Figs. 6,7). The smaller CB-peptides (subunit b: fract. II-IX, Fig. 6; subunit c: fract. III-IX, Fig. 7) were purified on HPLC and the sequence determined manually with the "DABITC"-reagent (15). Many peptides were sequenced up to the end, the C-terminal homoserine/homoserine lac-

182

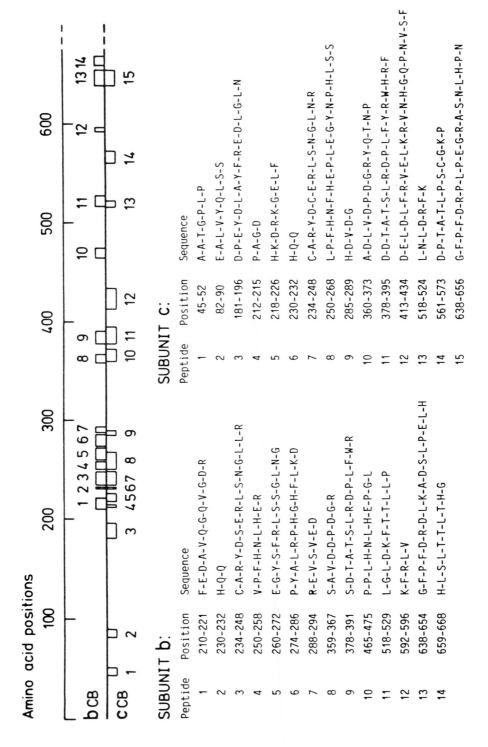

Fig. 8. Amino acid sequence and alignment of the CB-peptides of *Eurypelma* subunits b and c. Numbering of amino acids is according to (12). Leucine or isoleucine were not distinguished and are indicated with $\underline{L}$.

Amino acid positions

SUBUNIT b:

Peptide	Position	Sequence
1	210-221	F-E-D-A-V-Q-G-Q-V-G-D-R
2	230-232	H-Q-Q
3	234-248	C-A-R-Y-D-S-E-R-L-S-N-G-L-L-R
4	250-258	V-P-F-H-N-L-H-E-R
5	260-272	E-G-Y-S-F-R-L-S-S-G-L-N-G
6	274-286	P-Y-A-L-R-P-H-G-H-F-L-K-D
7	288-294	R-E-V-S-V-E-D
8	359-367	S-A-V-D-D-P-D-G-R
9	378-391	S-D-T-A-T-S-L-R-D-P-L-F-W-R
10	465-475	P-P-L-H-N-L-H-E-P-G-L
11	518-529	L-G-L-D-K-F-T-T-L-L-P
12	592-596	K-F-R-L-V
13	638-654	G-F-P-F-D-R-D-L-K-A-D-S-L-P-E-L-H
14	659-668	H-L-S-L-T-L-T-H-G

SUBUNIT c:

Peptide	Position	Sequence
1	45-52	A-A-T-G-P-L-P
2	82-90	E-A-L-V-Y-Q-L-S-S
3	181-196	D-P-E-Y-D-L-A-Y-F-R-E-D-L-G-L-N
4	212-215	P-A-G-D
5	218-226	H-K-D-R-K-G-E-L-F
6	230-232	H-Q-Q
7	234-248	C-A-R-Y-D-C-E-R-L-S-N-G-L-N-R
8	250-268	L-P-F-H-N-F-H-E-P-L-E-G-Y-N-P-H-L-S-S
9	285-289	H-D-V-D-G
10	360-373	A-D-L-V-D-P-D-G-R-Y-Q-T-N-P
11	378-395	D-D-T-A-T-S-L-R-D-P-L-F-Y-R-W-H-R-F
12	413-434	D-E-L-D-L-F-R-V-E-L-K-R-V-N-H-G-Q-P-N-V-S-F
13	518-524	L-N-L-D-R-F-K
14	561-573	D-P-T-A-T-L-P-S-C-G-K-P
15	638-656	G-F-P-F-D-R-P-L-P-E-G-R-A-S-N-L-H-P-N

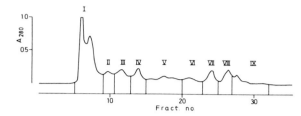

Fig. 6. Elution profile of the CB-peptides of subunit b on P 10 in 5 % acetic acid.

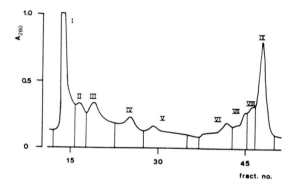

Fig. 7. Gel filtration of the CB-peptides of subunit c on Bio Gel P 10 in 5 % acetic acid.

tone was deduced from amino acid analysis. Since sequence homologies to **Eury a**, **d**, and **e** are rather high (ca. 45 %), arrangement of the CB-peptides of both sub-units within the whole chain was possible (Fig. 8). Up to now, sequencing of the smaller CB-peptides has revealed ca. 25 (subunit b) to 30 (subunit c) percent of the total structure. An interesting result is given by peptide **b CB 3**: Cys 239, common for all cheliceratan Hcs so far known and connected to Cys 234 in **Eury d** (13) by a disulfide bridge is replaced by a serine residue. Therefore, the previous conclusion that in all cheliceratan Hcs the secondary structure in this part of the chain should be the same (13) has possibly to be corrected. Even within one species, differences in the secondary and possibly tertiary structure of the Hc subunit have to be expected. Similarly, in subunit a of **Panulirus** Hc, there is a disulfide bridge from Cys 609 to Cys 562; in subunit c, Cys 609 is absent (19).

Acknowledgement. This work was supported by the Deutsche Forschungsgemeinschaft (Scha 317/3-3). We wish to thank Prof. Linzen for his critical discussions and reading of the MS, Mrs. H. Storz and Mrs. F. Althaus for drawing the figures, and Mrs. I. Krella for typing the MS.

184

Literature

1. Van Holde, K.E., and Miller, K.I., Q. Rev. Biophys. 15, 1-129 (1982).
2. Linzen, B., Life Chem. Rep., suppl. 1 (Wood, E.J., ed). Harwood Acad. Publ., Chur, London, New York, 27-38 (1983).
3. Markl, J., Markl, A., Schartau, W., and Linzen, B., J. Comp. Physiol. 130, 283-292 (1979).
4. Lamy, J., Sizaret, P.-Y., Lamy, J., Feldman, R., Bonaventura, J., and Bonaventura, C., Life Chem. Rep., suppl. 1 (Wood, E.J., ed) Harwood Acad. Publ., Chur, London, New York, 47-50 (1983).
5. Markl, J., Kempter, B., Linzen, B., Bijlholt, M.M.C., and van Bruggen, E.F.J., Hoppe-Seyler's Z. Physiol. Chem. 362, 1631-1641 (1981).
6. Schartau, W., Eyerle, F., Reisinger, P., Geisert, H., Storz, H., and Linzen, B., Hoppe Seyler's Z. Physiol. Chem. 364, 1383-1409 (1983).
7. Schneider, H.-J., Drexel, R., Feldmaier, G., Linzen, B., Lottspeich, F., and Henschen, A., Hoppe Seyler's Z. Physiol. Chem. 364, 1357-1381 (1983).
8. Nemoto, T., and Takagi, T., Report at the 56th annual meeting of the Japanese Biochemical Society, 29 September to 2 October 1983.
9. Yokota, E., and Riggs, A., J. Biol. Chem. 259, 4739-4747 (1984).
10. Bak, H.J. Soeter, N.M. Vereijken, J.M., Jekel, P.A., Neuteboom, B., and Beintema, J.J., this volume, pp. 149-152 (1986).
11. Gaykema, W.P.J., Hol, W.G.J., Vereijken, J.M., Soeter, N.M., Bak, H.J., and Beintema, J.J., Nature (London) 309, 23-29 (1984).
12. Linzen, B., Soeter, N.M., Riggs, A.F., Schneider, H.J., Schartau, W., Moore, M.D., Yokota, E., Behrens, P.Q., Nakashima, H., Takagi, T., Nemoto, T., Vereijken, J.M., Bak, H.J., Beintema, J.J., Volbeda , A., Gaykema, W.P.J., and Hol, W.G.J., Science 229, 519-524 (1985).
13. Eyerle, F., and Schartau, W., Biol. Chem. Hoppe-Seyler 366, 403-409 (1985).
14. Gross, W., and Witkop, B., J. Amer. Chem. Soc. 83, 1510-1511 (1961).
15. Chang, J.Y., Brauer, D., and Wittmann-Liebold, B., FEBS-Lett. 93, 205-214 (1978).
16. Markl, J., Decker, H., Linzen, B. Schutter, W.G., and van Bruggen, E.F.J., Hoppe-Seyler's Z. Physiol. Chem. 363, 73-87 (1982).
17. Schneider, H.-J., Markl, J., Schartau, W., and Linzen, B., Hoppe Seyler's Z. Physiol. Chem. 358, 1133-1141 (1978).
18. cf.: Markl, J., Stöcker, W., Runzler, R., and Precht, E., this volume, p. 281- 292 (1986).
19. Neuteboom, B., Beukeveld, G.J.J., and Beintema, J.J., this volume, p. 169-172 (1986).

MAPPING OF ANTIGENIC DETERMINANTS IN ANDROCTONUS AUSTRALIS HEMOCYANIN : PRELIMINARY RESULTS

J. N. Lamy, J. Lamy, P. Billiald, P.-Y . Sizaret, J.C. Taveau, and N. Boisset
University François Rabelais, Tours, France

J. Frank
New York State Department of Health, Albany, N.Y. 12201

G. Motta
University of Orléans, France

INTRODUCTION

During the past decade, the intramolecular localization of subunits within native hemocyanin has allowed the determination of the quaternary structure of the three most complex arthropod hemocyanins (1-3). The approach was a molecular immunoelectron microscopy (MIEM) using subunit-specific preparations of polyclonal Fab fragments. The main limitation of the method was the requirement for an antibody preparation highly specific for a small surface domain of the whole molecule. A simple solution to this essential requirement consisted of a subunit-specific antibody preparation obtained by giving highly purified subunits to rabbits and by removing the cross-reacting antibodies by immunoadsorption to the other subunits immobilized on DEAE-Sepharose. This limitation has been bypassed with the appearance of monoclonal antibodies. Indeed, the antibody clone is now directed against a single epitope and is perfectly specific for a small area of the external surface of the subunit. Therefore, the substitution of monoclonal for polyclonal antibodies considerably increases the resolution of the MIEM method, but in order to correctly interpret the results, it is necessary that the primary, the three-dimensional, and the quaternary structures of the molecule are known.

In chelicerate hemocyanins, no complete structure has been elucidated but the progress is so fast that we can anticipate that in the not too far distant future the structures of Androctonus australis and/or Eurypelma californicum and/or Limulus polyphemus will be completely resolved. Indeed, the partial or complete primary structures of several subunits have been recently determined (4-8), the three-dimensional structure of Panulirus interruptus hemocyanin is known at a 3.4 Å resolution (4) and the architecture of A. australis hemocyanin has been recently refined to a degree compatible with intramolecular localization of epitopes (9).

On the other hand, the preparation, the characterization of monoclonal antibodies (mAb), and the mapping of the corresponding epitopes on the external surface of the whole protein is a long term project which requires new technologies such as MIEM, image processing, cryoelectron microscopy, three-dimensional reconstruction, etc. By these methods epitopes will be circumscribed in very small areas of the hemocyanin external

Invertebrate Oxygen Carriers
Ed. by Bernt Linzen
© Springer–Verlag Berlin Heidelberg 1986

surface. When the three-dimensional structure of hemocyanin is known, the amino acid residues composing the monoclonal antibody binding area can then be immediately identified.

This paper presents the first results of intramolecular localization and mapping of antigenic epitopes in A. australis hemocyanin.

MATERIALS AND METHODS

Hemocyanin: All the work presented in this paper has been done on the hemocyanin of the tunisian scorpion Androctonus australis garzonii (10). Native 4x6-mers, dissociated hemocyanin and purified subunits have been prepared as previously reported (11).

Monoclonal antibodies: Immunocompetent cells were obtained from High Responder mice from Biozzi. The immunization procedure, the cell line and media, the protocol for fusion, cloning and subcloning, and the screening method for hybridomas have been recently described in detail (9).

MIEM: Monoclonal antibodies with high affinity and high specificity for a given hemocyanin subunit were selected and characterized as recently described (9). Soluble immunocomplexes between mAb's and A. australis were prepared and purified as previously reported for polyclonal antibodies (1). Specimens for electron microscopy were prepared by negative staining with 2 per cent uranyl acetate. In the microscope, the grid was oriented upside down with the carbon film facing the electron beam and the molecules facing the emulsion side of the photographic film. This procedure restores the correct orientation and handedness of the molecule on the print.

Image processing: The immunocomplexes, quantitatively analyzed by image processing, were scanned with a Perkin Elmer flat bed microdensitometer at 50 μ corresponding to a sampling resolution of 0.1 nm on the object scale. Selected images were subjected to a computer alignment (12). Correspondence analysis (13-15) was used to determine the presence of systematic trends in the interimage variation. Subsets of molecular images selected by this procedure were averaged and displayed with a limiting resolution of 1.7 nm.

Computer graphical techniques: All the models shown in this paper were drawn on a Apple IIe microcomputer using a program of three-dimensional representation developed by one of us (J.C.T). The program allows the rotation of three-dimensional objects composed of polygonal facets. Hidden facets may be or may be not visible on the picture. The program is quickly adaptable to any microcomputer accepting Pascal.

RESULTS

I - Architecture and quaternary structure of chelicerate hemocyanins: Reexamination and refinements

The final step of the intramolecular localization of an epitope within a protein by MIEM is always a comparison between E.M. views of the immunocomplex and a model of the architecture. Obviously this model must be as precise as possible. In the case of the 4x6-meric hemocyanin of Chelicerata, a controversy occurred at the EMBO meeting at Leeds in 1982 with respect to the enantiomorphic nature of the constituting 2x6-mers. Later, an important improvement in the knowledge of the subunit structure came from the determination by X-ray crystallography of the three-dimensional structure of P. interruptus hemocyanin (4). Therefore, previous models of architecture of 4x6-meric hemocyanin had to be reexamined and refined (9). The interpretation of MIEM requires a precise model of quaternary structure. The next paragraphs will describe in detail those refinements and their implications about the quaternary structure of the nx6-meric hemocyanins.

1-1 Shape of the subunits

It is considered that all the subunits composing the cheliceratan 4x6-mers and 8x6-mers share with P. interruptus subunits the following characters. A single polypeptide chain of approximately 670 residues is folded into a kidney-shaped particle composed of 3 domains of comparable size. As shown in figure 1a, the 3 domains are termed, with reference to P. interruptus hemocyanin, N-terminal (or domain 1), central (or domain 2), and C-terminal (or domain 3). Another extrapolation from P. interruptus hemocyanin is that the subunits possess one convex, one concave, and two kidney-shaped faces. Though the outlines of the two kidney-shaped faces are similar, they are not equivalent and may easily be discriminated by the presence of the N- and C-termini. One of the kidney-shaped faces is designated as the N-terminal face because in P. interruptus it contains the N-terminus while the opposite face contains the C-terminus. This nomenclature may have to be revised if, in other chelicerate hemocyanins, the N- and C-termini are located in different faces. There is no crystallographic data concerning cheliceratan subunits at a resolution comparable to that of P. interruptus hemocyanin. The only works of crystallography so far published are those of Magnus and Love (16) and Fearon et al. (17). However, both papers contain pieces of information in agreement with what we now know to be the structure of P. interruptus hemocyanin. These data, combined with the high degree of sequence homology between all the arthropodan subunits so far sequenced, make it very unlikely that the external shape of an arthropod subunit could be seriously different from those of the P. interruptus hemocyanin.

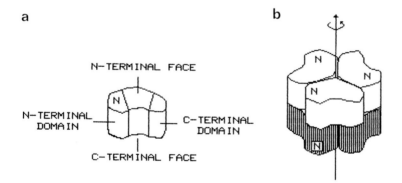

FIGURE 1 Schematic representation of the subunit (a) and of the hexameric building block (b) of arthropod hemocyanin. N: N-terminus in the N-terminal face; the arrows represent the pseudocrystallographic axis of the 1x6-mer.

1-2 Architecture of the hexamers

There is a long tradition of considering arthropod hemocyanins as composed of hexameric building blocks also called 1x6-mers. A discussion on whether or not this nomenclature is appropriate would be outside the scope of this paper. However, as it is widely used, we will describe the oligomers in terms of multiples of a "basic hexameric unit". This unit which has no physical reality, at least in cheliceratan hemocyanins, is considered to resemble the hexameric (stricto sensu) hemocyanin of P. interruptus and is shown in figure 1b. Notice that the subunits of the two trimeric layers have their C-terminal faces in contact meaning that only their N-terminal faces, their concave faces and the sides of their convex faces are accessible from outside. Furthermore, the two-fold axes around which the three-fold layers rotate are located such that the N-terminal domains in the two layers are superimposed. The same pattern of course occurs with the C-terminal domains. This configuration results in an alternation of superposed domain 1 and of superposed domain 3 when the molecule is viewed from its three-fold pseudocrystallographic axis. That this organization is also present in cheliceratan hemocyanins is supported by the E.M. views of the 2x6-meric fragments of L. polyphemus hemocyanin. Indeed, as shown in figure 2, image processing by the correspondence analysis method of van Heel and Frank (13) of negatively stained 2x6-mers repeats an alternation of domains with rectangular and triangular profiles exactly as in P. interruptus hemocyanin (4). Moreover, since the two faces of the 2x6-mer can be discriminated by the correspondence analysis, it is established that in L. polyphemus the rectangular domains and equivalently the triangular domains are superposed in the two trimeric layers (9). This disposition is an important direct argument in favor of a unitary organization of the "hexamers" throughout the arthropodan phylum.

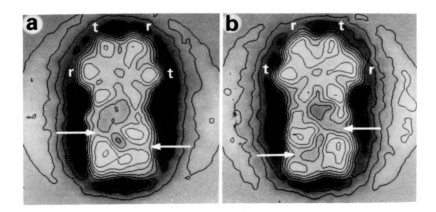

FIGURE 2 Average images of the two populations of 2x6-meric E.M. views of Limulus polyphemus hemocyanin separated by correspondence analysis. r and t : rectangular and triangular vertices; the arrows point to the cleft of minimal stain exclusion in the lower hexamer.

1-3   Architecture of the 2x6-mer

The 2x6-mer building block of chelicerate hemocyanins is more complex because of the so-called enantiomorphic convention. Let us suppose that, as shown in figure 3a, two hexamers are superimposed in such a way that the lower hexamer derives from the upper one by a translation. If the upper hexamer is rotated to the left or to the right, then a pair of isomers appears. By convention (19), the RIGHT and LEFT isomers result from 90° rotations of the upper hexamer to the right and to the left, respectively (Figs. 3b and 3c).

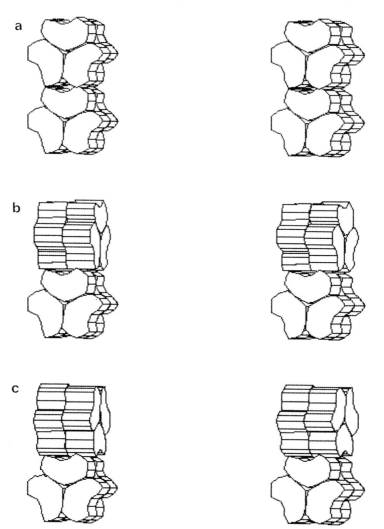

FIGURE 3 Stereopairs of models representing the formation of the right and left isomers of 2x6-meric hemocyanins according to Lamy et al. (1982). a) Two superposed hexamers; b) right isomer; c) left isomer.

In our previous reports, the N- and C-terminal domains and the N- and C-terminal kidney-shaped faces were not differentiated so that the model of the subunit had a two-fold axis passing through the center of the concave and convex faces. Therefore, the rotation of the upper hexamer to the left or to the right produced a pair of enantiomorphs. Because of the asymmetry resulting from the discrimination between the domains, the LEFT and RIGHT ISOMERS cannot be considered anymore as ENANTIOMERS and the terms LEFT and RIGHT ISOMERS will be prefered in the following. No crystallographic data being available for dodecameric hemocyanin, the choice between the LEFT and RIGHT ISOMERS had to be made on indirect arguments and led to a dispute between M. van Heel and J. Lamy at the EMBO meeting at Leeds in 1982. (20,3).

The publication of the three-dimensional structure of P. interruptus by Gaykema et al. (4) helped to settle the controversy leading to reexamination of the isomer choice in A. australis and L. polyphemus hemocyanins (9). Four sets of data, all obtained by electron microscopy and image processing, support the choice of the right isomer, as originally suggested by van Heel et al. (20). Three of them are indirect arguments obtained from the 4x6-mers and 8x6-mers which will be developed in the next section. They are the orientation of the rocking axis of the 4x6-mer, the pattern of stain exclusion of the 4x6-mer 45° view, and the outline of the symmetric pentagon of the 8x6-mer. The fourth argument is the position of the alternating rectangular and triangular vertices in the 2x6-mer (see figure 2). Indeed, in P. interruptus hemocyanin, the N-terminal domains protrude more outside the triangular envelope of the C-terminal domains than the C-terminal domains do outside the N-terminal domains (4). If this disposition is preserved in L. polyphemus hemocyanin, the rectangular vertex corresponds to the N-terminal domain, and the average images shown in figure 2 are the two faces of a 2x6-mer of the RIGHT isomeric type.

1-4   Architecture of the 4x6-mers

The construction of a model of the 4x6-mer requires that we know how the two dodecameric halves are associated. To answer this question, the following experimental data must be taken into account. First, the rocking axis of the molecule must join the lower right and the upper left hexamers when the longitudinal cleft of the molecule is oriented from top to bottom in the picture. This observation of van Heel and Frank (13) describes the fact that the 4x6-mer molecule in its top view does not stand on the support plane on four hexamers but only on three. Figure 4 presents a stereopicture of the model in its side view which describes the two extreme positions between which the molecule rocks. The second important feature of the 4x6-mer is its pattern of stain exclusion in the 45° view. This view described by van Heel et al. (20) results from a strong interaction of one of the dodecamers with the support film. The characteristic pattern of stain exclusion suggests that the molecule is standing on its left dodecamer when the longitudinal cleft is oriented from top to bottom and when the hexamers with hexagonal contours are located on the top of the figure. Figure 5 shows a 45° view and the model in the same orientation. Notice that in the E.M. view, the whiter half corresponds to the part of the molecule more deeply

a

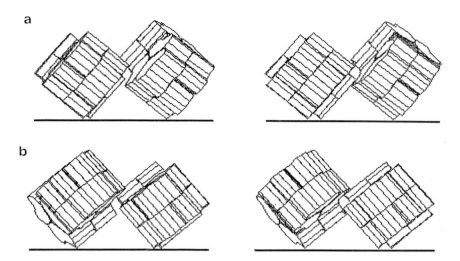

b

FIGURE 4 The rocking effect in 4x6-meric chelicerate hemocyanin. a) and b) Stereopairs of the extreme positions of the molecule in the rocking effect. The rocking axis passes through the front hexamer on the right side and the back hexamer on the left side.

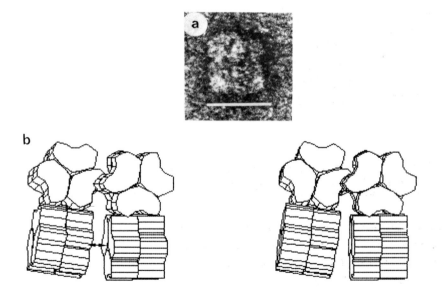

b

FIGURE 5 The 45° view of 4x6-meric chelicerate hemocyanin. a) E.M. view of Androctonus australis hemocyanin showing a strong pattern of stain exclusion on the right side. b) Stereopairs of the model in the same orientation. The double arrow materializes the bridge area (interdodecamer gap). The length of the bar is 25 nm.

192

immersed in the stain which produces the highest stain exclusion. The third important aspect of the 4x6-mer is the FLIP FLOP effect, a feature also discovered by Frank and van Heel (15) by correspondence analysis. The FLIP FLOP effect results from a slight shift of one dodecamer along the longitudinal cleft direction. As a result of this shift, the projection of the 4x6-mer in its top view is no longer a rectangle but a parallelogram with a long and a short diagonal. By correspondence analysis, van Heel and Frank distinguished two types of parallelogram with the longitudinal cleft oriented from top to bottom in the picture. In one type (the FLOP view), the long diagonal was oriented to the upper left, that is in the direction of the rocking axis, while in the other type (the FLIP view), the long diagonal was oriented to the upper right. The FLIP and FLOP views correspond to the two faces of the molecule that Sizaret et al. (19) proposed to designate as FLIP and FLOP faces to recall the origin of their discovery. By convention, the 4x6-mer stands on its FLOP face when the rocking axis is the long diagonal and on its FLIP face when the rocking axis is the short diagonal. (The character O in FLOP and LONG diagonal may be a mnemonic.) The representation of the Flip Flop effect is shown in figure 6. The last important aspect of the

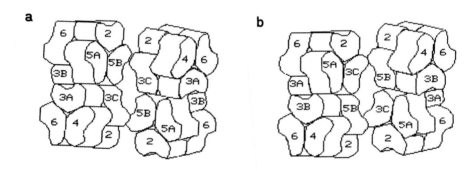

FIGURE 6   The FLIP and FLOP faces in the model of Androctonus australis hemocyanin (after Lamy et al., 1985). a) The molecule stands on its FLOP face (the FLIP face is exposed to the observer). b) The molecule stands on its FLIP face (the FLOP face is visible).

architectural model is the respective orientation of the dodecamers around their long axis. There are four different rectangular faces and four different edges on the dodecamer models in figure 3. Therefore, there are 16 different possibilities for associating two dodecamers side by side. In the model, the 1/1 edges are brought into contact. According to Lamy et al. (18), in the 1/1 edge, the two subunits located near the middle of the edge have their concave faces accessible from the outside. The best justification of this choice comes from the immunolabelling experiments of Lamy et al. (1) and Sizaret et al. (19) which demonstrate that subunits Aa 3C and Aa 5B, the components of the stable heterodimer, are located symmetrically apart from the middle of the dodecamer edge and that the dimeric subunit is the component of the bridge. The intramolecular location of the subunits recently reexamined by Lamy et al. (9) is also indicated in figure 6.

In the model of the 4x6-mer in figures 4, 5, and 6, no material is visible in the bridge area between the dodecameric halves, which seem independent from each other. This point was a matter of controversy. Indeed, originally Lamy et al. (18) proposed that the interdodecamer contacts occur between the flat faces of subunits Aa 3C and Aa 5B. In order to explain the orientation of the rocking axis, this assumption required that the 2x6-mer be of the LEFT isomeric type. As described above, it became evident that this choice was wrong and that the 2x6-mer is of the RIGHT type. However, this new choice requires that one rotate the two dodecamers away from each other to keep the correct rocking axis (see Lamy et al., 1985, for more detail). Thus, an imprecision still remains with respect to the nature of the material which fills the interdodecamer gaps. It could be an expansion of the C-terminal domain of subunit Aa 3C and/or Aa 5B or an insertion of a fourth domain. Therefore, as long as the precise nature of the bridges is unknown, we believe that it is best to keep gaps free of material in the models.

1-5   Architecture of the 8x6-mers

Under mild conditions the 8x6-meric hemocyanin of horseshoe crabs dissociates into 4x6-mers, then into dodecamers. The 4x6-mers resulting from the dissociation of L. polyphemus hemocyanin are morphologically very similar to the native 4x6-mers of A. australis or E. californicum hemocyanins. All three 4x6-mers exhibit the same rocking and FLIP-FLOP effects and by correspondence analysis they cannot be distinguished by the first two factors (21). Therefore, the 8x6-mers can reasonably be considered as composed of 4x6-mers of the above described type, and the determination of their architecture only requires that we understand how their halves are assembled to produce the 5 main E.M. views. These views catalogued by Lamy et al. (18) and designated as cross view, bow tie view, ring view, asymmetric pentagon and symmetric pentagon are shown in figure 7 together with the corresponding views of a model. The model of the 8x6-mer was built by assembling two copies of the 4x6-mer, as follows. First, one copy of the 4x6-mer was positioned on a support so that it presented its top view. On this was placed another copy of the 4x6-mer so that the longitudinal clefts between the dodecamers were superposed. The upper molecule was rotated in order to align its rocking axis with the cleft of the lower molecule. Finally, the upper 4x6-mer was translated down and slightly adjusted to optimize the inter-4x6-mer contacts and to fit the E.M. views. This model is in pretty good agreement with all the presently available data. Specifically, it perfectly fits the pentagonal views (symmetric and asymmetric) discriminated by correspondence analysis (18; van Heel, unpublished data quoted by J. Frank, 22). The model is also in excellent agreement with the immuno-complexes composed of cross-linked 8x6-mers and subunit-specific polyclonal Fab fragments (3).

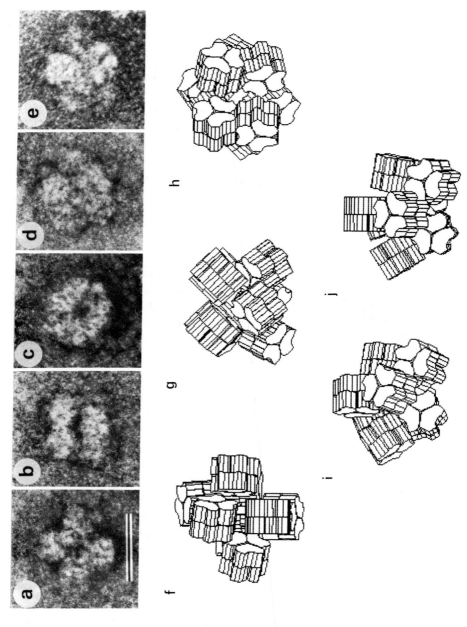

FIGURE 7 The 5 E.M. views of 8x6-meric hemocyanin and their translation in the model. a,f) Cross view; b,g) bow tie view; c,h) ring view; d,i) asymmetric pentagon, e,j) symmetric pentagon. The length of the bar is 25 nm.

Though no model of 4x6-meric or 8x6-meric hemocyanin has been published by M. van Heel, we must acknowledge that the short paper he published with his colleagues in the proceedings of the EMBO meeting held in Leeds in 1981 (20) was of considerable help for the building of the model shown in figure 7 and that their suggestion on the enantiomorphic structure of the constituting 2x6-mers were essentially correct.

II – Approach to the intramolecular localization of antigenic epitopes in Androctonus australis hemocyanin

Labelling an epitope within a protein (oligomeric or monomeric) with a monoclonal antibody is equivalent to labelling a subunit within an oligomer using polyclonal antibodies. Theoretically, if the antigenic particle possesses more than one copy of the epitope, a precipitation may occur. However, as shown below, monoclonal antibodies often produce ordered structures with high molecular weight easily visible in the electron microscope. The following examples show how such immunocomplexes can be used for localizing epitopes within subunits.

The first example reports the localization of the epitope of a monoclonal antibody (mAb6302) within the native 4x6-mer. This clone was demonstrated to have a high affinity and a high specificity for subunit Aa 2 either in the free state or when incorporated in the 4x6-mer (9). As shown in figure 8a, the immunocomplexes appear as long strings of 4x6-mers mostly in the side view and in the 45° view. The hemocyanin molecules do not come into contact and bridging material is clearly visible between them. The appearance of these immunocomplexes suggests that the epitope of mAb6302 is located near the top/bottom edges of the 4x6-mer. These linearly ordered immunocomplexes also confirm that the 4 copies of the epitope are distributed among the two faces of the molecule as expected from the previous localization of subunit Aa 2. To refine the localization, the experiment was repeated with monoclonal Fab fragments, producing the immunocomplexes shown in figure 8b. These structures are clearly composed of 4x6-mers in the side view bearing Fab fragments on the top edge of the left dodecamer. The reason why only one Fab fragment was usually visible on a single molecule, while there are four copies of subunit Aa 2 per 4x6-mer is explained by the mechanism of the negative stain. First, as shown in figure 8d, the two subunits in contact with the carbon grid completely exclude the stain while the 2 copies of subunit Aa 2 located on the top of the molecule are invisible because they emerge out of the stain layer. Second, on the side view, because of the rocking effect, one of two copies of subunit Aa 2 in contact with the carbon film projects itself near the contour line of the molecule (near the upper left edge on figure 8d) while the second copy is in the center of the second dodecamer projection (on the right in figure 8d) so that the Fab fragment does not appear on the contour line (arrow). The immunocomplexes of figure 8b were then submitted to an image processing which produced the average image of figure 8c. As predicted by the model, one Fab fragment is bound to the top of the upper left edge in an area corresponding to the top of the C-terminal domain of subunit Aa 2 and is oriented roughly perpendicular to the plane on the side of the longitudinal cleft.

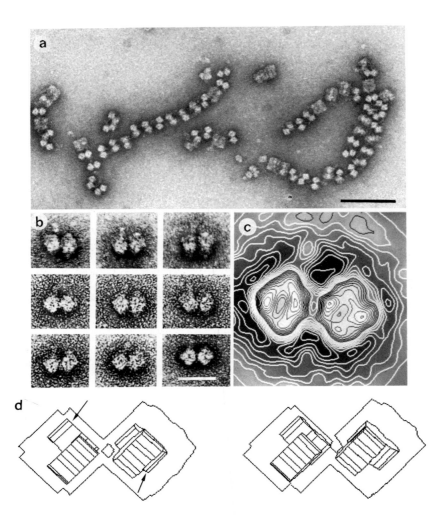

FIGURE 8 Localization of the epitope of mAb6302 in <u>Androctonus</u> <u>australis</u> hemocyanin. a) Immunocomplexes between whole hemocyanin and monoclonal IgG. b) Immunocomplexes between whole hemocyanin and Fab fragments. c) Average image of a subset of complexes shown in (b). d) Stereopairs of a model of intramolecular localization of mAb6302 epitope within subunit <u>Aa</u> 2. The arrows point to the top of C-terminal domain in the two copies in contact with the carbon film. The length of the bar is 100 nm in a) and 25 nm in b).

The second and third examples were both obtained with monoclonal antibodies having a high affinity for subunit <u>Aa</u> 6 and respectively termed mAb5701 and mAbL8 after the nomenclature of the laboratory. Figure 9a shows an E.M. field and selected views of immunocomplexes obtained with mAb5701 and subunit <u>Aa</u> 6. Obviously, the attachment point of the Fab fragment to the 4x6-mer is located in the corner of the hemocyanin molecule. The immunocomplexes shown in the galleries of selected views in the

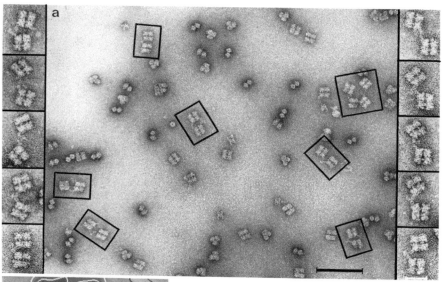

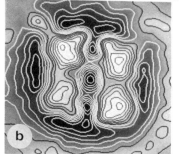

FIGURE 9 Localization of the epitope mAb5701. a) Immunocomplexes between whole hemocyanin and monclonal IgG. b) Average image of a subset of the immunocomplexes (the second hemocyanin molecule has been masked). c) Stereopairs of a model of intramolecular location of mAb5701 epitope within subunit Aa 6. The black area corresponds to the nonaccessible C- terminal face. The arrows point to the top of the C-terminal domain in the N-terminal face. The length of the bar is 100 nm and 25 nm in the enlargements.

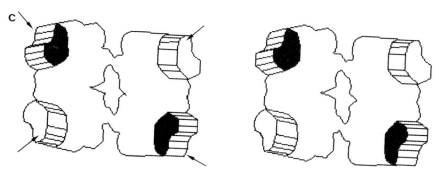

left and right parts of figure 9a differ by the rotation angle around the Fab arm. Actually, in the left gallery, the transition between complexes from top to bottom is easily explained by the segmental flexibility of the hinge. The same phenomenon also explains the transition in the right gallery. However, passing from complexes of the left to complexes of the right gallery requires a high degree of rotational flexibility around the Fab arm. Whether or not flexibility is forced by a strong interaction of the antibody and/or hemocyanin molecule with the carbon film is not yet established (9). However, similar phenomena also observed by negative stain have already been reported by Wrigley et al. (23,24) and Roux (25). Figure 9b shows an average obtained by correspondence analysis on a subpopulation of immuno-complexes similar to those shown in figure 9a. Clearly, the binding point of the Fab fragment to the 4x6-mer molecule is the corner corresponding to the domain of subunit Aa 6 equivalent to domain 3 (C-terminal) of P. interruptus hemocyanin. The stereo pair of the model of architecture shows the location of the epitope of mAb5701 within subunit Aa 6.

The immunocomplexes produced by mAbL8 and native 4x6-mers are shown in figure 10a. Here again, the attachment point is located in the corner of the molecule in an area corresponding to domain 3 and, in many molecules the four copies of subunit Aa 6 are labelled. This is a a posteriori verification of the correctness of the determination of the number of copies of subunit Aa 6 in the 4x6-mer molecule (26). Figure 10b presents an interpretation of the enlarged complex shown in the lower right part of figure

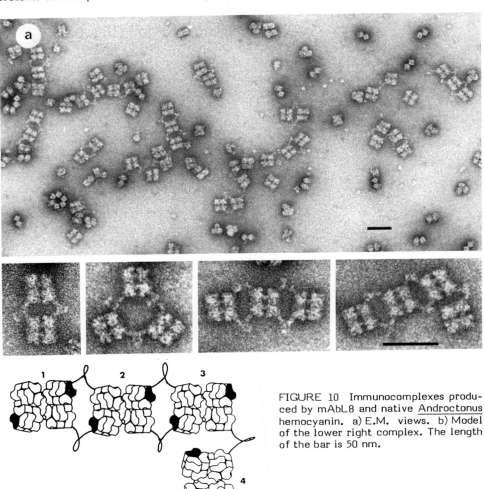

FIGURE 10 Immunocomplexes produced by mAbL8 and native Androctonus hemocyanin. a) E.M. views. b) Model of the lower right complex. The length of the bar is 50 nm.

10a. This complex is interesting because it contains two types of linkage which again demonstrate the high rotational flexibility of the Fab arm. Indeed, the complex is composed of a string of four 4x6-mer molecules labelled (1) through (4). In molecule (2), the four copies of subunit Aa 6 are exclusively linked with subunits of molecules (1) and (3). In the four bonds, an IgG molecule bridges two subunits having their flat accessible (N-terminal) face oppositely oriented. In contrast with this disposition, the IgG molecule bridging molecules (3) and (4) links two copies of subunit Aa 6 with the same orientation. The fact that both types of linkage coexist in the same complex suggests that the linkage type does not depend on the 4x6-mer molecule, but only on the rotation angle around the Fab arm.

Though an image processing of the complexes produced by mAbL8 has not yet been done, their resemblance with the complex produced by mAb5701 suggests that the locations of the two epitopes are similar. However, this method does not indicate whether the two monoclonal antibodies can bind simultaneously subunit Aa 6. If they do, this means that the two epitopes do not overlap and that the binding of one antibody to its epitope does not strongly change the three-dimensional structure of the second epitope and vice versa. To test this hypothesis, pure subunit Aa 6 was incubated with a mixture of the two monoclonal antibodies and the soluble immunocomplexes were purified. Figure 11a shows a gallery of annular immunocomplexes composed of two IgG molecules and two copies of subunit Aa 6 demonstrating that the two epitopes do not overlap and that they do not interfere with each other. In the map of figure 11b, the epitopes of mAb5701 and mAbL8 are located in the portion of the subunit surface accessible from outside. Both of them are located in domain 3 and they do not overlap. Other clones are presently being studied and it is expected that in the not too far distant future we will be capable of labelling almost all of the external surface of the A. australis hemocyanin molecule.

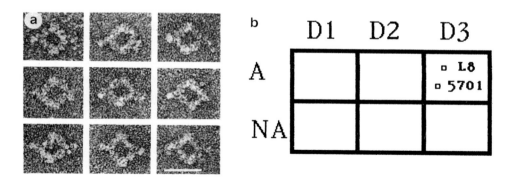

FIGURE 11 Mapping of mAb5701 and mAbL8. a) Soluble immunocomplexes produced by subunit Aa 6 and a mixture of mAb5701 and mAbL8. b) Schematic representation of the respective locations of mAbL8 and mAb5701.
A: Accessible faces, NA: non accessible faces from outside.
D1-3: Domain 1-3 of subunit Aa 6.
The length of the bar is 25 nm.

DISCUSSION AND CONCLUSION

The results presented in the previous sections open a series of new lines of research in the field of hemocyanin structure and function. Indeed, MIEM with monoclonal antibodies allows us to localize with precision epitopes within hemocyanin subunits. This is an interesting tool from a methodological point of view, but the question is "what can be done with this tool ?". In the present state of our knowledge of immunocomplex formation, the interest of this tool is multiple.

First, hemocyanin is in the progress of becoming one of the best models for studying the antigenicity of proteins. In fact, antigenic determinants are now considered as small hydrophilic and flexible surface domains coating a large part of the protein surface (27,28). It is generally accepted that most, if not all, antigenic sites are completely or partially composed of amino acid residues not contiguous in the sequence (29). Presently, more than 30 arthropodan hemocyanin subunits with important sequence homology are available in a pure and functional form (8 subunits in A. australis, 8 in L. polyphemus, 7 in T. tridentatus, 7 in E. californicum, at least 3 in P. interruptus hemocyanins). When their amino acid sequences and three-dimensional structures presently being studied are known, hemocyanin will probably be as good a model for studying the structure and localization of antigenic determinants as myoglobin, lysozyme or influenza virus hemagglutinin. Indeed, arthropod hemocyanins because of their huge size are easily visible in the electron microscope and the orientation of the external faces of the various subunits is accurately determined. Therefore, localizing the attachment point of the Fab fragment with respect to the boundary of the antigen allows one to decide in which surface domain of the subunit the epitope is located. When coupled to an image processing method, this direct approach to the localization of the epitope is much faster than the crystallographic approach which has until now been successful only with small proteins (30). On the other hand, the indirect localization by comparison of amino acid sequences of evolutionarily related proteins (the subunits in the case of arthropod hemocyanins) is difficult to use with discontinuous epitopes and can be considerably simplified by the MIEM approach.

A second major interest of immunolabelling hemocyanins with monoclonal antibodies is to study their structure-to-function relationships. It is well known that in chelicerate hemocyanin reversible oxygen binding involves the cooperation of many subunits. It would be surprising if such an important structural change could occur without conformational modification of the external surface of the subunits. From this point of view, monoclonal antibodies may become very powerful probes to detect minor conformational changes.

The third major interest of MIEM of hemocyanin with monoclonal antibodies is a better understanding of the antibody function. Indeed, it has already been shown that an important rotational flexibility around the Fab arm is a character of the immunoglobulin molecule. Hemocyanin appears to be a highly favorable material to study immunocomplexes with respect to one of the main characters of the immunoglobulin molecule, its flexibility.

ACKNOWLEDGMENT

This work has been financially supported by grants N° 518014 and INT-8313133 from NSF-CNRS Cooperative Program and by CNRS, RCP N° 080816.

REFERENCES

1. Lamy, J., Bijlholt, M.M.C., Sizaret, P.-Y., Lamy, J.N., and van Bruggen, E.F.J., Biochemistry 20, 1849-1856 (1981).
2. Markl, J., Kempter, B., Linzen, B., Bijlholt, M.M.C., and van Bruggen, E.F.J., Hoppe-Seyler's Z. Physiol. Chem. 362, 1631-1641 (1981).
3. Lamy, J., Lamy, J., Sizaret, P.-Y., Billiald, P., Jollès, P., Jollès, J., Feldmann, R.J., and Bonaventura, J., Biochemistry 22, 5573-5583 (1983).
4. Gaykema, W.P.G., Hol, W.G.J., Vereijken, J.M., Soeter, N.M., Bak, H.J., and Beintema, J.J., Nature 309, 23-29 (1984).
5. Schartau, W., Eyerle, F., Reisinger, P., Geisert, H., Storz, H., and Linzen, B., Hoppe-Seyler's Z. Physiol. Chem. 364, 1383-1409 (1983).
6. Schneider, H.J., Drexel, R., Feldmaier, G., Linzen, B., Lottspeich, F., and Henschen, A., Hoppe-Seyler's Z. Physiol. Chem. 364, 1357-1381 (1983).
7. Yokota, E., and Riggs, A.F., J. Biol. Chem. 259, 4739-4749 (1984).
8. Nemoto, T., and Takagi, T., Life Chemistry Reports, suppl. 1, 89-92 (1983).
9. Lamy, J., Lamy, J., Billiald, P., Sizaret, P.-Y., Cavé, G., Frank, J., and Motta, G., An approach to the direct intramolecular localization of antigenic determinants in Androctonus australis hemocyanin with monoclonal antibodies by molecular immunoelectron microscopy. Biochemistry (in press).
10. Goyffon, M., and Lamy, J., Bull. Soc. Zool. Fr. 98, 137-144 (1973).
11. Lamy, J., Lamy, J., and Weill, J., Arch. Biochem. Biophys. 193, 140-149 (1979).
12. Frank, J., Goldfarb, W., Eisenberg, D., Baker, T.S., Ultramicroscopy 3, 283-290 (1978).
13. van Heel, M., and Frank, J., Ultramicroscopy 6, 187-194 (1981).
14. Frank, J., Verschoor, A., and Boublik, M., J. Mol. Biol. 161, 107-133 (1982).
15. Frank, J., and van Heel, M., J. Mol. Biol. 161, 134-137 (1982).
16. Magnus, K.A., and Love, W.E., Life Chemistry Reports, suppl. 1, 61-64 (1983).
17. Fearon, E.R., Love, W.E., Magnus, K.A., Lamy, J., and Lamy, J., Life Chemistry Reports, suppl. 1, 65-68 (1983).
18. Lamy, J., Sizaret, P.Y., Frank, J., Verschoor, A., Feldmann, R., and Bonaventura, J., Biochemistry 21, 6825-6833 (1982).
19. Sizaret, P.Y., Frank, J., Lamy, J., Weill, J., and Lamy, J.N., Eur. J. Biochem. 127, 501-506 (1982).
20. van Heel, M., Keegstra, W., Schutter, W., and van Bruggen, E.F.J., Life Chemistry Reports, suppl. 1, 69-73 (1983).
21. Bijlholt, M.M.C., van Heel, M.G., and van Bruggen, E.F.J., J. Mol. Biol. 161, 139-153 (1982).
22. Frank, J., Ultramicroscopy 13, 153-164 (1984).
23. Wrigley, N.G., Brown, E.B., and Skehel, J.J., J. Mol. Biol. 169, 771-774 (1983).
24. Wrigley, N.G., Brown E.B., Daniels, R.S., Douglas, A.R., Skehel, J.J., and Wiley, D.C., Virology 131, 308-314 (1983).
25. Roux, K.H., Eur. J. Immunol. 14, 459-464 (1984).
26. Lamy, J., Lamy, J., Sizaret, P.Y., and Weill, J., in Invertebrate Oxygen Binding Proteins (Lamy, J., and Lamy, J., eds.), pp. 425-443, Marcel Dekker, New York (1981).
27. Westhof, E., Altschuh, D., Moras, D., Bloomer, A.C., Mondragon, A., Klug, A., and van Reggenmortel, M.H.V., Nature 311, 123-126 (1984).
28. Tainer, J.A., Getzoff, E.D., Alexander, H., Houghten, R.A., Olson, A.J., Lerner, R.A., and Hendrickson, W.A., Nature 312, 127-134 (1984).
29. Todd, P.E.E., East, I.J., and Leach, S.J., Trends Biochem. Sci. 7, 212-216 (1982).
30. Amit, A.G., Mariuzza, R.A., Phillips, S.E.V., and Poljak, R.J., Nature 313, 156-158 (1985).

CIRCULAR DICHROISM SPECTRA OF NATIVE 37 S HEMOCYANIN
FROM THE SPIDER EURYPELMA CALIFORNICUM AND ITS SUBUNITS

Peter Reisinger
Zoologisches Institut, Universität München,
Luisenstr. 14, 8000 München 2, F.R.G.
Present address: Biochemisches Laboratorium der Klinik und Poliklinik
für Hals-, Nasen- und Ohrenkranke der Universität München,
Marchioninistr. 15, 8000 München 70, F.R.G.

The 37 S hemocyanin of the tarantula Eurypelma californicum has a molecular mass of 1.8×10^6 Da and is built up from 24 subunits. Under alkaline conditions the molecule disaggregates into the monomers a, d, e, f, g and the dimers bc and ff (1). By reassembly experiments it was shown, that all seven subunits are necessary to reconstitute the native 37 S hemocyanin (2). Each subunit seems to play a specific role in the assembly of the whole molecule, and may therefore play an individual role in "molecular physiology" (i. e. Bohr effect and cooperativity). The heterogeneity of the immunologically distinct subunits seems to be based only on different amino acid sequences and not on post-translational modifications (1, 3).

The present work addresses the question, whether the differences of the subunits in primary structure are reflected in different secondary structures, orientation of amino acid side chains, and structure of the active center. To this end, CD spectra were recorded of tarantula hemocyanin and its isolated subunits, in their oxygenated and deoxygenated states. To my knowledge this is the first study of this kind.

37 S hemocyanin, the five monomers a, d, e, f, g and the heterodimer bc were isolated according to well established methods. The purity of the subunits was tested by crossed immunoelectrophoresis. Circular dichroism spectra were recorded at room temperature on a Jobin Yvon Modell Dichrograph V apparatus. Cylindrical quartz cells of 1.0 cm (for the 250 - 730 nm range) and 0.1 cm (for the 200 - 260 nm range) optical path length were used. Protein concentration (in 0.1 M Tris/HCl, pH 8.0) was adjusted appropiately. Deoxygenated samples were prepared by stirring in an argon atmosphere (200 - 260 nm range) or by addition of $NaHSO_3$ (250 - 730 nm range) (4). The calibration of the dichrograph was checked with a solution of epiandrosterone. The results are expressed in mean residue weight ellipticities, calculated from θ_{MRW} (degrees x cm^2 x decimole^{-1}) = 3300 x $\Delta\varepsilon$. MRW was assumed to be 116 (5).

The CD spectrum of native 37 S hemocyanin shows bands at 257 (in the deoxygenated state), 283, 340, 505 and 622 nm, which are typical for an arthropodan hemocyanin (Fig. 1). The position of the copper band (340 nm) is similar in the phylogenetically old Limulus hemocyanin, while it is at 335 nm in the Crustacea (6). The origin of these bands is briefly discussed by Ellerton et al. (7); the bands at 340, 505 and 622 nm originate from the copper-oxygen complex, while those at 257 and 283 nm are due to amino acid side chains. Upon deoxygenation the 340, 505 and 622 nm bands disappear, whereas the band at 283 nm doubles in intensity. In the range of 200 - 245 nm (peptide band) the spectrum changes little upon deoxygenation (Fig. 2) indicating the total α-helix content to remain nearly constant during binding or release of oxygen.

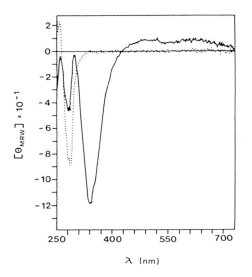 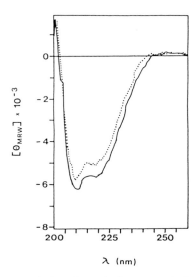

Figs. 1 (left) and 2 (right). CD spectra of oxygenated (———) and deoxygenated (---) Eurypelma hemocyanin.

The heterogeneity of the hemocyanin subunits can also be seen in their CD spectra. The spectral changes associated with oxygen binding vary significantly in extent. Subunit d shows the greatest difference, subunit g the second greatest, whereas the spectrum of subunit e is unaltered (Fig. 3).The average of the differences obtained with the subunits is similar to the difference between oxygenated and deoxygenated 37 S hemocyanin. Thus it seems very likely, that individual changes in secondary structure of different subunits are the same as those occuring in native 37 S hemocyanin. The changes in the peptide bands of subunits d and g are fully reversible, i. e. upon reoxygenation the CD signal reaches its earlier magnitude.

In the near ultraviolet and visible region, the CD spectra of the oxygenated subunits (Fig. 4) show considerable differences in band intensities (at 257, 283 and 340 nm). After deoxygenation the bands at 340, 505 and 622 nm disappear, while the 283 nm band increases (though to variable extent) reflecting changes in the active center and the geometry of amino acid side chains. These changes and those occuring in the 257 nm band provide further evidence for subunit heterogeneity. Clearly, the morphology and motion of individual parts associated with O_2 binding, must be distinct for each subunit, even if these differences should be subtle.

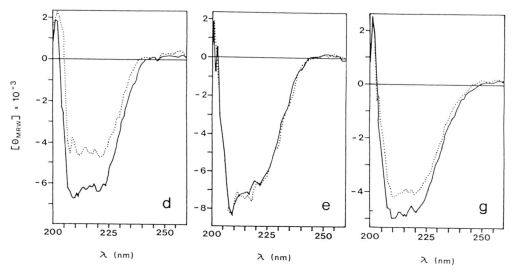

Figure 3. CD spectra (peptide band) of oxygenated (——) and deoxygenated (---) subunits d̲, e̲ and g̲.

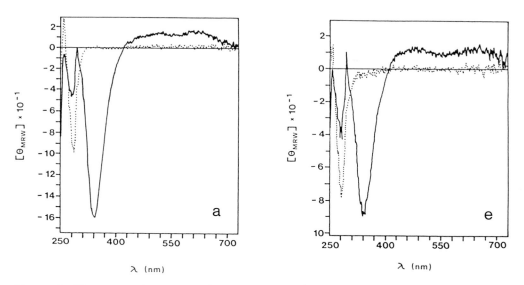

Figure 4. CD spectra (near UV and visible region) of oxygenated (——) and deoxygenated (---) subunits a̲, e̲ and heterodimer b̲c̲.

206

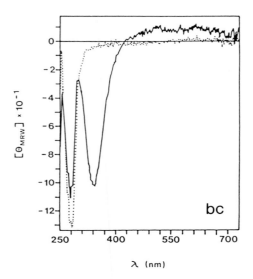

Figure 4 (continued)

Acknowledgements:

I wish to thank Prof. Dr. Hugo Scheer and Prof. Dr. Wolfhart Rüdiger, Botanisches Institut der Universität München, F.R.G., for the possibility to record the CD spectra. This work was supported by DFG grant to Prof. Dr. B. Linzen.

References:

1. Markl, J., Strych, W., Schartau, W., Schneider, H.-J., Schöberl, P., and Linzen, B., Hoppe Seyler's Z. Physiol. Chem. 360, 639-650 (1979).
2. Markl, J., Decker, H., Linzen, B., Schutter, W. G., and van Bruggen, E. F. J., Hoppe Seyler's Z. Physiol. Chem. 363, 73-87 (1982).
3. Linzen, B., Soeter, N. M., Riggs, A. F., Schneider, H.-J., Schartau, W., Moore, M. D., Yokota, E., Behrens, P. Q., Nakashima, H., Takagi, T., Nemoto, T., Vereijken, J. M., Bak, H. J., Beintema, J. J., Volbeda, A., Gaykema, W. P. J., Hol, W. G. J., Science, 229, 519-524 (1985).
4. Takesada, H., and Hamaguchi, K., J. Biochem. 63, 725-729 (1968).
5. Reisinger, P., Doctoral thesis, University of Munich (1985).
6. Nickerson, K. W., and van Holde, K. E., Comp. Biochem. Physiol. 39 B, 855-872 (1971).
7. Ellerton, H. D., Ellerton, N. F., and Robinson, H. A., Prog. Biophys. molec. Biol. 41, 143-248 (1983).

SUBUNIT HETEROGENEITY AND AGGREGATE FORMATION IN

Cherax destructor AND Jasus sp. HEMOCYANINS

G.B. Treacy and P.D. Jeffrey

John Curtin School of Medical Research

P.O. Box 334, Canberra City,

A.C.T. 2601, Australia

The hemolymphs of the two crustaceans, Cherax destructor a fresh water crayfish and a Jasus sp. salt water spiny lobster contain hemocyanin for oxygen carrying purposes.

The polyacrylamide gels in Fig. 1 show that at pH 7.8 Jasus hemolymph consists of hexameric hemocyanin, while Cherax hemolymph contains several hexameric bands, dodecamer and higher polymeric forms of hemocyanin, as previously reported (1).

Both hemocyanins are dissociated at pH 10 + EGTA (ethylene glycol bis (β aminoethyl ether) - N, N'-tetraacetic acid) and the polyacrylamide bands are shown in Fig. 1. Cherax hemocyanin dissociates to give three major components, M_1, M_2 and M_3' which can be separated (2). SDS gels show that the molecular weights of M_1 and M_2 are in the region of 75,000 and that of M_3' 150,000. On reduction with dithiothreitol, M_3' gives a band of M_3, also with molecular weight approximately 75,000 but this is different in amino acid composition from M_1 and M_2 (3 and 4). Jasus hemocyanin on dissociation gives two monomer bands which are very hard to separate, and no dimer band. The absence of dimer is also shown on SDS gels, and is consistent with the observation that the hemolymph contains no aggregates higher than hexamer.

Immunological Relationships

Using a rabbit antiserum raised to whole Jasus hemocyanin, it was found that no cross reaction was given with any component of Cherax hemocyanin or with whole Cherax hemolymph.

Formation of hexameric hybrids

Although unrelated immunologically, the monomers from the two hemocyanins can form hybrid hexamers by dialysis of varying mixtures of Cherax M_1 and dissociated Jasus hemocyanin at pH 10 into buffer at pH 7.8 containing 0.03M $CaCl_2$. Gels at pH 7.8 run on the resulting solutions are shown in Fig. 2.

Although discrete bands of the different hybrids cannot be seen, the bands obtained vary steadily in mobility from that of Jasus hexamer to that of $(M_1)_6$ as

Invertebrate Oxygen Carriers
Ed. by Bernt Linzen
© Springer–Verlag Berlin Heidelberg 1986

the proportion of Cherax M_1 is increased. The bands are not just mixtures of the two different homohexamers, as these give two bands, as shown in the Figure.

Formation of dodecameric hybrids, and higher MW polymers

The dimer obtained on dissociation of whole Cherax hemocyanin is the "glue" which is needed to link up 10 more monomer molecules to give the dodecamer molecule seen in Cherax hemolymph (4). To see if this dimer could be used to link Jasus monomers into dodecamers or higher polymers, solutions of each in buffer at pH 10 + EGTA were mixed in varying proportions, dialysed back to pH 7.8 + $CaCl_2$, and gels at pH 7.8 were run to investigate the products. Fig. 3 shows that many bands of higher polymers can be seen. A gel of whole Cherax hemocyanin is shown for comparison.

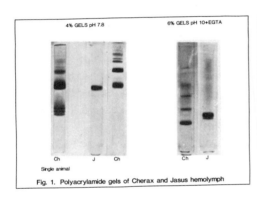

Fig. 1. Polyacrylamide gels of Cherax and Jasus hemolymph

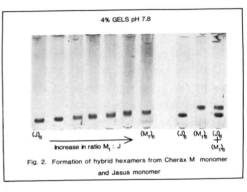

Fig. 2. Formation of hybrid hexamers from Cherax M monomer and Jasus monomer

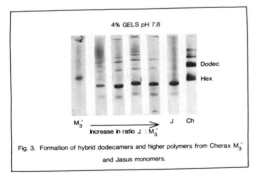

Fig. 3. Formation of hybrid dodecamers and higher polymers from Cherax M_3' and Jasus monomers.

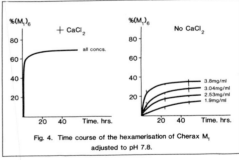

Fig. 4. Time course of the hexamerisation of Cherax M_1 adjusted to pH 7.8.

Hexamerisation of Cherax monomer M_1

By taking a solution of M_1 at pH 10 + EGTA and adjusting the pH instantaneously to pH 7.8, the rate of hexamer formation can be followed by removing samples for gel runs at various times. An estimate of the proportion of monomer to hexamer in the gels can be made by scanning them at 540nm. The results are shown in Fig. 4.

In the presence of $CaCl_2$, hexamerisation is almost instantaneous at all protein concentrations. At the same ionic strength, $I = 0.2$, in the absence of $CaCl_2$, the reaction is slower, and a set of curves for different protein concentrations is obtained.

As hemocyanin hexamer molecules are known to have a structure in which two sets of three monomers sit one above the other, it might be expected that dimers and then trimers would be formed before the hexamer when the conditions of pH are altered in these experiments. Attempts were made to detect dimers and trimers, the first being a meniscus depletion ultracentrifuge experiment on a sample of Cherax hemocyanin containing less than 15% of hexamer in the presence of the "monomer" in the absence of $CaCl_2$. Within experimental error, we found the presence of only one species besides the hexamer, and its molecular weight was found to be 75,400. This does not rule out the possibility that small polymers are present, but in amounts too small to be detected experimentally under these conditions.

In 1968, Moore, Henderson and Nichol studied Jasus hemocyanin in the ultracentrifuge (5) and measured the variation of the sedimentation coefficient of the 5S peak (that is the "monomer" peak) at pH 8.0 with protein concentration. Their results showed an initial positive slope to the curve, which is characteristic of systems which consist of a series of polymers in equilibrium, suggesting to them the contribution of dimers, trimers etc. to the 5S peak. Using whole Jasus hemolymph we were not able to repeat these results, because at pH 8.0 the 5S peak is too small to measure at concentrations less than 3g/100ml, where it is just detectable. The pre-treatment of the hemolymph by Moore et al, which included several exhaustive dialyses, had probably removed calcium ions, so that their system did not represent native hemocyanin.

Cross-linking experiments on equilibrium mixtures of Jasus monomer and hexamer

Glutaraldehyde has been used as a cross-linking reagent for Jasus hemocyanin (6). In the present experiments, Jasus hemocyanin was exposed to glycine buffer at pH 10 + EGTA for an hour, and then dialysed to pH 7.6 using phosphate buffer. At this stage the protein concentration was 0.145mg/ml, the buffer concentration 0.03M and gels at pH 7.8 showed the presence of "monomer" and hexamer bands, the former being the major component. To 18mls of solution were added 2mls of 1% glutaraldehyde in 0.1M phosphate pH 7.6, and the mixture was left overnight.

Any hemocyanin which had been present in dimer or trimer form would now be expected to be covalently linked by glutaraldehyde, so that SDS gels would indicate what molecular weight species were originally present. As small polymers are expected to be present in very small amounts the solution was dialysed exhaustively to remove glutaraldehyde and then concentrated to approx 1ml. The procedure was also carried out on Jasus hemocyanin at the same concentration which had not been exposed to high pH - i.e. hexameric hemocyanin -and on a sample of the same concentration which was exposed to pH 10 + EGTA and dialysed into borate buffer pH 9.0 before cross-linking (glycine buffer pH 10 cannot be used as it is itself cross-linked). This sample put onto gels at pH 10 + EGTA before cross-linking showed only monomer bands to be present. This latter experiment is the nearest control to cross-linking of monomer alone, for comparison with cross-linking the equilibrium mixture.

The results of SDS gels run on the cross-linked samples are shown in Fig. 5. Hexamer cross-linked gives only a band of molecular weight 6 times the monomer molecular weight. Monomer gives multiples of 2 and particularly 4 times the monomer molecular weight, while the equilibrium mixture gives no band with 4 times the monomer molecular weight, but some 3 and some 2 times the monomer molecular weight. These species cannot have arisen entirely as a consequence of cross-linking of monomer molecules, which would have produced a band of 4 times the monomer molecular weight. We may conclude that multiples of 2 and 3 were present in the original equilibrium mixture.

3.5 % SDS GELS

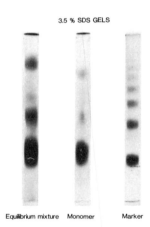

Equilibrium mixture   Monomer   Marker

Fig. 5. Cross-linking of Jasus samples

On this basis, we can say that a plausible model for hexamer formation is via dimer and trimer, possibly in reversible equilibrium with only small amounts of these two species, then association of two trimers to form hexamer with a very large equilibrium constant, so that the reaction is virtually irreversible. Presumably the process is the same in Cherax and Jasus as the monomeric subunits are apparently interchangeable. Also, there is no tendency in either animal for any further aggregation after hexamer formation, which must proceed via a different pathway since dimeric subunits have to be incorporated from the beginning for this to happen.

1. A.C. Murray and P.D. Jeffrey, Biochemistry 13: 3667 (1974).
2. P.D. Jeffrey and G.B. Treacy, Biochemistry 19: 5428 (1980).
3. P.D. Jeffrey, D.C. Shaw and G.B. Treacy, Biochemistry 15: 5527 (1976).
4. P.D. Jeffrey, D.C. Shaw and G.B. Treacy, Biochemistry 17: 3078 (1978).
5. C.H. Moore, R.W. Henderson and L.W. Nichol, Biochemistry 7: 4075 (1968).
6. M.J. Sculley, G.B. Treacy and P.D. Jeffrey, Biophysical Chemistry 19: 39 (1984).

ON THE ROLE OF INDIVIDUAL SUBUNITS
IN THE QUATERNARY STRUCTURE OF CRUSTACEAN HEMOCYANINS

Walter Stöcker, Uta Raeder, Martha M.C. Bijlholt*, Wilma G. Schutter*, Trijntje Wichertjes* and Jürgen Markl

Zoologisches Institut der Universität München
Luisenstrasse 14, 8000 München 2, F.R.G.
* Biochemisch Laboratorium, Rijksuniversiteit te Groningen
Nijenborg 16, 9747 AG Groningen, The Netherlands

INTRODUCTION

Among crustaceans, a widespread hemocyanin aggregate is the dodecamer (24S), composed of two hexameric halfs. According to computer correspondence analysis data (Bijlholt, unpublished), a one-point contact between the two hexamers is probable. A special subunit is required acting as inter-hexamer linker. In the dodecameric 24S hemocyanins of the crayfishes Cherax destructor and Astacus leptodactylus, a disulfide bridged dimer plays this role (1,2). Accordingly, heptameric intermediates were observed when Astacus dodecamers were stepwise dissociated into subunits. In contrast, dodecamers from the crab Cancer pagurus and the lobster Homarus americanus pass hexameric dissociation intermediates, and no dimers were found (3). One particular subunit, designated as alpha', is absent in native 16S hemocyanin of Cancer, and forms dimers under reassociation conditions (4). Apart from the bridging unit, additional subunit types form the two basic hexamers in 24S crustacean hemocyanins. Although X-ray data on the spiny lobster, Panulirus interruptus hemocyanin afforded a deep insight into the conformation of a 16S particle (5), the question remains whether or not those "hexamer-formers" are structurally equivalent and interchangeable. Recently, new data on subunit correspondencies of crustacean hemocyanins became available, resulting in a definition of the immunological subunit types alpha, beta, and gamma (6). This now enables a better interspecific comparison of the results of our earlier (4) and the more recent experiments about reassembly and immuno labeling. They were performed with hemocyanin subunits from the brachyuran crabs Cancer pagurus and Callinectes sapidus, and the astacuran crayfishes Homarus americanus and Astacus leptodactylus.

MATERIAL AND METHODS

After isolation of the 24S fraction, and alkaline dissociation of the hemocyanin into subunits, the subunit composition was analysed by alkaline polyacrylamide gel electrophoresis (PAGE), SDS-PAGE, and crossed immunoelectrophoresis (IE) as published previously (2,7). Each of the electrophoretically discernible subunits was isolated preparatively by ion exchange chromatography (DEAE Sepharose CL-6B, 0.05M Tris/HCl buffer pH 8.8, containing 1mM EDTA; 0.1-0.6M linear NaCl gradient), and subunit-specific rabbit antisera were raised; interspecific subunit correspondencies were analysed as described elsewhere (6). The subunit stoichiometry was determined by a combination of various methods published previously (8); moreover, defined amounts of purified subunits were mixed together until the native proportions were obtained as monitored by crossed immunoelectrophoresis. Reassembly experiments were performed by dialysing individual hemocyanin subunits, respectively mixtures of different subunits, at 4 degree Celsius for 7 days against 0.1 M Tris/HCl buffer of pH 7.5, containing 10 mM Calcium. Reassociation products were investigated by analytical ultracentrifugation, PAGE and, partially, electron microscopy. Immuno labeling experiments of native dodecamers with Fab fragments, respectively IgG molecules, specific for one subunit type, were carried out for Astacus, Homarus and Cancer, and the products investigated in the electron microscope; the methods are described elsewhere (9).

Invertebrate Oxygen Carriers
Ed. by Bernt Linzen
© Springer-Verlag Berlin Heidelberg 1986

RESULTS AND DISCUSSION

1. Subunit composition, stoichiometry, and interspecific correspondence

The dodecameric hemocyanins of Homarus and Astacus are composed of three immunologically distinct monomeric subunit types each, designated as alpha, beta, and gamma (Fig. 1a,b). In Astacus, additionally a fourth - dimeric - subunit type is present, which is immunologically closely related to alpha. Therefore, the designation alpha'-alpha' was chosen for this component. Astacus dodecamers are composed of 2 alpha ,2 alpha', 4 beta and 4 gamma subunits. In Homarus dodecamers, alpha, beta and gamma subunits are present in a molar ratio of 4 : 4 : 4.

In Cancer, electrophoretically four subunit components are discernible, corresponding to astacuran alpha, beta, and gamma (Fig. 1c). Gamma is antigenically deficient compared to alpha; beta is immunologically completely distinct (6,7). The alpha component consists of 2 immunologically identical subunits; one of them (designated as alpha') forms dimers under reassembly conditions and was strongly suggested as inter-hexamer bridge (4). Corresponding properties were detected in Callinectes, although electrophoretically this hemocyanin shows 6 bands (Fig. 1d). The beta component is represented by two immunologically identical subunits (bands 3 and 6). Despite its extreme cathodic electrophoretic position, band 6 is also monomeric as verified by the method of Hedrick and Smith (10). Bands 2, 4 and 5 are immunologically identical and correspond to alpha. Since 5 is present only in very minor quantities, and in SDS-PAGE shows the same molecular weight as component 2, we believe that it is a genetic variation of the latter rather than a new type of subunit. According to its electrophoretic position, band 4 should correspond to Cancer alpha'. At the anodic edge of the pattern migrates subunit gamma (band 1), which is immunologically related to alpha but, in contrast to Cancer gamma, shows no antigenic deficiency. Determination of the molar proportions of the four structural components of Cancer hemocyanin dodecamers yielded beta : alpha' : alpha/gamma = 4 : 2 : 6. Semiquantitative estimations indicated comparable values also for Callinectes. The proportion of alpha versus gamma seems to vary within the Brachyura; in a number of species, gamma has not been detected at all (6).

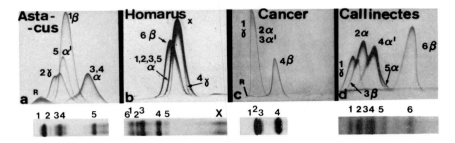

FIGURE 1:
Crossed immunoelectrophoresis patterns (above) of the subunits of 24S hemocyanins from Astacus, Homarus, Cancer, and Callinectes against their homologous antisera. The anode of the first dimension was on the left. R = undissociated hemocyanin; X = denatured hemocyanin. Subunits are designated with figures according to their electrophoretic mobility in alkaline PAGE (below). Additionally, immunologically corresponding subunits are identically labeled with Greek letters according to (6).

2. Hexamer and dodecamer formation in reassembly experiments

Purified alpha, beta, or gamma subunits from <u>Homarus</u> were capable to form regular 16S hexameric structures. For the formation of dodecamers, all three components were required. It could not be decided, however, which of them functions as inter-hexamer bridge. Interestingly, under the conditions used here, in this combination not only dodecamers, but also ringlike eight-hexamers, and four-hexameric half-rings were formed. <u>Homarus</u> hemocyanin provides no stable dimeric subunit and accordingly we observed heptameric intermediates neither in the present reassembly experiments nor in earlier dissociation experiments (3).

Also each of the three <u>Astacus</u> monomers was able to form homo-hexamers. Co-reassembly of <u>Astacus</u> alpha, beta and gamma yielded no dodecamers or heptamers, whereas hexamers were present in an 80% yield (Fig. 2a). Some heptamers and dodecamers, however, were found besides hexamers in reassembly products containing one of the three monomeric <u>Astacus</u> subunits combined with the alpha'-alpha' bridging unit. However, considerable amounts of heptamers and dodecamers were only formed, when all three monomers plus the dimer were present. Additionally, certain amounts of oligo-hexamers could be detected arranged in chains (Fig. 2b).

In <u>Cancer</u> and <u>Callinectes</u>, subunit beta was particularily capable to form homo-hexamers. Also alpha and gamma (the latter was only available in a pure form from <u>Callinectes</u>) formed homo-hexamers, but in much lower quantities. The proportion of hexamers clearly increased – and reached 80% – when alpha and gamma were reassembled as a mixture. Subunit alpha' in both species yielded a mixture of dimers and hexamers as reassembly products. The combination of alpha' with alpha and/or gamma yielded few heptamers in addition to many hexamers. However in both species, dodecamers were exclusively formed in the presence of all four subunit types; the best yield (20%) was obtained by combining the subunits in their native proportions.

Although the formation of hexamers is an inherent ability of all monomeric crustacean hemocyanin subunits investigated here, combinations of different subunit types work better in this respect. To assemble dodecamers in considerable yield, not only an "inter-hexamer linker unit" is required, but also the presence of the total set of different monomers in their native stoichiometry. Oligo-hexameric misfit products demonstrate how important it is to incorporate the linker unit correctly. Those structural constrains have probably stimulated the evolution of distinct types of monomeric subunits.

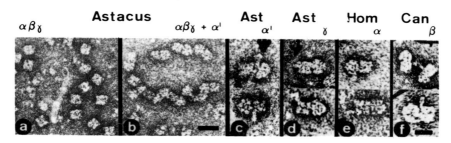

FIGURE 2:
Electron micrographs of negatively stained crustacean hemocyanin molecules.
(a) – (b): Reassembly products of Astacus hemocyanin subunits as indicated. The bar represents 20 nm.
(c) – (f): Immuno labeling of <u>Astacus</u>, <u>Homarus</u> and <u>Cancer</u> dodecamers. The subunit specificities of the Fab-fragments, respectively IgG molecules, are indicated. The bar represents 10 nm.

3. Immuno electron microscopy, and the model of quaternary structure

In the case of Astacus, distinct antisera could be raised against alpha and alpha', which was not possible for Homarus and Cancer due to the immunological identity of these components. Alpha' of Astacus is localized at the contact region between the two hexameric half-structures (Fig. 2c). In contrast, monomeric Astacus alpha occupies the peripheral short edges of the dodecamers. Experiments with Homarus alpha showed labeling of both positions (Fig. 2e).
In the case of Cancer, only an anti-alpha/alpha'/gamma antiserum was available, which labeled the dodecamer all over except for the long edge opposite to the bridge. This position is occupied by subunit beta as could be seen in Cancer (Fig. 2f), and also in the other species. Anti-gamma Fab fragments in Homarus and in Astacus clearly labeled the peripheral short edges of the dodecamer, but were also visible at the long edge opposite to the position of subunit beta (Fig. 2d). The resulting model is shown in Fig. 3.

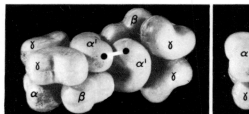

FIGURE 3:
Topologic model of the quaternary structure of 24S crustacean hemocyanins.
The different subunits are arranged symmetrically, with alpha'-alpha' as inter-hexamer bridge, a central cluster of the dimer and 4 beta units, 2 alpha units at the outer edges, and 4 gamma units to finish off the periphery. Although in the Brachyura partially gamma may be substituted by alpha, the structure is phylogenetically of a striking conservatism.

Acknowledgment: We thank Prof. Bernt Linzen and Prof. Ernst F.J. van Bruggen for support and discussions. Bruce Johnson (Beaufort, USA) provided subunit 6 of Callinectes hemocyanin. Klaas Gilissen and Wolfgang Forster prepared the photo documentation. Supported by DFG grants to J. Markl (Ma 843) and B. Linzen (Li 107), and by Dutch grants to E. van Bruggen (S.O.N. and Z.W.O.).

LITERATURE
 1. Jeffrey P.D. (1979): Biochemistry 12, 2508 - 2513.
 2. Markl J. & Kempter B. (1981): J.Comp.Physiol. 141, 594 - 502.
 3. Markl J., Decker H., Stöcker W., Savel A., Linzen B., Schutter W.G., &
    van Bruggen E.F.J. (1981): Hoppe-Seyler's Z.Physiol.Chem. 362, 185 - 188.
 4. Markl J., Stöcker W., Runzler R., Kempter B., Bijlholt M.M.C. & van
    Bruggen E.F.J. (1983): In: Structure and function of invertebrate oxygen
    binding proteins (Wood E.J., ed.). Life Chem. Reports 1, 39 -42.
 5. Gaykema W.P.J., Hol W.G.J., Vereijken J.M., Soeter N.M., Bak H.J.
    & Beintema J.J. (1984): Nature 309, 23 - 29.
 6. Markl J., Stöcker W., Runzler R. & Precht E. (1986): This volume,pp. 281-292.
 7. Markl J., Hofer A., Bauer G., Markl A., Kempter B., Brenzinger M. &
    Linzen B. (1979): J.Comp.Physiol. 133, 167 - 175.
 8. Markl J., Savel A. & Linzen B. (1980): Hoppe-Seyler's Z.Physiol.Chem. 361,
    649 - 660.
 9. Lamy J., Bijlholt M.M.C., Sizaret P.-Y., Lamy J. & van Bruggen E.F.J.
    (1981): Biochemistry 20, 1849 - 1856.
10. Hedrick J.L. & Smith A.J. (1968): Arch.Biochem.Biophys.. 126, 155 - 164.

STEM AND CRYO-TEM OF <u>LIMULUS</u> AND <u>KELLETIA</u> HEMOCYANIN

W.G. Schutter, W. Keegstra, F. Booy, J. Haker and E.F.J. van Bruggen
Biochemisch Laboratorium, Rijksuniversiteit
Groningen, The Netherlands

The image of the elastically scattered electrons (related to the mass-thickness of the object) can be obtained in the most efficient way using the ratio signal from Scanning Transmission Electron Microscopy (STEM). In this way we studied negatively stained (4x6)-meric molecules of <u>Limulus</u> Hc (hemocyanin). The results were analyzed by computer and compared with those from conventional bright field Transmission Electron Microscopy (TEM).

Cryo-EM has been developed recently to an applicable procedure to study biomacromolecules without fixation or staining close to their native state (1). Solutions of <u>Limulus</u> (8x6)-meric and <u>Kelletia</u> Hc were analyzed in this way. Use was made of focus dependent phase contrast between protein and amorphous ice. The results are compared with those from conventional negative staining.

Materials and methods

<u>Limulus</u> (4x6)-mers were prepared by dialysis of the blood against 20 mM Tris.HCl + 10 mM EDTA (pH 7.0). Specimens were made according to the single carbon layer technique of Valentine modified to a microscale (protein concentration 200 μg/ml; 1% uranyl acetate) (2). For STEM a Philips EM 400 T equipped with a field emission gun and a modified STEM attachment was used. We collected the dark field over bright field signal in the cold tip mode. Photographs were taken directly from the monitor. The negatives were digitized with a scan step of 5 Å at the specimen scale. The digitized images were processed using our IMAGIC software system on a NORD 10 minicomputer (3). Computer alignment was followed by correspondence analysis (CORAN) (4).

Specimens for cryo-EM were prepared by placing a drop of <u>Limulus</u> or <u>Kelletia</u> blood on a grid coated with a carbon supporting film. After adsorption for some 30 seconds the specimen was blotted partially dry with filter paper and quench-frozen in liquid ethane maintained close to its freezing point by liquid nitrogen. The frozen grid was transferred to the cooling holder (type Philips PW 6591) under liquid nitrogen, inserted into the Philips EM 400 T and examined at ca. -160°C (5). Electron micrographs were recorded in the 0 to 1 μm close-to-focus and 6 to 10 μm under focus range.

Invertebrate Oxygen Carriers
Ed. by Bernt Linzen
© Springer-Verlag Berlin Heidelberg 1986

Results and discussion

Fig.1(A) shows negatively stained (4x6)-meric molecules of <u>Limulus</u> Hc obtained
with the ratio signal in STEM. 206 molecules were selected, aligned and classified
by CORAN. The 4 most important classes of the CORAN aggregates were summed and are
reproduced in Fig.1(B-E). The results are similar to those obtained earlier from

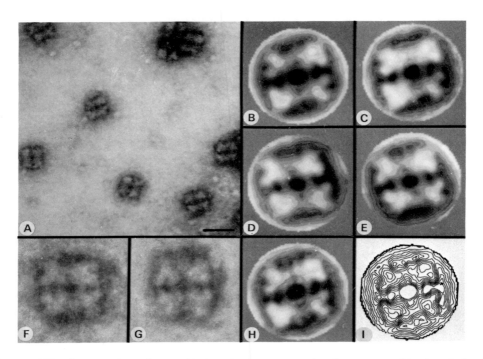

Fig.1: STEM (ratio signal) and CORAN of <u>Limulus</u> Hc (4x6)-mers negatively stained
with uranyl acetate. (A): STEM-survey; (B-E): 4 most important classes from CORAN,
B+C vs. D+E correspond with flip/flop, B+D vs. C+E show the rocking; (F-I):
molecules with 4 protein bridges, F and G as observed in STEM, H and I from CORAN
summation. Bar = 300 A.

bright field TEM (4). In STEM amplitude contrast only is observed, while in TEM we
are dealing with both amplitude and phase contrast. The similarity in our results
for STEM and TEM indicates that for TEM the eventual contribution from phase
contrast is lost during the computer averaging procedure. In STEM we did one
peculiar observation along the second CORAN-axis consisting of molecules with 4
instead of 2 protein bridges (Fig.1(H and I)). Upon closer look such molecules are
also visible on the original STEM graphs (Fig.1(F and G)). This might indicate,
that another type of (weaker?) interaction occurs between the (2x6)-mers. This was
mentioned earlier for <u>Eurypelma</u> Hc by the Munich group from biochemical data (6).

Fig.2(A and B) shows (8x6)-mers of the Limulus Hc in cryo-EM close-to-focus and under focus respectively. Conventional electron microscopy of the same solution after negative staining is demonstrated in fig.2(C). Some characteristic molecular profiles are compared in fig.2(D-F). The images are very similar. We expected more different orientations in the case of cryo-EM. Clearly complete adsorption to the

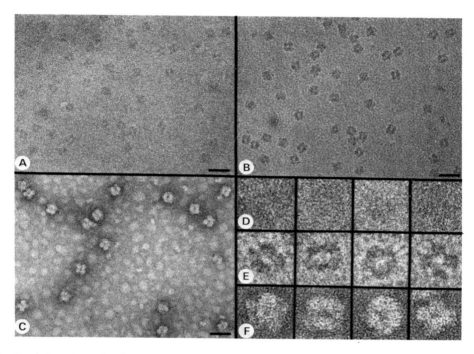

Fig.2: (A): close-to-focus cryo-EM of unstained Limulus Hc (8x6)-mers; (B): Idem under focus cryo-EM; (C): conventional EM of Limulus Hc (8x6)-mers negatively stained with uranyl acetate; (D): selected profiles cryo-EM close-to-focus unstained; (E): selected profiles cryo-EM under focus unstained; (F): selected profiles conventional EM negatively stained. Bar = 500 Å.

carbon film occurred before freezing of the solutions. The low contrast is a problem for computer analysis of these data.

Fig.3(A) is a cryo-electron micrograph of Kelletia Hc molecules. We do observe substructure although with low contrast and we can just distinguish domains. We see the well-known circular and rectangular projections of a cylindrical molecule. In addition we now observe intermediate projections (Fig.3(row B)). The dimensions of the cylinders are ca. 290 x 380 Å compared to 350 x 380 Å after negative staining. The difference is due to flattening.

We started experiments to detect differences between oxy- and deoxy-Hc by cryo-EM and cryo-electron diffraction.

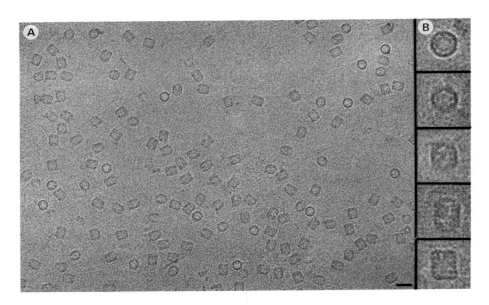

Fig.3: (A): under focus cryo-EM of unstained Kelletia Hc; (row B): selected profiles corresponding with different projections of a cylindrical molecule. Bar = 500 Å.

References

1. Adrian, M., Dubochet, J., Lepault, J., and Mc.Dowell, A., Nature, 308, 32-36 (1984).
2. Tichelaar, W., Schutter, W.G., Wichertjes, T., and Van Bruggen, E.F.J., Micron and Microsc.Act., 5, 195-203 (1984).
3. Van Heel, M., and Keegstra, W., Ultramicr., 7, 113-130 (1981).
4. Van Heel,M., and Frank, J., Ultramicr., 6, 187-194 (1981).
5. Booy, F.P., Ruigrok, R.W.H., and Van Bruggen, E.F.J., J. Mol. Biol. (1985) in press.
6. Markl, J., Kempter, B., Linzen, B., Bijlholt, M.M.C., and Van Bruggen, E.F.J., Hoppe-Seyler's Z. Physiol. Chem., 362, 1631-1641 (1981).

Acknowledgements

We thank K. Gilissen for printing and mounting of the photographs. This work was supported in part by the Netherlands Foundation for Chemical Research (SON) with financial aid from the Netherlands Organization for the Advancement of Pure Research (ZWO).

Subunit structure of molluscan hemocyanins

PRESENCE OF ONLY SEVEN FUNCTIONAL UNITS IN THE POLYPEPTIDE CHAIN OF THE HAEMOCYANIN OF THE CEPHALOPOD *OCTOPUS VULGARIS*

C. Gielens, C. Benoy, G. Préaux, and R. Lontie
Laboratorium voor Biochemie, Katholieke Universiteit te Leuven
Dekenstraat 6, B-3000 Leuven, Belgium

The subunits of the haemocyanins (Hc's) of gastropods (e.g. *Helix pomatia*, $M_r \simeq 450\ 000$) and of decapodan cephalopods (e.g. *Sepia officinalis*, $M_r \simeq 390\ 000$) are constituted of eight functional units (each containing a dioxygen-binding copper pair) with average M_r of respectively $\simeq 55\ 000$ and $\simeq 50\ 000$ (1). In order to check if the Hc subunits of octopodan cephalopods are built in the same way, the subunits of *Octopus vulgaris* Hc were submitted to limited proteolysis.

The haemolymph of *O. vulgaris* contained some agglutinin (2) and presumably traces of proteinase inhibitors as these were found in all organs investigated of the cephalopod *Loligo vulgaris* (3). In order to remove the contaminants the haemolymph was dialysed against Tris-HCl buffer, pH 8.2, I 50 mM, and chromatographed on DEAE-Sepharose with a linear gradient of NaCl (0 - 0.5 M) in the same buffer, whereby the agglutinin eluted before the Hc (4). In a more recent procedure the haemolymph was dialysed against Tris-HCl buffer, pH 7.0, I 0.1 M, 10 mM $CaCl_2$, and submitted to a preparative ultracentrifugation (Spinco Model L, rotor 40, 35 000 rev./min for 3 h), yielding a pellet of undissociated Hc molecules.

Crossed immunoelectrophoresis (CIE) in 1 % agarose at pH 8.8 in a 1/1 mixture of sodium barbital-barbital and Tris-glycine buffer, I 80 mM (5) with a rabbit antiserum against haemolymph showed for both purified Hc preparations (Hc mainly dissociated into its subunits) a single precipitation peak, pointing to the presence of only one type of subunit. By sodium dodecyl sulphate polyacrylamide gel electrophoresis (SDS-PAGE) on cylindrical gels (90 × 5 mm) of 5 % with a weight ratio acrylamide/N,N'-methylenebisacrylamide of 37:1 (6) in 0.1 M Tris-glycine buffer, pH 9.0, this subunit migrated slightly faster than that of the Hc of *S. officinalis* and distinctly faster than that of the Hc of *H. pomatia*. Electrophoresis of a mixture of these subunits confirmed this difference in mobility, and hence in size, as it yielded 3 bands (Fig. 1A). Using as references the subunit of β_c-Hc of *H. pomatia* and some of its proteolytic fragments, whose M_r values were determined by sedimentation equilibrium (7), a M_r of $\simeq 350\ 000$ was deduced for the

Invertebrate Oxygen Carriers
Ed. by Bernt Linzen
© Springer–Verlag Berlin Heidelberg 1986

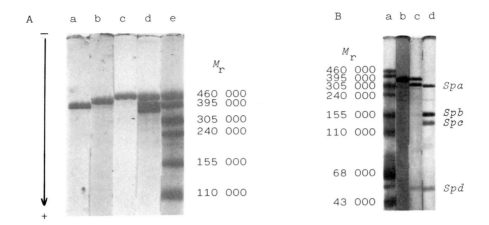

Fig. 1. SDS-PAGE on 5 % gels in 0.1 M Tris-glycine, pH 9.0, of
(A) the subunit of *O. vulgaris* Hc (a) as compared to that of *S. officinalis* Hc (b) and that of β_c-Hc of *H. pomatia* (c); a mixture of the
3 Hc's (d); β_c-Hc of *H. pomatia* and proteolytic fragments as references (e),
(B) hydrolysis products of the subunit of *O. vulgaris* Hc obtained by
limited proteolysis with *S. aureus* serine proteinase (E/S = 1/800,
w/w, at a Hc concentration of 1 %) at 20°C for 1 h (c) and 22 h
(d); calibration proteins (same as in gel Ae + bovine serum albumin
and ovalbumin) (a) and intact *O. vulgaris* Hc (b) as references.
All samples were treated with 1 % SDS in the presence of 1 % 2-mercap-
toethanol for 3 min at 100°C.

O. vulgaris Hc subunit (Table 1) from the quadratic relationship between log M_r and relative mobility (7). This value agreed fairly well
with that obtained by sedimentation equilibrium for the subunit of
Octopus dofleini Hc (8).

Limited proteolysis in borate-HCl buffer, pH 8.2, I 0.1 M, 0.02 %
NaN_3, of the Hc subunits (most probably in equilibrium with 1/5-mole-
cules as indicated by $s_{20,w}$ = 13.4 S at 5 mg/ml) with *Staphylococcus
aureus* (V8) serine proteinase (EC 3.4.21.19; Miles, lot no. 0019)
yielded on SDS-PAGE both in the presence and in the absence of 2-mer-
captoethanol two main fragments: *Spa*, M_r ≃ 300 000, and *Spd*, M_r ≃
50 000. On increasing the hydrolysis time a cleavage of the former
fragment was observed into fragments *Spb* and *Spc*, both with M_r ≃
150 000 (Fig. 1B). Chromatography on Ultrogel AcA34 in 0.1 M NH_4HCO_3
of a hydrolysate, obtained by incubation of a 1 % Hc solution in the
pH 8.2 borate buffer for 5 h at 20°C at an E/S ratio of 1/800 (w/w),
yielded 3 peaks respectively corresponding to *Spa*, *Spb* + *Spc*, and *Spd*.

Fragments *Spb* and *Spc*, present in only minor amounts, were not separated from each other due to a similar M_r. On repeating the chromatography after hydrolysis for 8 h at a Hc concentration of 5.5 % (with
the other conditions kept constant) the first peak was markedly decreased in favour of the second one, as a result of a further cleavage
of *Spa* into *Spb* and *Spc*. On chromatography of fraction (*Spb* + *Spc*),
concentrated by ultrafiltration on a Diaflo PM-10 membrane, on DEAE-
Sepharose in Tris-HCl buffer, pH 8.2, I 50 mM, with a linear gradient
of NaCl, *Spc* eluted before *Spb*. For each chromatography the fragments
were located in the elution curves by SDS-PAGE and fused rocket immunoelectrophoresis (5).

All fragments showed besides the absorption band at 278 nm, due to
the aromatic amino acids, the characteristic copper band with maximum
near 346 nm, proving that on proteolysis the functional groups were
preserved. The specific absorption coefficient at 278 nm (a_{278}) was
determined in 50 mM borax (pH 9.2) on solutions for which the protein
concentration (0.2 - 0.6 %) was obtained by differential refractom-
etry at 546 nm, using for the specific refraction increment the value
determined for β_c-Hc of *H. pomatia* (1.944×10^{-3} dl/g) (7). The a_{278}
values were fairly constant (Table 1), pointing to a similarity in
aromatic amino-acid content. The specific absorption coefficient at
346 nm (a_{346}) was obtained in 50 mM borax after regeneration of the
copper band with H_2O_2 ($H_2O_2/Cu = 10$). The observed values (Table 1)
seem to indicate differences at the active sites of the functional
units. The M_r values (averages of 15-30 measurements) were determined
by SDS-PAGE (in the presence of 2-mercaptoethanol) on 5 % gels from
the quadratic ($M_r > 10^5$) or linear ($M_r < 10^5$) relationship between
$\log M_r$ and relative mobility (7). Copper was determined from the atomic
absorption at 324.7 nm of solutions (protein concentration $\approx$ 0.5 g/l)

Table 1. Characteristics of the subunit and fragments of *O. vulgaris* Hc

| | a_{278} | a_{346} | M_r | | Copper | | N-Terminal Amino Acid |
	($1 \ g^{-1}cm^{-1}$)				%	Pairs	
Subunit	1.45	0.360	346 ± 21	(30)	0.25_7	7.0	Asn
Spa	1.43	0.358	298 ± 22	(24)	0.25_7	6.0	Asn
Spb	1.40	0.285	151 ± 11	(22)	0.22_9	2.7	Asn
Spc	1.48	0.388	140 ± 15	(24)	0.23_9	2.6	(Arg)
Spd	1.45	0.428	55 ± 3.2	(15)	0.25_9	1.1	Val

which were rendered 10 mM in EDTA and thereafter extensively dialysed against borate-HCl, pH 8.2, I 0.1 M, 0.02 % NaN_3. From the M_r values and the percentage of copper in the subunit and the fragments (Table 1) it could be concluded that the subunit contained 7 functional units, Spa 6, Spb and Spc both 3, and that Spd corresponded to a single functional unit. N-terminal amino acid analysis with 4-NN-dimethylamino-azobenzene 4'-isothiocyanate (9) allowed to locate the fragments in the polypeptide chain: ^{N}Spa-Spd^{C} and ^{N}Spb-Spc-Spd^{C}. Tandem CIE (5) showed that Spb and Spc were immunologically different. Their 3-unit structure was further confirmed by a limited trypsin treatment followed by SDS-PAGE: Spb was cleaved into fragments with $M_r \simeq$ 110 000 and 40 000 (held together by a disulphide bridge) and Spc into fragments with $M_r \simeq$ 90 000 and 50 000.

The results of the limited proteolysis with $S.$ $aureus$ serine proteinase at pH 8.2 thus proved that the subunit of $O.$ $vulgaris$ Hc is constituted of only seven functional units.

Acknowledgement. We wish to thank Dr. G. Nardi, Stazione Zoologica, Naples, Italy, for the haemolymph of $O.$ $vulgaris$. We are grateful to the Fonds voor Collectief Fundamenteel Onderzoek for research grants.

REFERENCES

1. Préaux, G., and Gielens, C., *in* Copper Proteins and Copper Enzymes (Lontie, R., Ed.), Vol. II, pp. 159-205, CRC Press, Boca Raton, FL, (1984).
2. Renwrantz, L., and Uhlenbruck, G., Z. Immunitätsforsch. *148*, 16-22 (1974).
3. Tschesche, H., and von Rücker, A., Hoppe-Seyler's Z. Physiol. Chem. *354*, 1510-1512 (1973).
4. Préaux, G., Gielens, C., and Lontie, R., *in* Metalloproteins, Structure, Molecular Function and Clinical Aspects (Weser, U., Ed.), pp. 73-80, Georg Thieme Verlag, Stuttgart (1979).
5. Axelsen, N.H., Krøll, J., and Weeke, B., Scand. J. Immunol. *2*, Suppl. 1, 1-167 (1973).
6. Weber, K., Pringle, J.R., and Osborn, M., Methods Enzymol. *26*, 3-27 (1973).
7. Wood, E.J., Chaplin, M.F., Gielens, C., De Sadeleer, J., Préaux, G., and Lontie, R., Comp. Biochem. Physiol., in press.
8. Miller, K.I., and Van Holde, K.E., Comp. Biochem. Physiol. *73B*, 1013-1018 (1982).
9. Chang, J.Y., Brauer, D., and Wittmann-Liebold, B., FEBS Lett. *93*, 205-214 (1978).

Immuno-EM of <u>Sepia officinalis</u> hemocyanin.

T. Wichertjes[1], C. Gielens[2], W.G. Schutter[1], G. Préaux[2], R. Lontie[2] and
E.F.J. van Bruggen[1]
1. Biochemisch Laboratorium, Rijksuniversiteit, Groningen, The Netherlands
2. Laboratorium voor Biochemie, Katholieke Universiteit, Leuven, Belgium.

Negatively stained molecules of <u>Sepia officinalis</u> hemocyanin (Hc) (M_r 3.8×10^6)
have the form of a partly hollow cylinder with 350 Å diameter and 170 Å height.
The circular profiles show material inside the cylinder: the so-called collar
which has 5-fold (or possibly 10-fold) symmetry (1). At higher pH values this Hc
dissociates into five subunits, each consisting of 2x8 domains (<u>a-h</u>) with average
M_r of 50,000 (2). Most probably this dissociation is similar to the dissociation
of half <u>Helix pomatia</u> α-Hc molecules, i.e. along helical grooves in the cylinder
wall (3).
We made antibody complexes of whole Hc-molecules with IgG anti-<u>abc</u>, IgG anti-<u>de</u>
and IgG anti-<u>gh</u> in order to determine which domains form the "collar" and whether
the ten polypeptide chains <u>a-h</u> are all running parallel to each other (i.e.
parallel to the grooves) or are alternating.
The antibodies were isolated from antiserum against whole <u>Sepia</u> Hc by exhaustion
with respectively fragment <u>d-h</u> (for IgG anti-<u>abc</u>), fragment <u>a-f</u> (for IgG anti-<u>gh</u>)
and fragments <u>abc</u>, <u>f</u> and <u>gh</u> (for IgG anti-<u>de</u>). Complexes were made (7-19 hr, 4°C),
purified over an Ultrogel A4 column (135x2 cm) and prepared for electron
microscopy using the spray-droplet technique (3) with uranyl acetate as negative
stain.
Sedimentation analysis of the fractions indicated, that the faster moving
fractions contain complexes with two or more Hc molecules, while the slower
fractions contain single molecules and/or dissociation products of Hc.
The complexes on the electron micrographs were classified in several groups after
which the relative amount of each group was calculated. Then a comparison was made
between the data for complexes with different IgG's.
Hc+IgG anti-<u>abc</u> gives a high percentage of Hc molecules with IgG attached to the
edge of the cylinder-formed molecules (fig.1A). In the first fractions many pairs
of Hc molecules are found with relatively large distances; quite often the two Hc
molecules are not parallel to each other (fig.1A1). With IgG anti-<u>gh</u> similar
results are obtained, the main difference with Hc+IgG anti-<u>abc</u> being the shape of
the pairs of Hc molecules, present in the first fractions; with IgG anti-<u>gh</u> the Hc
molecules are situated more parallel and closer to each other than with IgG
anti-<u>abc</u> (fig.1B). Hc+IgG anti-<u>de</u> also gives pairs of Hc molecules in the first
fractions; now the Hc molecules are sitting quite close to each other and parallel
(fig.1C1). Another characteristic of Hc+IgG anti-<u>de</u> is the presence of pairs of

Invertebrate Oxygen Carriers
Ed. by Bernt Linzen
© Springer–Verlag Berlin Heidelberg 1986

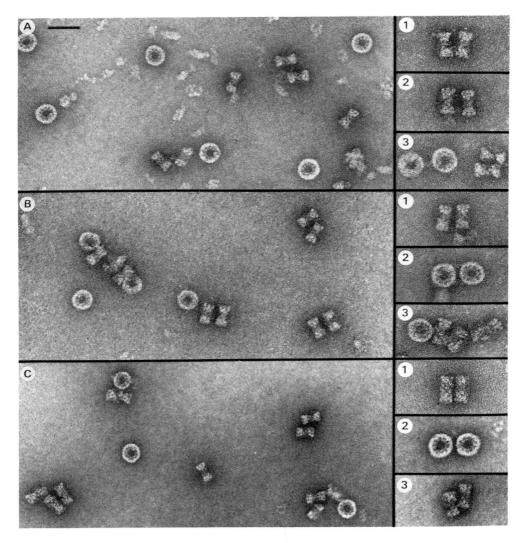

Figure 1. Complexes of S.officinalis hemocyanin with IgG (scale bar = 500 Å).
1A. With IgG anti-abc; 1B. With IgG anti-gh; 1B3. With IgG anti-gh,
treated with IgG anti-abc; 1C. With IgG anti-de.

molecules, in which one Hc molecule is pulled into the hollow inside of the other
molecule (fig.1C3). In all cases complexes are observed formed by two circular
views of Hc molecules connected by one or more IgG molecule(s) (fig.1A3, 1B2 and
1C2).

From these observations qualitative conclusions can be drawn:

. abc and gh are entirely or partly situated near the edge of the Hc molecule

. de is situated in the middle of the molecule

. abc, gh and de are entirely or partly situated in the wall, because in all cases
  IgG is seen attached to the outside of the cylinder wall.

In order to get more quantitative results we classified all complexes dependent on the position of the antigenic site on the Hc molecule. The upper and lower third part of the Hc molecule were called region A, the remaining part of the Hc the B-D region, B being the outside and D the inside. Further we assumed that pairs of molecules with distance 0-20 Å are connected by one or more IgG's attached to the B or D region, while all distances larger than 20 Å are assumed to correspond to attachment sites in the A region. The results of this classification are shown in table 1.

Table 1.

	Hc+IgG anti-abc	Hc+IgG anti-gh	Hc+IgG anti-de
A	52.5	50.5	32.5
A or B	17	9	13
B	0	2	1.5
D	24.5	13	24
B or D	6	19	17.5
A or D	0	3	9
non-classifiable	0	2	3

High percentages for attachment site A for IgG anti-abc and IgG anti-gh point at abc and gh sitting entirely or partly near the edge of the Hc molecule, while the lower percentage for Hc+IgG anti-de means that de is situated rather more in the center of the Hc molecule. Low percentages for site B for all cases are in accordance with the expectation, because of the cylindrical shape of the Hc molecule: the chance to see IgG attached to the middle of the cylinder wall with the Hc molecule in its rectangular view is very small. The low percentage for region D found for complexes with IgG anti-gh means that gh is sitting closer to the edge of the molecule, while abc must sit partly near the middle region.
These results are in perfect agreement with the model for Sepia Hc we derived from the shapes of compact 1/5 molecules (fig.2).
In this model the domains b, c, d, e, f and g are forming the wall, while domains a and h are sitting inside, thus forming two collars: one in the upper part of the cylinder and one in the lower part.
The question whether the ten polypeptide chains a-h are all running parallel to each other or are alternating is not easy to answer. In order to solve this problem we added IgG anti-abc to Sepia Hc molecules, complexed with IgG anti-gh.

We found several complexes of three or four Hc molecules, connected by IgG, which could represent these mixed complexes (fig.1B3). We plan to repeat this experiment in order to get more evidence for or against alternating chains.

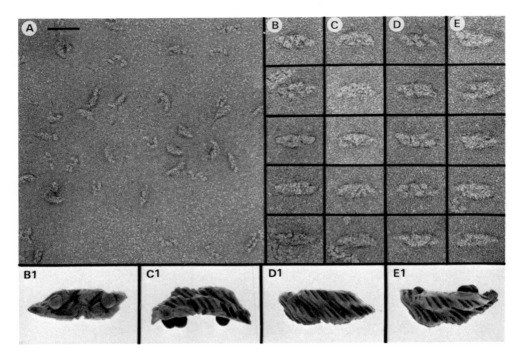

Figure 2A. Subunits of S.officinalis Hc in 0.05 M Tris-HCl, pH 7.4, ionic
strength 0.5, 1 mM EDTA (scale bar = 500 Å).
B,C,D,E. Idem, classified in four categories.
B1,C1,D1,E1. Model for compact 1/5 molecules.

Acknowledgements:

We thank the Station marine, Wimereux, France for the gift of Sepia hemolymphe,
Mr. K. Gilissen for photographic assistance and Dr. W. Keegstra for text
processing. This work was partly supported by the Netherlands Foundation for
Chemical Research (S.O.N.) with financial aid from the Netherlands Organization
for the Advancement of Pure Research (Z.W.O.)(E.v.B.) and Fonds voor Collectief
Fundamenteel Onderzoek, Belgium (R.L.).

References:

1. Bruggen, E.F.J. van, Wiebenga, E.H. and Gruber, M., J.Mol.Biol. 4, 8-9 (1962).
2. Gielens, C., Bosman, F., Préaux, G. and Lontie, R. in Structure and Function of
   Invertebrate Respiratory Proteins, ed. E.J. Wood, Harwood Academic Publishers,
   Chur, London, New York, 1983, p. 121-124.
3. Siezen, R.J. and Bruggen, E.F.J. van, J.Mol.Biol. 90, 77-89 (1974).

PRELIMINARY RESULTS ON THE STRUCTURE OF OCTOPUS DOFLEINI HEMOCYANIN

J. Lamy, J.N. Lamy, M. Leclerc, and S. Compin
François Rabelais University, Tours, France

K.I. Miller and K.E. van Holde
Oregon State University, Corvallis, Oregon, U.S.A.

INTRODUCTION

The hemocyanin of Octopus dofleini has provided an excellent model for the study of the behavior of multi-subunit proteins (1-3). Sedimentation experiments indicated that the native molecule (Mr = 3.6×10^6, $S_{20,w}$ = 51S) is a decamer of polypeptide chains (Mr = 3.6×10^5, $S_{20,w}$ = 11.1S). Preliminary electrophoretic studies suggested that there might be only a single kind of polypeptide chain, making this structure simpler than that of most molluscan hemocyanins. Furthermore, the molecular weight of the subunit is smaller than that of gastropod hemocyanins, suggesting that it might contain fewer than eight oxygen binding domains. We report here on our investigations of the structure of O. dofleini hemocyanin.

MATERIALS AND METHODS

Hemocyanin: Hemocyanin was prepared by gel filtration of whole hemolymph on Biogel A-5m as described previously (1). Whole molecules were dissociated into subunits by dialysis against a buffer containing 10 mM EDTA at pH 8.0 or above.
Antisera: Rabbit polyclonal antisera were prepared as previously reported (4). All the antisera were used as such without any saturation step.
Immunoelectrophoreses: Crossed immunoelectrophoreses (CIE) and crossed-line immuno-electrophoresis (CLIE) were carried out according to Weeke (5) and Kroll (6), respectively.
Sedimentation behavior: Experiments were conducted with the Beckman model E analytical ultracentrifuge equipped with photoelectric scanner.
Electon microscopy: Specimens for electron microscopy were negatively stained with 2 per cent uranyl acetate.
Proteolysis experiments: Bovine pancreas trypsin (Sigma T1005), subtilisin (Boehringer Mannheim 165905), serine protease from Staphylococcus aureus strain V8 (Sigma P8400) and bovine pancreas α-chymotrypsin (Serva 17160) were used for limited proteolysis experiments with various incubation times and enzyme/substrate ratio less than 1/50. Zymofren[@] was used to stop the incubations of trypsin and chymotrypsin and PMSF the incubations of subtilisin and staphylococcal serine protease.
Polyacrylamide gel electrophoreses were conducted in the presence and absence of SDS according to the methods of Davis (7) and Laemmli (8).

RESULTS AND DISCUSSION

1 - The native molecule contains a single type of polypeptide chain

Two arguments are in favor of a single type of polypeptide chain. First, dissociated hemocyanin is homogeneous in molecular weight by SDS gel electrophoresis and sedimentation equilibrium (1). Furthermore, when one calculates the integral distribution of sedimentation coefficients by the method of van Holde and Weischet (9), one finds complete homogeneity of sedimentation behavior.

Invertebrate Oxygen Carriers
Ed. by Bernt Linzen
© Springer–Verlag Berlin Heidelberg 1986

232

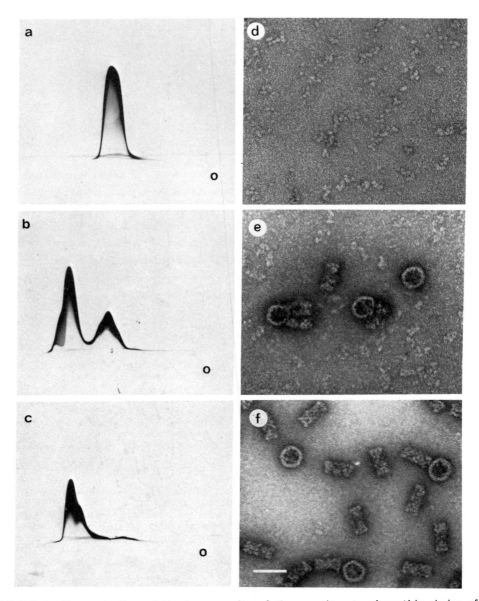

FIGURE 1 Demonstration of the homogeneity of the constituent polypeptide chains of Octopus dofleini hemocyanin by crossed immunoelectrophoresis and electron microscopy. a, b, c: CIE of whole O. dofleini hemocyanin. The first dimension occurred a) at pH 8.6 in barbital buffer, b) in 50 mM Tris-HCl buffer at pH 7.5, 2.5 mM calcium lactate; c) in 50 mM Tris-HCl buffer at pH 7.5, 5 mM calcium lactate. d, e, and f are electron micrographs of hemocyanin separated as in Figs. a, b, c, respectively.

Second, under the standard conditions of crossed immunoelectrophoresis (barbital buffer pH 8.6), the whole molecule dissociated (Fig. 1d) and produced a single homogeneous precipitation peak against an antiserum specific for dissociated hemocyanin (Fig. 1a). When the first dimension occurred in a 50 mM Tris-HCl buffer, pH 7.5, containing 2.5 mM calcium lactate, hemocyanin partially dissociated producing two precipitation peaks corresponding to the native molecule and to the polypeptide chain, respectively (Figs. 1b and 1e). In the presence of 5 mM calcium lactate, no dissociation occurred and only an asymmetric peak corresponding to whole hemocyanin was found (Figs. 1c and 1f).

2 -   The polypeptide chains are composed of 7 globular domains

Figure 2 shows a gallery of selected views of polypeptide chains resulting from the dissociation of the native molecule by dialysis against a I 0.1 Tris-HCl buffer pH 8.0, 10 mM EDTA. Seven globular domains are clearly visible on the micrographs. This result is in fairly good agreement with the molecular weight of 350 kDa of the whole chain determined by sedimentation equilibrium (1).

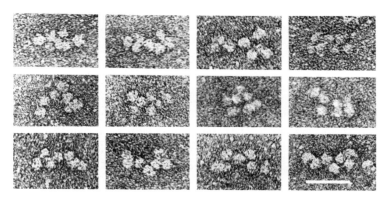

FIGURE 2   Selected E.M. views of multidomain polypeptide chains. The length of the bar is 25 nm.

3 -   The proteolysis of the subunit produces seven antigenically distinct 50 kDa domains

A 20 hour trypsinolysis of dissociated hemocyanin with a 1/50 enzyme/substrate ratio produced four major 1-domain fragments and one major 2-domain fragment characterized by thin-layer gel filtration (TLG) for the molecular weight and by CIE. The four 1-domain fractions did not cross-react or cross-reacted at most very weakly. They were arbitrarily termed domains 1, 2, 3, and 4. The 2-domain fragment was clearly composed of domain 2 and of another domain called domain 5. In addition, two minor fractions were subsequently identified as domain 6 and domain 7. The reason why domain 6 only produced a small peak with trypsin while it produced a high precipitation peak with subtilisin is not yet clear.

Domains 1, 2, and 6 were purified to a high degree of purity by preparative polyacrylamide gel, and were injected into rabbits. This procedure allowed the preparation of perfectly monospecific antisera as judged in CIE. An antiserum specific for domain 1 + 3 was also prepared.

4 - Limited proteolysis experiments allow partial determination of the sequence of the domains

Four proteases were studied at various enzyme/substrate ratios and at various incubation times. The fragments purified by preparative polyacrylamide gel electrophoresis were studied by CIE and CLIE for the domain composition. The results of these experiments are shown in Table I.

TABLE I

Enzyme	6 domains	4 domains	3 domains	2 domains
		Fragments		
Subtilisin		(1,2,5,6); (3,4,6,7);	(1,2,5); (3,4,7);	(2,5);(4,7);
Staphylococcal Serine protease	(2,3,4,5,6,7);		(2,5,6); (3,4,7);	(2,5);
α chymotrypsin		(1,2,5,6)	(2,5,6); (3,4,7);	
Trypsin	(2,3,4,5,6,7)	(3,4,6,7)	(1,2,5); (3,4,7);	(2,5)
Elastase		(1,2,5,6)	(3,4,7),	

CONCLUSION

We have demonstrated that Octopus dofleini hemocyanin consists of a single polypeptide chain, containing seven immunologically distinct globular oxygen binding domains. The sequence of these domains has been determined with some remaining uncertainty of position to be $\left\{ 1 - (2,5) - 6 - [3, (4,7)] \right\}$. There appears to be a labile zone in the middle of the subunit. Hydrolysis of this zone by all proteases so far studied produced a 4-domain fragment $[1 - (2,5) - 6]$ and a 3-domain fragment $[3, (4,7)]$. That this particular susceptibility to proteases is related to the architecture of the molecule is not yet established.

REFERENCES

1. Miller, K., and van Holde, K.E., Comp. Biochem. Physiol. 73B, 1013-1018 (1982).
2. van Holde, K.E., and Miller, K., Biochemistry (in press) (1985).
3. Miller, K., Biochemistry (in press) (1985).
4. Lamy, J., Lamy, J., and Weill, J., Arch. Biochem. Biophys. 193, 140-149 (1979).
5. Weeke, B., Scand. J. Immunol. 2 (suppl. 1), 47-56 (1973).
6. Kroll, J., Scand. J. Immunol. 2 (Suppl. 1), 79-81 (1973).
7. Davis, B., Ann. N. Y. Acad. Sci. 121, 404-427 (1964).
8. Laemmli, U.K., Nature 227, 680-685 (1970).
9. van Holde, K.E., and Weischet, W., Biopolymers 17: 1387-1404 (1978).

FUNCTIONAL AND STRUCTURAL PROPERTIES OF THE 50,000 D SUBUNIT OF Octopus vulgaris HEMOCYANIN

F. Ricchelli, B. Filippi[*], S. Gobbo, E. Simoni[§], L. Tallandini[§] and P. Zatta

C.N.R. Centre of Hemocyanins and other metalloproteins, * Dept. of Organic Chemistry and § Dept. of Biology

University of Padova, Padova (Italy)

Although the functional and structural properties of hemocyanins (Hc) have been well characterized (1,2), the interpretation of the results is often difficult owing to the quite complicated quaternary structure of these copper proteins. Under physiological conditions, Octopus vulgaris Hc exists as a large aggregate with a sedimentation coefficient $s^o_{20,w}$ = 50 S (MW=2.7x10^6). The smallest component which can be obtained without splitting covalent bonds (by increasing the pH or in the presence of 3 M urea) has $s^o_{20,w}$ = 11 S (MW=250,000)(3). A fully functional component with s = 5 S (MW=50,000), corresponding to the theoretical minimal subunit, has been obtained from O. vulgaris Hc after proteolysis with trypsin under controlled conditions. The functional and conformational properties of 5 S subunit have been compared with those of the protein in the two main aggregation states (11S, 50S). A definite picture of the tryptophan (Trp) residues localization can be obtained with no interference of the quaternary structure.

MATERIALS AND METHODS

11S and 50S components of O. vulgaris Hc were prepared as described elsewhere (3). 5S subunit was obtained by hydrolysis of 50S Hc with trypsin (Hc/trypsin = 800/1) at 37°C for 24 h in 0.1 M bicarbonate/carbonate buffer, pH= 8.2. The full procedure for separating the different fractions of the digested protein has been previously reported (4). From the last chromatography on an ionic exchange gel, two 50,000 molecular weight components were obtained. Since the two 5S components showed only limited differences in the O_2-affinity and in the conformational stability, only the 5S subunit which was eluted the last was used for this study. 5S, 11S, and 50S apoHcs were prepared by dialysis against 25 mM KCN. Oxygen equilibrium experiments were carried out according to Brouwer et al., 1977 (5). Cations were removed from 5S Hc by dialysis against 10 mM EDTA. Fluorescence quenching experiments were performed using acrylamide, Cs^+ and I^- as quenchers. The quenching data were analyzed according to the Stern-Volmer relationship (2):

$F°/F = 1 + K_Q [X]$, where F° and F are the fluorescence intensities in the absence and in the presence of the quencher, respectively, $[X]$ is the quencher concentration and K_Q the quenching constant. When the fluorescence was heterogeneous a modified Stern-Volmer equation was used (2), which allows to calculate the fraction of fluorescence accessible to the quencher (f_a) and the quenching constant associable to this fraction (K_Q).

RESULTS AND DISCUSSION

Like 50S and 11S components, 5S Hc binds O_2 reversibly. The maximal O_2-affinity is shown at pH= 9.0 (A_{345}/A_{280} = 0.25). In fig. 1 the data of the oxygen affinity as a function of pH are reported for the 50S Hc (both in the presence and in the absence of Ca^{++}) and for the 5S component. The positive Bohr effect observed in the whole protein is strongly reduced in 5S subunit. The disruption of quaternary structure leads to the disappearance of cooperative effects, as deduced by Hill plots (data not shown). For the entire pH range examined, n_{max} = 1. Fluorescence quantum yields (Q) for all Hc samples, as calculated by using N-Acetyltryptophanamide as reference standard (Q= 0.13), are reported in Table 1. The fluorescence ratio F_{deoxy}/F_{oxy} is 2.6 for 5S Hc. Clearly, the relationship between the fluorescence quantum yields of oxy- deoxy- and apo- forms of 5S Hc (1:2.6:4.8) is nearly the same as found in most Hcs (1:2.8:5). The pH profiles of the fluorescence quantum yields (in the pH range 5.0-9.5) of both 50S and 5S native Hc show two transitions, around pH 6.7 and 8.0 (fig. 2). This is due to the ionization of specific groups, since no denaturation of the proteins occurs, as indicated by the constancy of the fluorescence

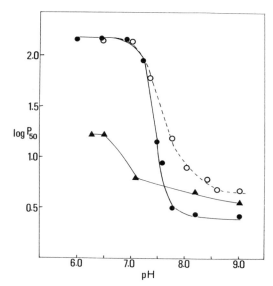

Fig. 1: O$_2$-affinity as a function of pH for 50S Hc in the presence of 10 mM Ca^{2+} (●-●), or 0.25 M NaCl (O--O) and for 5S Hc (▲-▲). The buffer was Tris/HCl I= 0.1.

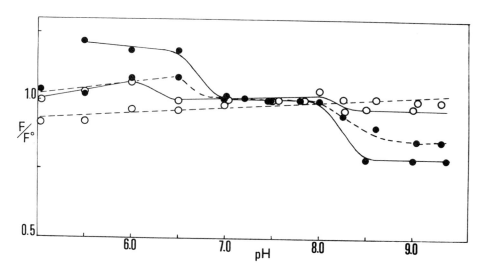

Fig. 2: Fluorescence quantum yields as a function of pH for 50S native Hc (●-●), 50S apoHc (O-O), 5S native Hc (●--●) and 5S apoHc (O--O). All fluorescence intensities were normalized to that of Hc at pH 7.5.

TABLE 1

Fluorescence emission properties of O. vulgaris Hc

Hc at pH= 7.5	Q	Emission maximum(nm)[a]	Acrylamide quenching	
			f_a	$K_Q(M^{-1})$
50S oxyHc[b]	0.011	330	1.0	0.38
50S apoHc[b]	0.053	331	1.0	0.62
11S oxyHc[b]	0.01	333	0.63	0.72
11S apoHc[b]	0.049	335	0.55	2.9
5S oxyHc	0.02	333	0.35	10.4
5S apoHc	0.096	335	0.45	10.1
5S oxyHc (with 0.7 M Cs$^+$)[c]			1.0	0.9
5S apoHc (with 0.7 M Cs$^+$)			0.75	3.2

a) The excitation wavelength was 295 nm

b) Ref. 2

c) The quenching experiments were carried out at pH 9.0 where
$A_{345}/A_{278} = 0.25$

emission maximum and circular dichroism experiments. These findings
clearly indicate that the main physico-chemical properties of Hc are
preserved in 5S subunit. Some conformational variations, however, are
observed. It is known that, for the maintenance of the conformational
stability, subunit interactions are more important than the presence
of copper ions in the active site (2). This can be evidenced in 5S Hc
from the lower value of the α-helix content (13% for both native and
apoHc), as compared with that of 11S and 50S Hc (17%). Moreover,
quenching experiments (Table 1) indicate a greater exposure of the
Trp residues to the solvent. In 5S Hc, 35% of the fluorescence is associated with a class of fully exposed Trps, as indicated by the values
of K_Q which are typical of Trps completely accessible to the quencher.
On the contrary, the K_Q values, obtained for 11S and 50S Hc, are

indicative of a deep burial of the fluorophores, due to the quaternary organization. A limit case is represented by the 50S Hc where the great compactness of the protein structure leads to an homogeneity of all Trp environments ($f_a = 1$). These results indicate that the exposed Trps are localized, within the quaternary structure, near to the intersubunit contact sites. Cs^+ and I^- selectively quench the most exposed class of fluorophores ($f_a = 0.4$)(data not reported). The combined use of ionic species and acrylamide as quenchers allow to understand definitely the emissive properties and the localization of the Trp residues in Hc. When the most exposed Trps were previously quenched with Cs^+ (or I^-), native 5S Hc shows only one residual class of Trps susceptible to acrylamide quenching ($f_a = 1.0$, $K_Q = 0.9$ M^{-1}, very buried Trps). The same experiment on 5S apoHc reveals a further class of partially ex-posed Trps ($f_a = 0.75$, $K_Q = 3.2$ M^{-1}). These results confirm that three classes of Trps are present in Hc, one of which is not fluorescent in the native protein (6).

REFERENCES

1. Van Holde, K.E., and Miller, K.I., Quarterly Reviews of Biophysics 15, 1-129 (1982).
2. Ricchelli, F., Jori, G., Tallandini, L., Zatta, P., Beltramini, M., and Salvato, B., Arch.Biochem.Biophys. 235, 461-469 (1984).
3. Salvato, B., Jori, G., Ghiretti-Magaldi, A., and Ghiretti, F., Biochemistry 18, 2731-2736 (1979).
4. Ricchelli, F., and Zatta, P., Médicine Biologie Environnement 13 105-108 (1985).
5. Brouwer, M., Bonaventura, C., and Bonaventura, J., Biochemistry, 15, 2618-2623 (1977).
6. Shaklai, N., Gafni, A., and Daniel, E., Biochemistry 17, 4438-4442 (1978).

IMMUNOCHEMICAL RELATIONSHIPS AND SUBUNIT COMPOSITION OF SELECTED MOLLUSCAN HEMOCYANINS

Michael Brenowitz, Kathy Munger, Celia Bonaventura, and Joseph Bonaventura, Department of Physiology and Marine Biomedical Center, Duke University Marine Laboratory, Beaufort, NC 28516 (USA)

Among the molluscan hemocyanins that have been studied in this laboratory, the constituent subunits of a gastropod, the giant atlantic murex, Murex fulvescens, and a cephalopod, the chambered nautilus, Nautilus pompilius, have been isolated and characterized (1,2). In the first case, two electrophoretically-distinct subunits having different structural and functional properties were observed. In the latter case, three subunits with distinct oxygen binding characteristics were isolated. In this study we have used immunoelectrophoresis to determine the subunit composition of these hemocyanins and others whose self-assembly and oxygen binding properties have been studied. We have also investigated whether the immunochemical relationships follow phylogenetic relationships or some other structural or functional criteria.

MATERIALS AND METHODS

Hemocyanins were purified essentially as described (1,2). Helix pomatia hemocyanin was a gift from Dr. E.F.J. van Bruggen. Except for Nautilus hemocyanin (Phillipines), Megathura crenulata and Octopus dofleini (Newport, Oregon) all hemocyanins were obtained from specimens collected in the vicinity of Beaufort, N.C. Antisera were prepared as has been described (3) and were stored in liquid nitrogen until use.

Immunochemical relationships were determined using the Ouchterlony diffusion technique (4). One millimeter thick, 1% agarose plates made up in 0.05M Tris/HCl, 10 mM EDTA buffer at pH 8.9 were employed. Hemocyanin samples were dissociated by dialysis versus this buffer. Diffusion was allowed to proceed at room temperature for 48 hours. Subunit studies were carried out by crossed (CIE) and crossed-line immunoelectrophoresis (CLIE) (3).

RESULTS

Immunochemical Relationships

Ouchterlony double diffusion experiments were performed using antisera raised against dissociated and unfractionated Nautilus, Octopus, Loligo pealei, Murex, and Buscyon carica hemocyanins (Fig.1, Table I). Dissociated hemocyanins from these species and a number of other molluscs were used as antigens. All the hemocyanins tested against Murex antiserum except

TABLE 1: Ouchterlony Diffusion Experiments

ANTIGENS	ANTISERA				
Gastropoda	Murex	Busycon	Nautilus	Octopus	Loligo
Murex	T	P	P	N	N
B. carica	P	T	N	N	N
B. contrarium	P	T	N		
B. canaliculatum		T			
Ilyanassa	P	P	N	N	N
Littorina	P	P	N	N	N
Fasciolaria	P	P	N	N	N
Helix	FP	N	N	N	N
Megathura	N	N	N	N	N
Cephalopoda					
Nautilus	N	N	T	N	N
O. dofleini	FP	N	FP	T	FP
O. vulgaris				T	
Loligo	FP	N	FP	FP	T

T = Total Identity
P = Partial Identity
N = No Identity
FP = Faint precipitate

Invertebrate Oxygen Carriers
Ed. by Bernt Linzen
© Springer–Verlag Berlin Heidelberg 1986

Megathura and Nautilus showed partial identity. While Octopus and Loligo hemocyanins react with Murex antiserum, this reaction was non-reciprocal in that neither cephalopod antisera reacted with Murex hemocyanin. Total identity is observed among all three species of Busycon tested against Busycon carica antiserum. The Busycon anti-serum did not react with Helix, Megathura or with any of the cephalopod hemocyanins. The antisera raised against the cephalopod hemocyanins did not cross-react with any gastropod hemocyanins with the exception of a non-reciprocal partial reaction of Murex hemocyanin and Nautilus antiserum.

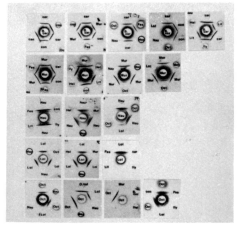

Figure 1.   Ouchterlony double dif-fusion experiments with the anti-serum in the center well.   Row 1, Busycon carica (B. car) antiserum; Row 2, Murex fulvescens (Mur) antiserum;   Row 3, Nautilus pompilius (Nau) antiserum;   Row 4, Loligo pealei (Lol) antiserum. Row 5, Octopus dofleini (Oct) antiserum. Abbreviation for the antigens are Ilyanassa Ily; Helix pomatia, Hel; Littorina littoria, Lit; Megathura crenulata, Meg; Fasciolaria, Fas; B. carica. car; B. canacuniculatum, can; B. contrarium, con.

Subunit composition: Cephalopod hemocyanins.   Upon CIE Octopus hemocyanin shows a single symmetrical peak (Fig. 2D).   Loligo hemocyanin shows a single major peak, with minor components which are immunochemically deficient relative to the main peak, suggesting that they are undissociated or denatured hemocyanin (Fig. 2E).

Nautilus hemocyanin shows four distinct precipitation lines (Fig. 2A). To identify these peaks an analysis was performed on zones A, B, and C isolated from Nautilus hemocyanin (2) and on a fourth fraction (zone D) which was subsequently isolated from zone A (5).   Zone A shows a antigenically-deficient leading shoulder indicating that some degradation of the subunit may have occurred (Fig. 3A). Only zone B shows any appreciable contamination by the other subunits (Fig.B). Crossed-line immunoelectrophoresis was used to identify each of the peaks of the whole subunit mixture (Figs. 3E-H).   The shoulder of peak A (Fig. 2A) is identical to zone A (Fig. 3E). CLIE shows that zone D is immunochemically deficient to peak A (Fig. 3H) which strongly suggests that zone D is a degradation product of zone A.   Zone C is identical with the most cathodal peak in the dissociated hemocyanin (Fig. 3G) and with the leading shoulder of the peak (see Fig. 3E where the obscuring peak is removed).

Figure 2. CIE of dissociated hemo- cyanins. First dimension migration is from left (anode) to right (cathode). A. Nautilus; B. Murex; C. Murex (another sample); D. Octopus; E. Loligo; F. Busycon.

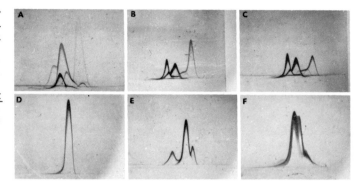

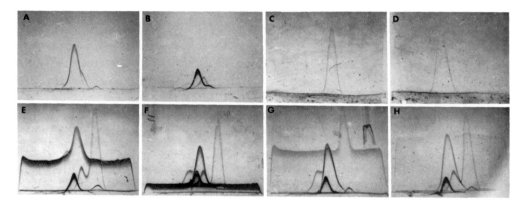

Figure 3. Identification of the isolated Nautilus subunits. The top row (A-D) shows CIE with purified Nautilus zones A, B, C, D. The lower row (E-H) shows CLIE of unfractionated and dissociated hemo- cyanin (in the hole) versus zones A through D in the line.

Gastropod hemocyanins   Upon, CIE Busycon carica hemocyanin shows a single peak with an antigenically deficient leading shoulder (Fig. 2F). A second precipitation line is also apparent under the leading edge of the peak. Three peaks are apparent upon CIE of Murex hemocyanin. When different samples of Murex hemocyanin were tested the most cathodal peak varied in height (Fig. 2B, C). Crossed immunoelectrophoresis of the two subunits isolated from Murex hemocyanin by (1) shows both subunit preparations to be homogeneous (Fig. 4A and B). Upon CLIE, subunit A fuses with the anodal peak; although, unexpectedly the peak is antigenically deficient to the line (Fig. 4C). The line containing subunit B completely fuses with the middle peak (Fig. 4D). The cathodal peak was immunochemically deficient relative to both the purified subunits.

Figure 4. Identification of the
isolated _Murex_ subunits. Plate A,
CIE of isolated subunit A; plate B,
isolated subunit B, plate C, CLIE
of unfractionated hemocyanin (in
the hole) versus subunit A (in the
line); plate D, versus subunit B.

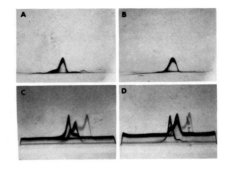

CONCLUSIONS

The pattern of cross-reactivity is that expected from the phylogenetic relationships. Species of the same genus are immunochemically
identical. Partial identity is observed within some of the other
superfamilies of Gastropoda and Cephalopoda while there is almost no
cross-reaction between the hemocyanins of the two classes. The non-
reciprocal reactions which occur between _Murex_ and the cephalopod
hemocyanins suggests certain antigenic determinants were accessible
for antibody development in some hemocyanins but not others. It is
possible that partial denaturation as a result of dissociation of the
hemocyanins could expose more determinants during formation of the
antibodies (6). There may be two sets of determinants, external
surfaces that are unique to each species and possibly internal
determinants that are common to all molluscan hemocyanins.

Of the five hemocyanins investigated three appear to have a
single kind of polypeptide chain. _Murex_ has two and _Nautilus_ three
distinct subunits. The extent of polypeptide chain heterogeneity of
the molluscan hemocyanins is much less than in arthropod hemocyanins
of comparable molecular weight.The absence of any clear difference in
the subunit composition of the hemocyanins of the two classes, which
differ in size and morphology, suggests that there is not a general
relationship between the number of kinds of subunits and the
aggregation state of native molluscan hemocyanins as has been
demonstrated for arthropod hemocyanins (7).

REFERENCES

1. Brouwer, M., Ryan, M., Bonaventura, C., Bonaventura, J.
   Biochemistry 17: 2810-2815 (1978).
2. Bonaventura, C., Bonaventura, J., Miller, K., Van Holde, K.E.
   Arch. Biochem. Biophys. 211: 589-598 (1981).
3. Lamy, J., Lamy,J., Weill,J., Bonaventura, J., Bonaventura, C.,
   Brenowitz, M., _Arch. Biochem. Biophys._ 196: 324-339 (1979).
4. Ouchterlony, O. _Prog. Allergy_ 6: 30-154 (1962).
5. Gerald Godette, personal communication.
6. Maurer, P.H. and Callahan, M. _Meth. Enzymol._ 70: 49-69 (1980).
7. Markl, J. & Kempter, B. _J. Comp. Physiol._ B141: 495-502 (1981).

KINETICS AND EQUILIBRIA OF OCTOPUS HEMOCYANIN ASSOCIATION

K.E. van Holde and Karen I. Miller
Oregon State University, Corvallis, OR 97331 USA

Molluscan hemocyanins are constructed from decamers of large polypeptide chains (see van Holde and Miller, 1982, Ellerton et al., 1983). Although complete dissociation of these molecules is readily effected, quantitative reassociation has not been possible in most cases where it has been tried (see, for example, van Holde and Cohen, 1964; Siezen, 1974; Brouwer et al., 1978). However, as shown by Miller and van Holde (1982), dissociation of the hemocyanin of Octopus dofleini is completely reversible. Addition of sufficient quantities of a divalent cation (Ca^{2+}, Mg^{2+}) results in the quantitative reassociation of subunits into functional decameric molecules, indistinguishable from the native hemocyanin by physical techniques (van Holde and Miller, 1985). Because of this behavior, we felt that this system would be an excellent candidate for studies of the kinetics of reassociation of a molluscan hemocyanin. Virtually nothing is known of such processes.

Experimental Methods

Hemocyanin was purified from O. dofleini hemolymph as described (van Holde and Miller, 1982). All experiments used 0.1 I Tris buffers, at pH 8.0, containing either EDTA or Mg^{2+}, as required. All studies were conducted at 20°±2°. Sedimentation experiments were carried out in a Beckman Model E ultracentrifuge equipped with scanner optics. Light scattering experiments utilized either a fluorimeter constructed in the laboratory of Dr. I. Isenberg, or a Dionex D100 stopped-flow apparatus, in the laboratory of Dr. M. Schimerlik.

Results and Conclusions

Initial studies of the reassociation reaction utilized sedimentation velocity experiments begun at various times after adding Mg^{2+} to a solution of hemocyanin subunits. Figure 1 depicts two important points: (1) Under the conditions chosen the reaction is relatively slow, requiring many hours for completion. (2) As the scans show, 11S subunits and 51S decamers are the principal components at all times; there is very little evidence for the existence of intermediates.
In such an event, it is possible to use a simple technique like light scattering to follow the reaction over the entire time span. The mixture initially contains only monomers (with scattering intensity i_0). If the reaction proceeds by a mechanism in which monomer and decamer are the only species present, to eventually quantitatively convert to decamers (with scattering intensity i_∞), then the fraction of decamers can be shown to be

$$f_D(t) = \frac{i(t)-i_0}{i_\infty-i_0} \tag{1}$$

where i(t) is the intensity at time t. Experiments were conducted as follows:
A solution of monomers (in 5 mM EDTA) was placed in the fluorimeter, and the 90°

Invertebrate Oxygen Carriers
Ed. by Bernt Linzen
© Springer–Verlag Berlin Heidelberg 1986

scattering at 384 nm recorded. A small volume of 4 M MgCl$_2$ was added at t=o, so as to compensate for the EDTA and yield a known excess of Mg^{2+}. Scattering was recorded for about 30 minutes. After 12-24 hours, a final value (i$_\infty$) was measured. At the same time, the sample was checked in the analytical ultra-centrifuge to determine that reassociation was complete.

A representative experiment is shown in Figure 2. The data are deficient only in the fact that we could not obtain a first data point sooner than about 30 sec. after mixing. As Figure 2 shows, the scattering curve does not appear to pass through the origin.

Fig. 1   Fig. 2

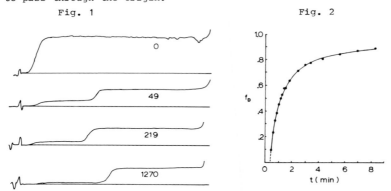

Figure 1. The top panel shows a sedimentation velocity experiment with <u>Octopus</u> hemocyanin subunits in 10 mM EDTA, pH 8.0. The material sediments as a homogeneous 11S boundary. Aliquots of a concentrated stock of this solution were diluted into buffer containing sufficient MgCl$_2$ to yield a final concentration of free Mg^{2+} of 8 mM. At times of 49', 219', and 1270' after mixing, sedimentation velocity runs were started, with results indicated. The 1270' scan probably represents a near-equilibrium state at the Mg^{2+} and hemocyanin concentrations used here.

Figure 2. The fraction of decamer (f$_D$) as a function of time from light scattering. At t=o, a solution of subunits was made 40 mM in Mg^{2+}. The final scattering value was measured after 24 hours.

We decided to test the data against various kinetic models. The simplest reasonable model will involve a second-order reaction between monomers as the rate-limiting step. If this were the case, the fraction of decamers at time t should be given by either of the two expressions below:

$$\frac{f_D}{1-f_D} = kC_o t \tag{2}$$

or
$$\frac{1}{f_D} = 1 + \frac{1}{kC_o t} \tag{3}$$

Where k is a second-order rate constant and C$_o$ is the initial concentration. Figure 3 depicts a plot according to Equation (2). All data gave similar linear

graphs, but in all cases the line did not pass through the origin. In other
words, the system behaves like a second-order reaction with a "lag" period. If
one subtracts the observed lag (δ) as determined from graphs like Figure 3, and
graphs according to equation (3), with $t'=t-\delta$, excellent straight lines are again
obtained (Figure 4). This test allows us to examine the behavior at very long
times. Graphs like (3) and (4) demonstrate that the data (when corrected for the
lag period) accurately follow second order kinetics over a very wide time span.

Another test of second-order kinetics can be carried out by seeing if the
quantity kC_0 (obtained from the slopes of graphs like Fig. 3 and Fig. 4) is in
fact proportional to C_0. This asks, in effect, if k is really a constant.
Figure 5 shows that this is so. The rate-limiting step in the association reaction is second order in monomer.

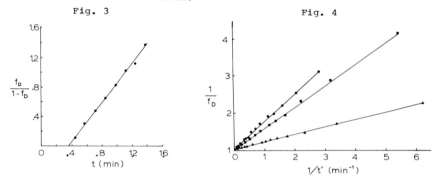

Fig. 3   Fig. 4

Figure 3. A graph of the data in Figure 2 according to Equation (2).

Figure 4. Graphs of hemocyanin reassociation kinetics according to equation (3).
The data cover virtually the whole time span. In each case, the time t' has been
calculated by subtracting the "lag time" δ, determined from graphs like Figure 3.

While these experiments imply the existence of a lag period, they do not
allow us to examine behavior in this time range. Accordingly, we have turned to
stopped-flow methods, again employing light scattering. In these experiments, a
dissociated hemocyanin solution in one syringe was mixed with a magnesium-
containing buffer in another syringe. Figure 6 shows the results of a typical
experiment. The lag period is demonstrated by the initial upward turn in the
data. We wondered if this could be the consequence of some unimolecular
rearrangement, and accordingly examined the tryptophan fluorescence of the hemocyanin by stopped-flow after addition of $MgCl_2$. A rapid reaction, first order in
hemocyanin was observed (data not shown). However, the half-time of this reaction
is only about 50 msec., so it cannot in itself explain the lag time.

This reaction may be only part of a series of unimolecular events leading to
the formation of an activated monomer, M^*.

On the basis of these observations, we propose the following tentative model
for the reaction:

    (a) $M \rightarrow M^*$: unimolecular; possibly a series of processes

    (b) $2M^* \rightarrow M_2^*$: rate limiting; bimolecular

    (c) $M_2^* + M_2^* + \ldots \rightarrow$ decamer; fast; details unknown

Fig. 5

Fig. 6

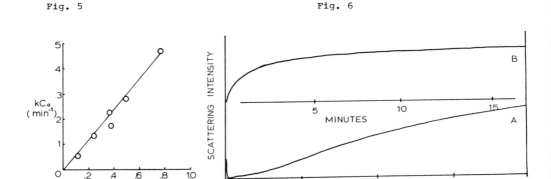

Figure 5. The quantity "kC_0" has been determined as either the slope of a graph like Figure 3, or the inverse of the slope of graph like Figure 4 (results are virtually identical. Graphing this versus C_0 (the known initial monomer concentration) gives a straight line, indicating that k is constant.

Figure 6. Representative scattering data from stopped-flow experiments showing lag period.

Preliminary studies of the effect of magnesium ion concentration on the bimolecular reaction and the lag time indicate that the former is insensitive to Mg^{2+}, but that lag time increases with decreasing magnesium. This would suggest that reaction (a) is the step at which divalent cations are involved. It may be of changes in the subunit conformation, triggered by binding of divalent cations. Clearly, further studies are required.

Acknowledgments

This research was supported by a National Science Foundation grant (PCM82 12347). We wish to acknowledge the assistance of Dr. L. Libertini in the light scattering studies, and Dr. M. Schimerlik in the stopped-flow experiments.

References

Brouwer, M., Ryan, M., Bonaventura, J. and Bonaventura, C. (1978) Biochemistry 17, 2810-2815.
Ellerton, H.D., Ellerton, N.F., and Robinson, H. (1983) Prog. Biophys. Mol. Biol. 41, 143-248.
Miller, K.I., and van Holde, K.E. (1982) Comp. Biochem. Physiol. 73B, 1013-1018.
Siezen , R.J. (1974) J. Molec. Biol. 90, 103-113.
van Holde, K.E. and Cohen, L.B. (1974) Brookhaven Symp. Biol. 17, 184-193.
van Holde, K.E. and Miller, K.I. (1982) Quart. Rev. Biophys. 15, 1-129.
van Holde, K.E. and Miller, K.I. (1985) Biochemistry (In Press).

SUBUNIT STRUCTURE OF HEMOCYANIN FROM THE GASTROPOD Levantina hierosolima

I. Avissar, E. Daniel, D. Banin and E. Ilan[1]
Department of Biochemistry, Tel-Aviv University, Tel-Aviv 69978
[1]Department of Bio-medical Engineering, Technion, Haifa 32000
Israel

Introduction

Electron microscopy and hydrodynamic studies have shown that gastropod hemocyanin can exist in a number of discrete states of aggregation 100, 60, 20 and 11 S
that have been identified as 1 (whole), 1/2, 1/10 and 1/20 molecules (1,2).   The
11 S species, which corresponds to the fully dissociated polypeptide chain, has in
turn been shown to consist of 8 functional units, similar in size, each carrying a
binuclear copper centre (3,4).   The molecular weight of the 100 S native molecule
has been determined to be $\sim 9 \times 10^6$ (Archachatina marginata, 9.08×10^6 (5); Busycon
canaliculatum, 8.8×10^6 (6); Helix pomatia, 8.95×10^6 (7), Helix pomatia α, 8.91×10^6
(5), Helix pomatia β, 8.95×10^6 (5), 9.02×10^6 (8), 7.55×10^6 (9)).   A molecular
weight of 55 000 has been determined for Helix pomatia functional unit (3,10).
From these findings, a molecular weight of ~ 450 000 has been predicted for the polypeptide chain of gastropod hemocyanin $((1/20) \times 9 \times 10^6$ or 8x55 000).   However, the
molecular weight determined for the polypeptide chain is ~ 350 000 (Busycon canaliculatum 300 000 (6); Helix pomatia α, 365 000 (11) and 330 000 (12), Helix pomatia β,
350 000 (8)).   The reason for the discrepancy could, a priori, be in the molecular
weight determination of the functional unit, the polypeptide chain or the native
molecule; it could also involve an error in the number of functional units per polypeptide chain or the number of polypeptide chains per 100 S molecule.   The solution
to this problem provided the motivation for this study.

Experimental

Preparation of hemocyanin.   Hemocyanin was prepared from the snail Levantina hierosolima.   Emphasis was placed on rapid isolation.   The purification of hemocyanin
involved: (a) Rapid bleeding of the snail (~ 5 min); (b) Gel filtration of the hemolymph over G75 Sephadex (~ 5 min); (c) Two precipitations and redissolutions in a
preparative ultracentrifuge.   All solutions contained 0.1 mM phenylmethylsulfonylfluoride.

Copper content.   Copper was determined with 2,2'biquinoline after reduction with
cysteine hydrochloride (13) and by atomic absorption spectroscopy.

Invertebrate Oxygen Carriers
Ed. by Bernt Linzen
© Springer–Verlag Berlin Heidelberg 1986

Ultracentrifugation. Molecular weights were determined by meniscus depletion sedimentation equilibrium. A partial specific volume of hemocyanin in water, 0.729
ml/g, and an isopotential specific volume in 6 M guanidinium chloride, 0.713 ml/g,
were measured in a digital density meter. For details, see elsewhere (14).

Results

Molecular weight of the native 100 S molecule. This was determined by meniscus
depletion sedimentation equilibrium. The molecular weight was obtained by extrapolation to zero concentration of a plot of reciprocal point-by-point weight average
molecular weight, $1/\bar{M}_{w,r}$, versus protein concentration, c. A molecular weight of
10.4×10^6 was obtained.

SDS gel electrophoresis. The electrophoretic pattern of Levantina hierosolima
hemocyanin in SDS polyacrylamide gel electrophoresis shows a single band with a
mobility slightly lower than that of thyroglobulin (Fig. 1). The same pattern

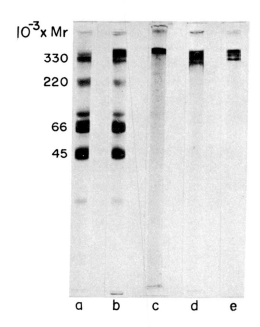

Fig. 1. SDS polyacrylamide gel electrophoresis of Levantina hemocyanin.
(a) Mixture of protein markers (thyroglobulin, 330 000; ferritin, 220 000; bovine
serum albumin 66 000; ovalbumin, 45 000); (b) Purified hemocyanin + protein markers;
(c) Purified hemocyanin; (d) Levantina hemolymph + thyroglobulin; (e) Hemocyanin
purified by rapid (∿5 min) passage through G75 Sephadex + thyroglobulin

was obtained with hemolymph from a live animal and with hemocyanin that was prepared by rapid gel filtration.   From the band mobility, a molecular weight of 360 000 was estimated for the polypeptide chain.

Molecular weight of the polypeptide chain.   This was determined by meniscus depletion sedimentation equilibrium in 6 M guanidinium chloride.   Extrapolation of a plot of $1/\bar{M}_{w,r}$ versus protein concentration c to c= 0 gave a value of 334 000 for the molecular weight of the polypeptide chain.

Copper content.   Determination of the copper content with 2,2' biquinoline gave 0.226 %.   Values of 0.23, 0.229 and 0.232 % were obtained by atomic absorption spectroscopy using three spectrometers, Varian Techtron AA5, Perkin Elmer 403 and Perkin Elmer 5000 respectively.   From the average value obtained for the copper content, 0.229 %, a value of 55 000 was calculated for the molecular weight per 2 copper atoms.

Discussion

The molecular weight determined in this study for the native Levantina 100 S hemocyanin, 10.4×10^6, is higher than any reported for 100 S gastropod hemocyanin. Our value for the molecular weight per 2 copper atoms is in very good agreement with the size of the functional unit determined for Helix pomatia hemocyanin (3,10). The size of the polypeptide chain, 334 000, is not very different from the corresponding values reported for other gastropod hemocyanins.   The identity of electrophoretic mobility of purified hemocyanin with that of the hemocyanin band in the hemolymph rules out the possibility of proteolysis and indicates unequivocally that we are dealing with intact polypeptide chains.

From the molecular weights determined in this study, the number of polypeptide chains per 100 S molecule is $10.4 \times 10^6/334\ 000 = 31.1$.   Taking into consideration the 5-fold symmetry seen in the electron microscope, an acceptable value is a multiple of five, and hence 30.   The ratio of the molecular weight of a polypeptide chain to that of a functional unit is 334 000/55 000 = 6.07.   The number of functional units per polypeptide chain is the nearest integer, 6.

On the basis of our findings, we can arrive at a definition of the various states of aggregation of gastropod hemocyanins in terms of the numbers of polypeptide chains and functional units they contain:

Aggregation state	Relation to whole 100 S molecule	Number of polypeptide chains	Number of functional units
100 S	1	30	180
60 S	1/2	15	90
20 S	1/10	3	18
11 S	1/30	1	6

Concluding remark

In the current model of gastropod hemocyanin
(4), two polypeptide chains are involved in the
formation of a 1/10 molecule.   Each polypeptide
chain contains 8 functional units so that, in all,
there are 16 functional units, 12 in the wall and
4 in the collar.   According to our findings, a
1/10 molecule contains 3 polypeptide chains.
Each polypeptide chain contains 6 functional units,
18 functional units in all.   If the wall is still
to contain 12 functional units, as suggested from
the 3-dimensional image reconstruction of Melema
and Klug (15), the remaining 6 functional units
must go to the collar.   A possible arrangement
is shown in the accompanying sketch.

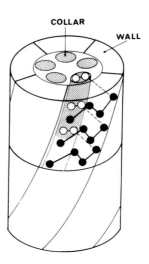

References

1. Van Holde, K.E., and Miller, K.I. , Quart. Rev. Biophys. 15, 1-129 (1982).
2. Ellerton, H.D., Ellerton, N.F., and Robinson, H.A., Prog. Biophys. Molec. Biol.
   41, 143-248 (1983).
3. Lontie, R., Life Chem. Rep. Suppl. 1, 109-120 (1983).
4. Van Bruggen, E.F.J., Life Chem. Rep. Suppl. 1, 1-14 (1983).
5. Wood, E.J., Bannister, W.H., Oliver, C.J., Lontie, R., and Witters, R., Comp.
   Biochem. Physiol. 40B, 19-24 (1971).
6. Quitter, S., Watts, L.A., Crosby, C., and Roxby, R., J. Biol. Chem. 253,
   525-530 (1978).
7. Pilz, I., Kratky, O., and Moring-Claesson, I., Z. Naturforsch 25B, 600-606 (1970).
8. Berger, J., Pilz, I., Witters, R., and Lontie, R., Eur. J. Biochem. 80, 79-82
   (1977).
9. Herskovits, T.T., and Russel, M.W., Biochemistry 23, 2812-2819 (1984).
10. Van der Laan, J.M., Torensma, R., and Van Bruggen, E.F.J., in Invertebrate
    Oxygen Binding Proteins (Lamy, J., and Lamy, J., eds.), Marcel Dekker, New York,
    739-744 (1981).
11. Siezen, R.J., and Van Bruggen, E.F.J., J. Mol. Biol. 90, 77-89 (1974).
12. Pilz, I., Walder, K., and Siezen, R., Z. Naturforsch 29C, 116-121 (1974).
13. Felsenfeld, G., Arch. Biochem. Biophys. 87, 247-251 (1960).
14. Ilan, E., and Daniel, E., Biochem. J. 183, 325-330 (1979).
15. Melema, J.E., and Klug, A., Nature 239, 146-150 (1972).

Evolution of hemocyanins

PARTIAL PRIMARY STRUCTURE OF THE HELIX POMATIA β_C-HEMOCYANIN FUNCTIONAL UNIT D

R. Drexel, H.-J. Schneider, S. Sigmund, B. Linzen
Zoologisches Institut, Luisenstr. 14, D-8000 München 2, F.R.G.

C. Gielens, R. Lontie, G. Préaux
Laboratorium voor Biochemie, Dekenstraat 6, B-3000 Leuven, Belgium

F. Lottspeich, and A. Henschen
MPI für Biochemie, am Klopferspitz, D-8033 Martinsried, F.R.G.

Introduction

Hemocyanins occur in two different phyla: molluscs and arthropods. Although their physiological role and some spectroscopic properties are similar, their quaternary structure and arrangement of oxygen binding units show great differences. The evolutionary relationship between both hemocyanin classes can only be solved by comparing their amino acid sequences. The first two sequences of arthropodan subunits were published in 1983 (1,2) and more data have since become available (3,4). Here we report the nearly complete amino acid sequence of domain d of Helix pomatia β_C hemocyanin, which allows for the first time a direct comparison of molluscan and arthropodan hemocyanin structures.

Material and Methods

The isolation of the functional unit β_C-d from Helix pomatia hemolymph was carried out as described previously (5). Thereby one obtains a mixture of the structurally intact unit d (=T3) and of two large fragments (ca. 43 and 10 kDa, called T3a and T3b, respectively) which are connected by a disulphide bridge. The whole mixture was reduced and carboxymethylated and separated by gel filtration on Sephacryl S200. Cleavage with CNBr was performed according to Gross and Witkop (6) in 70 % formic acid for 24 h at room temperature. Enzymatic digestion with trypsin, Staphylococcus aureus protease, chymotrypsin and Lys-C protease from Lysobacter enzymogenes was performed in volatile buffers for appropriate time periods at 37 °C. The resulting peptides were isolated by gel filtration and reversed phase HPLC. Sequence determination was done either manually according to Chang et al. (7) or with a liquid or gas phase sequenator. The amino acid analyses were kindly performed by Dr. W. Schartau.

Results and Discussion

The mixture of domain d and its two large fragments was separated by gel filtration (Fig. 1). The N-termini of fractions I (intact domain) and II (43 kDa) were identical, leaving fraction III (10 kDa) for the C-terminus.

Invertebrate Oxygen Carriers
Ed. by Bernt Linzen
© Springer–Verlag Berlin Heidelberg 1986

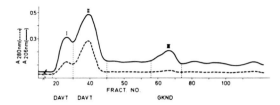

Fig. 1. Separation of domain d (I) and its fragments T3a (II) and T3b (III) by gel filtration. The N-terminal sequences are indicated below each peak.

The two fragments, T3a and T3b, were subjected to conventional cleavage methods, the resulting peptides separated by gel filtration and HPLC, and sequenced. The result is shown in Fig. 2.

One tryptic peptide (pos. 253-268) showed abnormal behavior: during gel filtration it was eluted at a position expected for a peptide of 50 residues, and during HPLC it seemed to be very hydrophilic. The first degradation step yielded a blank while the 15 following steps (right through the C-terminus) gave clear results. From the amino acid analysis the first amino acid was Asx, and since the third one was found to be threonine, we conclude that Asn-253 is the carbohydrate attachment site.

All in all, 402 positions have been determined so far. If to the corresponding mol. weight 6.0 % carbohydrate (8) are added, one arrives at ca. 50,000, close to the mol. weight of the entire functional unit.

```
1                  10                  20                  30                  40                  50
D A V T V A S H V R K D L D T L T A G E I E S I R S A F L D I Q Q D H T Y E N I A S F H G K P G L C
51                 60                  70                  80                  90                  100
Q H E G H K V A C S V S G M P T F P S W H R L Y V E Q V E E A L L D H G S S V A V P Y F D W I S P I
101                110                 120                 130                 140                 150
Q K L P D L I S K A T Y Y N S R E Q R F D P N P F F S G K V A G E D A V T T R D P Q P E L F N N N Y
151                160                 170       *       *180              190                 200
F Y E Q A L Y A L E Q D N F D D F E I Q F E V L H N A L H S W L G G H A K Y S F S S L D Y T A F D P
201          *     210                 220                 230                 240                 250
V F F L H H A N T D R L W A I W Q E L Q R Y R G L P Y N E A D C A I N L M R K P L Q P F Q D K K L N
251  <CHO>          260                 270                 280                 290                 300
P R N I T N I Y S R P A D T F D Y R N H F H Y E Y D T L E L N H Q T V P Q L E N L L K R R Q E Y G R
301                310                 319
V F A G F L I H N N G L S A D V T V Y
1                  10                  20                  30                  40                  50
G K N D C N H K A G V F S V L G G E L E M P F T F D R L Y K L Q I T D T I K Q L G L K V N N A A S Y
51                 60                  70                  80    83
Q L K V E I K A V P G T L L D P H I L P D P S I I F E P G T K E R
```

Fig. 2. Amino acid sequence of Helix pomatia ß_C hemocyanin functional subunit d. CHO marks the attachment site of the carbohydrate side chain. The presumptive copper binding histidines of the 'Copper B' site are marked by asterisks.

Fig. 3. Sequence comparison of a section of the primary structures of P. interruptus chain a, E. californicum chain e, H. pomatia ß$_C$ d and N. crassa tyrosinase. The boxed amino acids are identical in all compared sequences or in tyrosinase and Helix hemocyanin. Asterisks denote the three histidines identified as copper ligands in the 3-D structure, the + marking the active site His in tyrosinase (9). The α-helices were identified in the 3-D structure of P. interruptus hemocyanin (10).

To our great surprise there are no similarities with arthropodan hemocyanins for most of the sequence. We found only one region (Fig. 3) where the Helix sequence could be aligned with arthropodan hemocyanins. This same region shows also many residues identical to Neurospora tyrosinase (9). This region has been identified in Panulirus interruptus hemocyanin as the 'Copper B' binding site and comprises the helices 2.5, 2.6 and 2.7. It is noteworthy that of the eight residues identical in the four hemocyanins which are compared in Fig. 3, five are arranged so that they stick out more or less on the same side of the helix. In addition, this region shows many residues with isofunctional side chains.

Helix hemocyanin and Neurospora tyrosinase show a much greater similarity in this section, 25 out of 46 amino acids being identical. With arthropodan hemocyanins further alignments are not possible. In particular the 'Copper A' site with its characteristic His-Trp cluster was not found. In contrast, it is possible to align Helix hemocyanin and Neurospora tyrosinase in a few other regions. The majority of these is shown in Fig. 4. In order to achieve these alignments, several great deletions had to be assumed, especially in the Helix sequence. A curious feature is the identity of the heptapeptide YEQALYA which is C-terminal of the (also identical) peptide PYFDW in Helix hemocyanin, but located N-terminal in

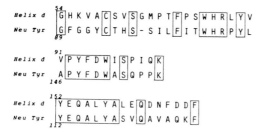

Fig. 4. Sequence comparison of three sections of Helix d and N. crassa tyrosinase. Boxed amino acids are identical. Note that these three sections are far apart in both sequences.

the tyrosinase. This indicates translocation of a small piece of a gene.

The search for possible 'Copper A' ligands has led to only one conserved histidine, in the sequence WHR flanked by predominantly hydrophobic residues (Fig. 4). Note that in the arthropodan hemocyanins, Trp is frequently associated with the copper binding sites (3); Trp (pos. 181 and 216) is also in proximity of the presumed 'Copper B' site of Helix hemocyanin. The histidines assumed to belong to the active site in Neurospora tyrosinase (9) do not occur in the corresponding region of the Helix sequence. Other residues must take over this function.

We conclude that tyrosinase and all hemocyanins have a common ancestor for the region of the 'Copper B' binding site, but that the 'Copper A' site has evolved independently in tyrosinase, mollusc hemocyanin, and arthropod hemocyanin. However, mollusc hemocyanins are much closer to tyrosinase than arthropod hemocyanins, and this would support the hypothesis put forward by van Holde and Miller (11) that the two classes of hemocyanins evolved by branching off from tyrosinase on two different occasions.

1. Schneider, H.-J., Drexel, R., Feldmaier, G., Linzen, B., Lottspeich, F., and Henschen, A., Hoppe-Seyler's Z. Physiol. Chem. 364, 1357-1381 (1983).
2. Schartau, W., Eyerle, F., Reisinger, P., Geisert, H., Storz, H., and Linzen, B., ibidem 364, 1383-1409 (1983).
3. Linzen, B., Soeter, N.M., Riggs, A.F., Schneider, H.-J., Schartau, W. et al., Science 229, 519-524 (1985).
4. Bak, H.J., Soeter, N.M., Vereijken, J.M., Jekel, P.A., Neuteboom, B., and Beintema, J.J., this volume, p. 149-152 (1986).
5. Gielens, C., Verschueren, L.J., Préaux, G., and Lontie, R., Eur. J. Biochem. 103, 463-470 (1980).
6. Gross, E. and Witkop, B., J. Am. Chem. Soc. 83, 1510-1511 (1961).
7. Chang, J.Y., Brauer, D., and Wittmann-Liebold, B., FEBS Let. 93, 205-214 (1978).
8. Wood, E.J., Chaplin, M.F., Gielens, C., De Sadeleer, J., Préaux, G., and Lontie, R., Comp. Biochem. Physiol. 82B, 179-186 (1985).
9. Lerch, K., Proc. Nat. Acad. Sci. USA 75, 3635-3639 (1978).
10. Gaykema, W.P.J., Hol, W.G.J, Vereijken, J.M., Soeter, N.M., Bak, H.J., and Beintema, J.J., Nature 309, 23-29 (1984).
11. Van Holde, K.E., and Miller, K.I., Quart. Rev. Biophys. 15, 1-129 (1982).

AMINO ACID SEQUENCE OF THE C-TERMINAL DOMAIN OF OCTOPUS
(Paroctopus dofleini dofleini) HEMOCYANIN

Takashi Takagi

Biological Institute, Faculty of Science, Tohoku University,
Sendai, Japan 980

Recently the structural work on arthropod hemocyanins has greatly progressed.The primary structures of several species have been determined and the three dimensional structure of Panulirus interruptus hemocyanin has been established at the 3.2 Å level(1, 2). On the other hand, we have very little information on the primary structure of mollusc hemocyanin. The native structures observed by electron microscopy, and subunit structures of arthropod and mollusc hemocyanins are greatly different. Thus to elucidate the relation of arthropod and mollusc hemocyanins at the primary structure level, we have started the amino acid sequence determination of mollusc hemocyanin.

MATERIALS AND METHODS

Hemocyanin: Octopus (Paroctopus dofleini dofleini) hemolymph was kindly supplied by Dr. S. Hirai (Marine laboratory at Asamushi, Tohoku Univ., deceased 1983). It was kept at -80°C until use. Hemolymph was dialyzed against 25 mM Tris-HCl buffer containing 1 mM EDTA, pH 8.5 and purified with DEAE-cellulose column chromatography by increasing the concentration of NaCl from 0.1 M to 0.5 M linearly. Fractions containing hemocyanin were pooled and concentrated with ultrafiltration. It showed a single band by SDS PAGE.

Tryptic Digestion:To hemocyanin in 25 mM Tris-HCl containing 1 mM EDTA, pH 7.9 or 8.5 trypsin was added (TPCK treated, Worthington) at a ratio of substrate to enzyme =100 : 1 (w/w). After 2 h at 20°C, the reaction was stopped by addition of soybean trypsin inhibitor. The reaction mixture was applied to a column (3 x 90 cm) of Sephacryl S-200. The domains of 100k, 50k and 45k were eluted in the same fractions. These domains were separated by a column chromatography on DEAE-cellulose by increasing the concentration of NaCl linearly. Fractions containing 100k, 50k and 45k domains were pooled separately and concentrated by ultrafiltration and further purified by gel filtration with Sephadex G-100. The 45k domain was used for the sequence determination.

Sequence Determination Fifty nmol of reduced and carboxymethylated 45k domain in 170 µl of 0.1 M NH_4HCO_3, pH 8.5 were digested with 25 µl of trypsin (1mg/ml) or Staphylococcus aureus V8 protease (Miles, 1 mg/ml) at 37°C for 4 h and the reaction mixture was subjected to HPLC. The column (4 x 250 mm, Lichrosorb RP-8, Merck) was equili-

Invertebrate Oxygen Carriers
Ed. by Bernt Linzen
© Springer–Verlag Berlin Heidelberg 1986

brated with 98 % of solvent A (50 mM ammonium formate, pH 6.8) and 2
% solvent B (90 % acetonitrile in 50 mM ammonium acetate), and
eluted by increasing the concentration of solvent B linearly. Some
of the peptides were further purified by Asahipak GS-320 column (7.5
x 500 mm, Asahi Chemical Ind., Kawasaki) with isocratic elution of
50 mM ammonium formate for hydrophylic peptides or by a small column
(4 x 150 mm, Lichrosorb RP-18) for the more hydrophobic peptides
employing 0.1 % trifluoroacetic acid instead of ammonium formate
in solvents A and B. The peptide (1 - 3 nmol) was hydrolyzed with
50 μl of trifluoroacetic acid/conc. HCl (1 : 2, v/v) containing 0.02
% phenol in evacuated, sealed tubes at 170° C for 20 min and the
hydrolyzate was analyzed by a Hitachi 835-50 amino acid analyzer.
The amino acid sequence of the peptides was determined by the manual
Edman method according to Tarr(3) and modified as in(4).

RESULTS AND DISCUSSION

Only leucine was detected in the <u>native</u> hemocyanin by the manual
Edman method; by carboxypeptidase P digestion at 37°C for 1 h,
Thr, Lys and His were identified in the ratio of 1.5 : 0.9 : 1.0.
The 45k domain was first treated with CNBr, however, due to difficulties in purification of CNBr peptides, we could not get so much
information by this procedure. Thus the domain was digested with
trypsin and the reaction mixture was directly subjected to HPLC as
shown in Fig. 1. Some of the peptides were further purified as
described in MATERIALS AND METHODS. Thirty four tryptic peptides

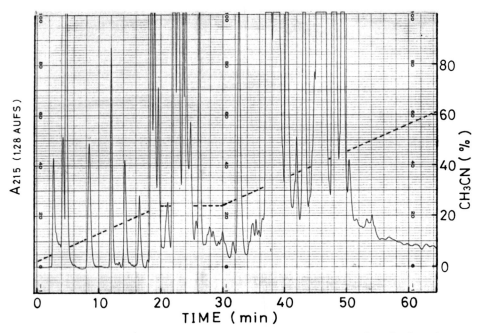

Fig. 1 HPLC separation of tryptic peptides of 45k domain.

were sequenced   (shown in Fig. 2). Larger peptides were further
digested with chymotrypsin, thermolysin or staphylococcal protease.
About 280 residues were sequenced; and the sequences of about 100
residues have not been established yet. The peptide T5 in Fig. 2 was a
tripeptide and its amino acid sequence was Thr-Thr-His. Considering
the amino acid composition released by carboxypeptidase P digestion
of entire hemocyanin, the peptide T5 seems to be not only the C-
terminal peptide of 45k domain but also that of entire hemocyanin.
By carboxypeptidase P digestion of 45k domain at 37°C for 10 min,
Thr, Lys and His were released in a ratio of 3.1 : 2.2 : 1.7.   Thus
45k domain is derived from the C-terminus of octopus hemocyanin.
The peptide T39 contained carbohydrate.   The sequence of T39 was
established by using its chymotryptic and staphylococcal protease
peptides.   In the sequence determination of T39, 11th serine, 15th
asparagine and 22nd glycine were not identified.   Thus, carbohydrate is
attached to 15th asparagine and there is a possibility that 11th
serine is also glycosylated.   The reason of 22nd glycine was not

```
T5        T T H
T6  a     E I Q H E D R
    b     D A M A K
    c     E L R
    d     S P R
T9        Y R
T13       V Q A D T S D D G Y Q K
T15       H L R
T17       V T I K
T19       L D A H D P I A V D T D F K
T20       A Q L F D D P D K G K
T21       Q M E D A L V A K
T22       L L T K
T23 a     A H S T G A T S F D Y H K
    b     G L P Y N T A N C E I K
T24 a     T V G S A I P S D D R L K
    b     T I G Q V A D D V C T S N F K
T25       G I D G H V L S D K
T26       D N S F H H A V I D T T R
T27       L F K
T29       Y D I T T S L K
T33       S F F Y R
T38 a     L G Y D (        ?           )
    b     M F W A W D F R
                        (*)        *
T39       I A S Y H G I P L C S H Y E N G T A Y A C G Q M V
          T F P N H R
T40       L V K P L K P F D L D (    ?     )
T42       Y L S V P T P F L A P A K
T44       I G H D (             ?             )
T47       G S H V I P Y W D W T T W F A N L P V L V T E E K
T48       V F A G F L L R
T49       I W S V W Q A L Q K
```

Fig. 2 Amino acid sequence of tryptic peptides of 45k domain.
Peptide numbers show the elution time in Fig. 1. * : carbohydrate.

```
1            10            20            30
E A V R G T I I R K N V N S L T P S I K E L R D A M A K V Q
R N V I T T F I K K N E L E L T H G I D F G I D F - - - - G
           40            50 (*)        *           60
A D T S D D G - Y Q K I A S Y H G I P L C S H Y E N G T A Y
T T N S V K V K Y P H L E H E P F S F Q I T V - E N T S G A
           70            80            90
A C G Q M V T F P N H R S T K H Y T S Y P L F L L T K Q M E
K K D A T V R I F L A P K L D E L G N Q L P A N I Q R L F I
          100           110           120
D A L V A K G S H V G I P D Y W D W T T W F A N L P V L V T
E L D K F H K E L I S G Q N I I T H N A A D S S V T V S T L
          130           140           150
E E K D N S F H H A V I D T T R S P R A Q L F D D P D K G K
R T F E D L K A G K G V S E D A T E Y C S C G W P Q N M L I
```

Fig. 3 Sequence comparison of octopus C-terminal domain (upper) and
Tachypleus hemocyanin alpha chain (lower). The residue number of
octopus hemocyanin is tentative. The sequence of Tachypleus alpha
chain starts from 399th residue. - , deletion; * , carbohydrate;
identical and isofunctional residues are shown in bold face and
underlined, respectively.

identified is not clear. Tryptic peptides were overlapped by using
staphylococcal protease peptides. Although the sequence is not
completed yet, a partial sequence (149 residues) of 45k domain was
compared with Tachypleus hemocyanin sequence (Fig. 3). The partial
sequence of 45k domain showed the maximum homology with the sequence
from 399th residue of Tachypleus alpha chain. The first 25 residues
showed 40 % identity, however, after 26th residue significant homology was not observed. The corresponding sequence of Tachypleus
hemocyanin to carbohydrate attachment site of 45k domain also shows
Asn-Thr-Ser, although no carbohydrate was found in Tachypleus hemocyanin. The sequence around the copper binding sites of 45k domain
was not determined yet. But the typical copper binding site sequence, His-X-Y-Z-His has not been found in tryptic peptides. Thus
there is a possibility that mollusc hemocyanin binds copper in a
different way than arthropod hemocyanin.

REFERENCES

(1) Gaykema, W.P.J., Hol, W.G.J., Vereijken, J.M., Soeter, N.M.,
    Bak, H.J., Beintema, J.J. Nature (London) 309, 23-29 (1984)
(2) Linzen, B., Soeter, N.M., Riggs, A.F., Schneider, H.-J.,
    Scharatau, W., Moore, M.D., Yokota, E., Behrens, P.Q.,
    Nakashima, H., Takagi, T., Nemoto, T., Vereijken, J.M., Bak,
    H.J., Beitema, J.J., Volbeda, A., Gaykema, W.P.J., Hol, W.G.L.
    Science 229, 519-524 (1985).
(3) Tarr, G.E. Methods Enzymol. 47, 335-357 (1977)
(4) Takagi, T., Konishi, K. J. Biochem. (Tokyo) 96, 59-67 (1984)

SEQUENCE HOMOLOGIES OF PAROCTOPUS DOFLEINI AND HELIX POMATIA HEMOCYANIN

R. Drexel[1], T. Takagi[2] and B. Linzen[1]

[1]Zoologisches Institut der Universität München, Luisenstr. 14, D-8000 München 2, F.R.G.

[2]Biological Institute, Faculty of Science, Tohoku University, Sendai 980, Japan

The meeting at Tutzing provided the opportunity to compare the amino acid sequences obtained so far for the domain d of Helix pomatia β_c hemocyanin (1) and the C-terminal 45 K domain of Paroctopus dofleini hemocyanin (2). 25 out of 32 tryptic peptides of the octopus hemocyanin could be easily aligned (Fig. 1). For the best fit, several short (1-3 residues) deletions/insertions and a major one (pos. 117-132 of the Helix sequence) had to be made. 278 positions could be compared. 107 (=38 %) of these had identical residues, another 37 (=13 %) amino

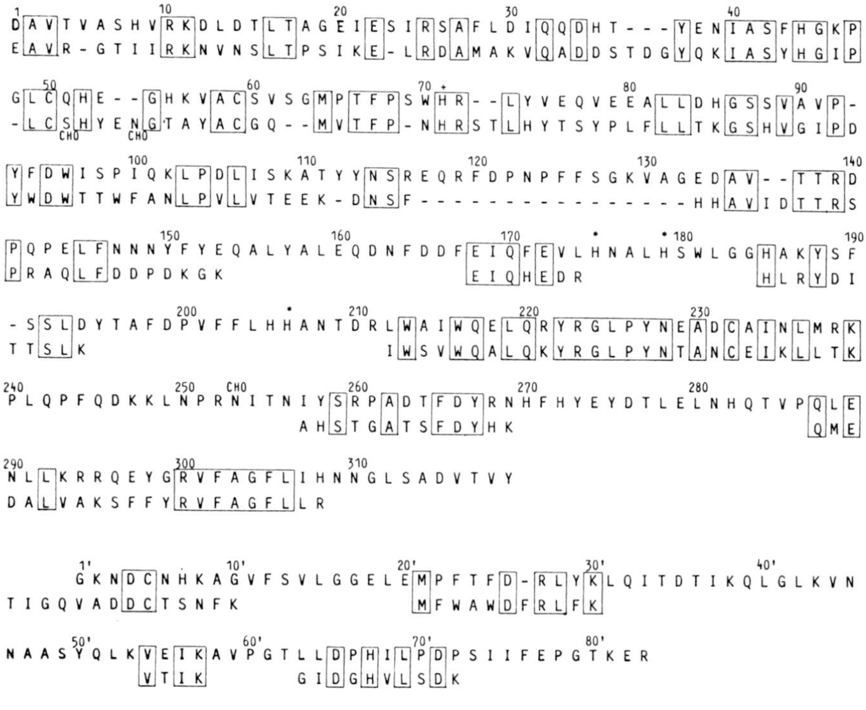

Fig. 1. Alignment of partial sequences from Helix ß$_C$ domain d (upper line) and Paroctopus 45 K C-terminal domain (lower line) hemocyanins. Identical amino acids are boxed. Deletions/insertions(-) have been introduced to obtain maximum homology. * indicates histidines of the presumptive Copper B site, + indicates the histidine which is invariant in Helix and Paroctopus hemocyanin, and in Neurospora and Streptomyces tyrosinase. The numbering is that used for the Helix hemocyanin (1).

Invertebrate Oxygen Carriers
Ed. by Bernt Linzen
© Springer–Verlag Berlin Heidelberg 1986

acids with similar side chains, raising the degree of homology to ca. 50 %. As the two domains which are compared here, are not only derived from species of two classes of molluscs, but also occupy different positions in the whole subunit - the Paroctopus 45 K domain being C-terminal and the Helix β_c domain d being in the middle (3) - it is tentatively concluded that all mollusc hemocyanin domains are derived from a common ancestral gene. A much higher degree of homology is to be expected if functional domains with the same position in the 8(7) - domain subunit are compared.

Unfortunately, for the Paroctopus hemocyanin the sequence around the "Copper B" site (pos. 175-185 and 195-210) has not yet been established. However, it is already clear that, like Helix β_c domain d, Paroctopus hemocyanin lacks the structure which is typical for "Copper A" in the arthropodan hemocyanins (4). It is tempting to speculate that His-71 represents one of the "Copper A" ligands as it is invariant also in comparison to the tyrosinases sequenced so far (5). In this region there are only two other conserved histidines, in position 44 and 52; only one of them, His-44, occurs also in Streptomyces, but not in Neurospora tyrosinase. It should be noted that all cysteine residues identified so far, are in identical positions in both hemocyanins.

1. Drexel, R., Schneider, H.-J., Sigmund, S., Linzen, B., Gielens, C., Lontie, R., Préaux, G., Lottspeich, F., and Henschen, A., this volume, p. 255-258 (1986).
2. Takagi, T., this volume, p. 259-662 (1986).
3. Gielens, C. Verschueren, L.J., Préaux, G., and Lontie, R., Eur. J. Biochem. 103, 463-470 (1980).
4. Linzen, B., et al., Science 229, 519-524 (1985).
5. Huber, M., and Lerch, K., this volume, p. 265-276 (1986).

ACTIVE-SITE AND PROTEIN STRUCTURE OF TYROSINASE: COMPARISON TO HEMOCYANIN

M. Huber and K. Lerch
Biochemisches Institut der Universität
Zürich, Winterthurerstrasse 190
CH-8057 Zürich, Switzerland

INTRODUCTION

Tyrosinase is a copper-containing monooxygenase catalyzing both the o-hydroxylation of monophenols (cresolase activity) and the oxidation of o-diphenols to o-quinones (catecholase activity) (1). The enzyme is widely distributed in microorganisms, plants and animals, where it is involved in the formation of melanins and other polyphenolic compounds (2).

$$\text{monophenol} \; + \; O_2 \; \longrightarrow \; \text{o-quinone} \; + \; H_2O \quad \text{(cresolase activity)}$$

$$2 \; \text{o-diphenol} \; + \; O_2 \; \longrightarrow \; 2 \; \text{o-quinone} \; + \; 2 \; H_2O \quad \text{(catecholase activity)}$$

Tyrosinases from the common mushroom Agaricus bisporus and Neurospora crassa were shown to contain an antiferromagnetically coupled copper pair which in the reduced state binds molecular oxygen reversibly (3,4). This binuclear copper complex can be obtained in a number of different forms (met-, halfmet-, deoxy- and oxytyrosinase), depending on the oxidation state of the metal ions (5,6). The chemical and the spectroscopic properties of the various forms were shown to be remarkably similar to those reported for the oxygen-transporting hemocyanins (7). However, distinct differences between the active site structures of the two copper proteins are apparent from peroxide displacement and binding studies of tyrosinase sub-

Invertebrate Oxygen Carriers
Ed. by Bernt Linzen
© Springer–Verlag Berlin Heidelberg 1986

strate analogues (8), relating to the different biological functions of tyrosinase (monooxygenase) and hemocyanin (oxygen-transporting).

The striking similarity of the binuclear copper active site structures of tyrosinase and hemocyanin is also born out by comparison of their primary structures. All proteins sequenced so far contain a highly conserved region with three invariant histidine residues shown to be metal ligands to Cu(B) in Panulirus interruptus hemocyanin by X-ray cristallography (9). However, only limited sequence homology is apparent for the region containing the three histidine residues which bind to Cu(A).

In the present paper the chemical and spectroscopic properties of two tyrosinases as well as their primary structures will be compared to those of various hemocyanins. To obtain further insight into the ligand environment of Cu(A) in tyrosinase, the structural gene of Streptomyces glaucescens tyrosinase was modified by site-directed mutagenesis. Replacement of histidine residue 62 by asparagine resulted in a completely inactive enzyme strongly suggesting that this residue constitutes a ligand to Cu(A).

MATERIALS AND METHODS

Tyrosinase from Streptomyces glaucescens was isolated from liquid cultures as described previously (10). The enzyme had a specific activity of 1400 U/mg using L-3,4-dihydroxyphenylalanine as substrate (11).

Oligonucleotides were synthesized on a manual solid phase DNA synthesizer (BACHEMGENTECH) according to the phosphotriester method (12). The fully deprotected oligomers were purified by polyacrylamide gel electrophoresis in 8 M urea and chromatography on NACS-52 (BRL).

Site-directed mutagenesis was carried out by the double primer approach (13) using the M13 universal primer and the 15-mer 5'-CTG-GAACCGCAGATA-3' as the mutagenic primer. The phage M13 mp18 containing the complete tyrosinase gene from S. glaucescens was used as a template.

Protoplasts of the tyrosinase deficient strain GLA 205 (14) were prepared and transformed as described previously (15). Transformants were selected on thiostrepton (Squibb & Sons) and plated on a medium

(15) supplemented with L-tyrosine, L-methionine, L-leucine, and copper sulfate to indicate melanin formation. The mutants were further
characterized by immunological techniques.

Antibodies were raised in rabbits against SDS denatured purified
enzyme. Antisera were tested for antibodies with the Ouchterlony
double diffusion test.

RESULTS AND DISCUSSION

Chemical and spectroscopic properties of tyrosinase and hemocyanin

Structural similarities between the copper active sites of tyrosinase and hemocyanin were first noticed more than 4 decades ago by
Kubowitz when he studied the binding of carbon monoxide to the two
copper proteins (16). In the meantime, a great number of chemical
and spectroscopic studies have led to a detailed picture of the geo-
metric and electronic structure of the binuclear copper sites in
tyrosinase and hemocyanin (7). Table I summarizes and compares the
most important properties of the binuclear copper complexes. In the
reduced state (deoxy form) both proteins bind molecular oxygen reversibly leading to the corresponding oxy forms (7). Resonance Raman spectroscopy clearly revealed that the oxygen in the oxy form is
bound as peroxide (17,18). This assignment is also in agreement with
the interpretation of the intense absorption band in the UV as a peroxide to Cu(II) charge transfer transition (5,7). The deoxy forms
moreover bind CO reversibly yielding a characteristic luminescence
with a large Stokes shift (19,20). The met- as well as the oxy forms
contain two tetragonal Cu(II) ions antiferromagnetically coupled
through an endogeneous bridge (R) resulting in a EPR silent state.
The Cu-Cu distances determined by EXAFS-spectroscopy are again remarkably similar for the two proteins (21,22). Both proteins can be
half-reduced yielding half-met-derivatives with characteristic EPR
spectra depending on the nature of the exogenous ligand bound to the
site. Finally, in the presence of NO the deoxy forms undergo a
change leading to the dimer forms, which are characterized by a
broad EPR spectrum in the $g \sim 2$ region and a 7 line spectrum in the

Form	Structural site	Tyrosinase	Hemocyanin	References
Deoxy	$\overset{N}{\underset{N}{>}}Cu(I)\ Cu(I)\overset{<N}{\underset{<N}{}}$	colorless, EPR-silent	colorless, EPR-silent	3,5,7
CO complex		λ_{max} of emission : 550 nm	λ_{max} of emission : 540 nm	19,20
Oxy	$\overset{N}{\underset{N}{>}}Cu(II)\overset{O-O}{\underset{R}{<}}Cu(II)\overset{<N}{\underset{<N}{}}$	blue, EPR-silent$\\\varepsilon 345 : 17'200\ M^{-1},\ cm^{-1}\\$Cu-Cu distance : 3.63 Å$\\$O-O stretch frequency :$\\755\ cm^{-1}$	blue, EPR-silent$\\\varepsilon 345 : 20'000\ M^{-1},\ cm^{-1}\\$Cu-Cu distance : 3.58-3.66 Å$\\$O-O stretch frequency :$\\744 - 749\ cm^{-1}$	5,7$\\$4,7$\\$21,22$\\$17,18
Met	$\overset{N}{\underset{N}{>}}Cu(II)\overset{L}{\underset{R}{<}}Cu(II)\overset{<N}{\underset{<N}{}}$	green, EPR-silent$\\\varepsilon 700 : 260\ M^{-1},\ cm^{-1}\\$Cu-Cu distance : 3.39 Å	green, EPR-silent$\\\varepsilon 680 : 120\ M^{-1},\ cm^{-1}\\$Cu-Cu distance : 3.39-3.45 Å	5,7$\\$6,33$\\$21,22
Halfmet	$\overset{N}{\underset{N}{>}}Cu(II)\overset{L}{\underset{R}{<}}Cu(I)\overset{<N}{\underset{<N}{}}$	EPR parameters for$\\$Halfmet-NO_2^- :$\\$g$\parallel$: 2.296$\\$g$\perp$: 2.078$\\$A$\parallel$: 131 x $10^{-4}\ cm^{-1}$	EPR parameters for$\\$halfmet-NO_2^- (Busycon)$\\$g$\parallel$: 2.302$\\$g$\perp$: 2.096$\\$A$\parallel$: 125 x $10^{-4}\ cm^{-1}$	5
Dimer	$\overset{N}{\underset{N}{>}}Cu(II)\overset{L}{\underset{R}{<}}Cu(II)\overset{<N}{\underset{<N}{}}$	EPR parameters :$\\$g~2 broad, g~4, 7 lines$\\$Cu-Cu distance : 5.9 Å	EPR parameters (Cancer):$\\$g~2 broad, g~4, 7 lines$\\$Cu-Cu distance : 5.7 Å	23,24

Table I

Comparison of the chemical and spectroscopic properties of the different forms of tyrosinase and hemocyanin

$g \sim 4$ region, typical for dipole-dipole coupled Cu(II) pairs (23).
Calculations based on computer simulations of the spectra resulted
in a Cu-Cu distance of approximatively 6 Å (24). Taken together,
these and other data convincingly show, that the binuclear copper
active sites of tyrosinase and hemocyanin are very similar. On the
other hand, it is well known that the two proteins serve different
biological functions (oxygen-transport vs. monooxygenase activity).
Distinct differences between the binuclear copper centers of tyrosinase and hemocyanin are observed when the peroxide displacement by
different ligands from the oxy forms are studied. As shown in Table
II, the pseudo-first-order rates for peroxide displacement vary considerably for the small ligand azide and the larger one L-mimosine,
a substrate analogue. These data unambiguously demonstrate, that the
tyrosinase copper active site is much more accessible to large ligands such as organic substrates than that of hemocyanin. Tyrosinase
can therefore be viewed in simplified terms as a hemocyanin with an
exposed binuclear copper complex. This idea is consistent with the
results from the three-dimensional structure of P. interruptus hemocyanin showing a highly buried copper site (9).

Ligand	Tyrosinase	Hemocyanin		
		Limulus	Cancer	Busycon
Azide	0.95	$\ll 10^{-4}$	0.04	0.002
L-mimosine	162	$\ll 10^{-4}$	$\ll 10^{-4}$	$\ll 10^{-4}$

Table II

Peroxide displacement by azide and L-mimosine from oxytyrosinase and
different oxyhemocyanins (5). The pseudo-first-order rates are
shown (h^{-1}).

Comparison of the primary structure of tyrosinase and hemocyanin

In the preceeding section dealing with the chemical and spectroscopic properties of tyrosinase and hemocyanin it was shown that the
active site structures of these binuclear copper proteins are strikingly similar. This similarity is also demonstrated by a comparison
of their primary structures. For arthropod hemocyanins numerous amino acid sequences have become available recently (25,26). Furthermore, the X-ray structure of the arthropod P. interruptus hemocyanin
has been determined, identifying two clusters (designated as Cu(A)
and Cu(B)) comprising three histidyl residues each as ligands to the
binuclear active site copper (9). Not surprisingly, these two regions are highly conserved in all arthropod hemocyanins sequenced so
far. In the case of tyrosinase, however, only the amino acid sequences of the proteins from the fungus Neurospora crassa (27) and the
bacterium Streptomyces glaucescens (28) are known. A comparison of
the two primary structures surprisingly yields an overall sequence
homology of only 24.2 % (Fig. 1). This seems to be a rather low number for two proteins performing the same enzymatic function. The
bacterial enzyme is much smaller and deletions have occured both at
the N- and C-terminal part of the protein (Fig. 1). However, there
are two regions showing a sequence homology of more than 50 %, including some invariant histidyl residues (Fig. 1). One of these regions corresponds to Cu(B), as demonstrated by the sequence comparison of N. crassa and S. glaucescens tyrosinase and E. californicum
hemocyanin (25, Fig. 2). The three histidyl residues of S. glaucescens and N. crassa tyrosinase are located exactly in the same positions as those of E. californicum hemocyanin. Moreover, the flanking sequences of the third histidyl residue show a very high degree
of homology (80 % between N. crassa and S. glaucescens tyrosinase;
66 % between the two tyrosinases and E. californicum hemocyanin)
(Fig. 2). The involvement of the third histidyl residue as a copper
ligand is also supported by protein modification experiments with
N. crassa tyrosinase showing that histidyl residue 306 is specifically destroyed by a mechanism-based inactivation process with the
concomitant loss of one copper atom (26). Fig. 2 shows also that
the distances between the first two histidyl residues and the third
one in the three proteins compared vary considerably. From the data
of the X-ray structural analysis of P. interruptus hemocyanin it is

N.c.   Ac-S T D I K F A I T G V P T T P S S N G A V P L R R E L R D L Q Q N Y

N.c.   P E Q F N L Y L L G L R D F Q G L D E A K L D S Y Y Q V A G I H G M
S.g.   T V R K N Q A T L T A D E K R R F V A A V L E L K R S G R Y D E F V

P F K P W A G V P S D T D W S Q P G S S G F G G Y C T H S S I L F I
T T H N A F I I G D T D   A G E R T G H R S P S F L

T W H R P Y L A L Y E Q A L Y A S V Q A V A Q K F P V E G G L R A K
P W H R R Y L L E F E R A L Q S V D A

Y V A A A K D F R A P Y F D W A S Q P P K G T L A F P E S L S S R T
        S V A L P Y W D W   S

I Q V V D V D G K T K S I N N P L H R F T F H P V N P S P G N F S
A D R T A R A S L W A P D F L G G T G R S L D G R V M D G P   F A

A A W S R Y P S T V R Y P N R L T G A S R D E R I A P I L A D E L A
A S A G N W P I N V R V D G R A Y L R R S L G T A V R   E L

S L R N N V S L L L L S Y K D F D A F S Y N R W D P N T N P G D F G
P T R A E V G S V L G M A T Y D T A P W N S A S D G F R N H L E G W

S L E A V H N E I H D R T G G N G H M S S L E V S A F D P L F W L H
R G V N L H N R V H V W V G G R M A T G M S P N D P V F W L H

H V N V D R L W S I W Q D L N P N S F M T P R P A P Y S T F V A Q
H A Y V D K L W A E W Q R R H P G S G Y L P A A G T P D V V D L N D

N.c.   E G E S Q S K S T P L E P F W D K S A A N F W I S E Q V K D S I T F
S.g.   R M K P W N D T S P A D L L D H T A H Y T F D T D

N.c.   G Y A Y P E T Q K W K Y S S V K E Y Q A A I R K S V T A L Y G S N V F

Fig. 1   Amino acid sequence comparison of S. glaucescens (S.g.) and
N. crassa (N.c.) tyrosinases (27,28). Identical residues are indicated by boxes.

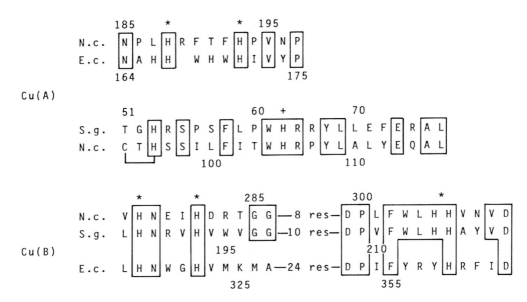

Fig. 2   Amino acid sequence comparison of regions containing residues serving as ligands to the active site copper of S. glaucescens (S.g.) and N. crassa (N.c.) tyrosinases and E. californicum d (E.c.) hemocyanin (25). Identical residues are indicated by boxes. The asterisks denote histidyl residues involved as copper ligands. The plus indicates His 62 of S. glaucescens tyrosinase changed to asparagine by site-directed mutagenesis (Fig. 3). Cys 94 is covalently linked to His 96 in N. crassa tyrosinase (32).

evident that these histidyl residues occur in two α-helices separated by a loop structure. Such loops connecting adjacent elements of secondary structure have been shown to vary in length quite often in different proteins (30).

Very recently, the primary structure of the functional unit d of H. pomatia hemocyanin β_C has been determined (31). This fragment was shown to contain a Cu(B) site very similar to those found in arthropod hemocyanins and tyrosinases.

As shown in Fig. 2 the region involving the Cu(A) site of arthropod hemocyanins shows only limited sequence homology to that of N. crassa tyrosinase. Two histidyl residues are found in a sequence which is similar in N. crassa tyrosinase and E. californicum hemocyanin, although the distance in between differs by one amino acid (Fig. 2). Concerning the third histidyl residue there is no similarity between N. crassa tyrosinase and arthropod hemocyanins. In the fungal enzyme this residue is located about 100 amino acids towards the C-terminus in the Cu(B) site (Fig. 1 and 2). In arthropod hemo-

cyanins the third histidyl residue of Cu(A) is separated by 26
amino acids. In N. crassa tyrosinase the histidyl residues of Cu(A)
(His 188, 193 and 289, Fig. 2) were identified as copper ligands by
photoinactivation studies (32). Surprisingly, these three histidines
are absent in S. glaucescens tyrosinase. However, the two tyrosinase molecules share another region with a high degree of sequence
homology (Fig.2). Furthermore, this stretch contains two histidines
which are separated by 8 amino acids. It is therefore likely, that
His 53 and 62 in S. glaucescens tyrosinase are involved as copper
ligands thus representing a new type of a Cu(A) site. In N. crassa
tyrosinase His 96 (corresponding to His 53 in S. glaucescens tyrosinase) is covalently linked via a thioether to Cys 94 (33). Hence,
one could speculate that His 96 and 105 were originally ligands to
Cu(A) which were then replaced during evolution by His 188 and 193
(Fig. 2). In this context it should be mentioned that the amino acid
sequence involving the Cu(A) site of arthropod hemocyanins is completely different from that of the functional unit d of H. pomatia
hemocyanin β_C (31). However, substantial sequence homology of the
molluscan hemocyanin to N. crassa and S. glaucescens tyrosinase is
observed in the region involving the histidyl residues proposed to
be ligands to Cu(A) in the bacterial enzyme.

Replacement of His 62 in S. glaucescens tyrosinase by site-directed
mutagenesis

_____

As suggested above, His 62 is a likely candidate as a copper ligand
to Cu(A) in S. glaucescens tyrosinase. To test this hypothesis, this
residue was changed to asparagine by site-directed mutagenesis
(Fig. 3). For this purpose the cloned tyrosinase gene of S. glaucescens (28) was subcloned in the phage M13 mp18. Using the double
primer technique (13) the single mutation His 62 ⟶ Asn was achieved
(Fig. 3). The mutant tyrosinase gene was subcloned into plasmid pIJ
364 (15), a multicopy broad-host range vector derivative of S. livi-
dans. The resulting plasmid was introduced by protoplast transformation into a S. glaucescens strain devoid of the tyrosinase structural gene. On indicator plates for melanin formation only white
colonies could be detected, whereas the wild-type gene typically
yields black colonies. The actual expression of the mutant tyrosi-

nase gene was demonstrated by immunodiffusion techniques. Crude extracts of the mutant were completely devoid of tyrosinase activity.

```
        58                  62              65
      Phe Leu Pro Trp His Arg Arg Tyr    wild type
      Phe Leu Pro Trp Asn Arg Arg Tyr    mutant

              5'-C TGG AAC CGC AGA TA-3'  mutagenic primer
                        *
  ...TTC CTG CCC TGG CAC CGC AGA TAC...  template
```

Fig. 3   Part of the amino acid sequence of S. glaucescens tyrosinase and the corresponding nucleotide sequence. The box indicates the amino acid exchange (His 62 → Asn) by site-directed mutagenesis using the mutagenic primer shown below. The asterisk denotes the corresponding nucleotide exchange.

The fact that the replacement of His 62 to Asn leads to a completely inactive enzyme strongly argues for the participation of this histidyl residue in the binding of Cu(A). Alternatively, this amino acid substitution may result in a conformational change of the enzyme molecule thus rendering it inactive. Studies are currently in progress to further characterize this mutant.

ACKNOWLEDGEMENTS

We would like to express our thanks to Squibb and Sons for the generous gift of thiostrepton and G. Hintermann for valuable discussions. This work was supported by Schweizerischer Nationalfonds Nr. 3.285-0.82 and the Kanton of Zürich.

REFERENCES

1. Mason, H.S., Annu. Rev. Biochem. 34, 595-634 (1965).

2. Lerch, K. in Metal Ions in Biological Systems (Sigel, H. ed), vol. 13, pp. 143-186, Marcel Dekker, New York (1981).

3. Jolley, R.L., Jr., Evans, L.H., Makino, N., and Mason, H.S., J. Biol. Chem. 249, 335-345 (1974).

4. Lerch, K., FEBS Lett. 69, 157-160 (1976).

5. Himmelwright, R.S., Eickman, N.C., LuBien, C.D., Lerch, K., and Solomon, E.I., J. Am. Chem. Soc. 102, 7339-7344 (1980).

6. Lerch, K., M. Cell. Biochem. 52, 125-138 (1983).

7. Solomon, E.I. in Copper Proteins (Spiro, T.G., ed.), pp. 41-108, Wiley and Sons, Inc., New York (1981).

8. Winkler, M.E., Lerch, K., and Solomon, E.I., J. Am. Chem. Soc. 103, 7001-7001 (1981).

9. Gaykema, W.P.J, Hol, W.G.J., Vereijken, N.M., Soeter, M.N., Bak, H.J., and Beintema, J.J., Nature (London) 309, 23-29 (1984).

10. Lerch, K., and Ettlinger, L., Eur. J. Biochem. 31, 427-437 (1972).

11. Fling, M., Horowitz, N.H., and Heineman, S.F., J. Biol. Chem. 238, 2045-2053 (1963).

12. Miyoshi, K., Miyaka, T., Hozumi, T., and Itakura, K., Nucleic Acids Res. 8, 5473-5505 (1980).

13. Norris, K., Norris, F., Christiansen, L., and Fiil, N., Nucleic Acids Res. 11, 5103-5112 (1983).

14. Crameri, R., Hintermann, G., Hütter, R., and Kieser, T., Can. J. Microbiol. 30, 1058-1067 (1984).

15. Kieser, T., Hopwood, D.A., Wright, H.M., and Thompson, C.J., Mol. Gen. Genet. 185, 223-238 (1982).

16. Kubowitz, F., Biochem. Z. 299, 32-57 (1938).

17. Freedman, T.B., Loehr, J.S., and Loehr, T.M., J. Am. Chem. Soc. 98, 2809-2815 (1976).

18. Eickman, N.C., Solomon, E.I., Larrabee, J.A., Spiro, T.G., and Lerch, K., J. Am. Chem. Soc. 100, 6529-6531 (1978).

19. Kuiper, H.A., Finazzi-Agrò, A., Antonini, E., and Brunori, M., Proc. Natl. Acad. Sci., USA 77, 2387-2389 (1980).

20. Kuiper, H.A., Lerch, K., Brunori, M., and Finazzi-Agrò, A., FEBS Lett. 111, 232-234 (1980).

21. Woolerey, G.L., Powers, L., Winkler, M., Solomon, E.I.,and Spiro, T.G., J. Am. Chem. Soc. 106, 86-92 (1984).

22. Woolerey, G.L., Powers, L., Winkler, M., Solomon, E.I., Lerch, K., and Spiro, T.G., Biochim. Biophys. Acta 788, 155-161 (1984).

23. Schoot Uiterkamp, A.J.M., and Mason, H.S., Proc. Natl. Acad. Sci., USA 70, 993-996 (1973).

24. Schoot Uiterkamp., A.J.M, Van der Deen, H., Berendsen, H.C.J., and Boas, J.F., Biochim. Biophys. Acta 372, 407-425 (1974).

25. Schneider, H.-J., Drexel, R., Feldmaier, G., Linzen, B., Lottspeich, F., and Henschen, A., Hoppe-Seyler's Z. Physiol. Chem. 364, 1357-1381 (1983).

26. Yokota, E., and Riggs, A.F., J. Biol. Chem. 259, 4739-4749 (1984).

27. Lerch, K., J. Biol. Chem. 257, 6414-6419 (1982).

28. Huber, M., Hintermann, G., and Lerch, K., Biochemistry, in press (1985).

29. Dietler, C., and Lerch, K., in Oxidases and Related Redox Systems (King, T.E., Mason, H.S., and Morrison, M., eds.), pp. 305-317, Pergamon Press, New York (1982).

30. Stroud, R.M., Kay, L.M., and Dickerson, R.E., Cold Spring Harbor Symp. Quant. Biol. 36, 125-140 (1971).

31. Drexel, R., Schneider, H.-J., Sigmund, S., Linzen, B., Gielens, C., Lontie, R., Préaux, G., Lottspeich, F., and Henschen A., in Invertebrate Oxygen Carriers (Linzen, B., ed.) pp. 255-258, Springer-Verlag, Berlin-Heidelberg-New York (1986).

32. Lerch, K., Methods Enzymol. 106, 355-359 (1984).

33. Makino, N., Van der Deen. H., MacMahill, P., Gould, D.S., Moss, T.H., Simo, C., and Mason, H.S., Biochim. Biophys. Acta 532, 315-326 (1978).

THE HEMOCYANIN OF THE UNIRAMOUS ARTHROPODS

C.P. Mangum and G. Godette
Department of Biology, College of William & Mary, Williamsburg VA 23185 USA
and
Duke University Biomedical Center, Beaufort NC 28516 USA

INTRODUCTION

Recently we described a hemocyanin (Hc) found in the uniramous arthropods, the group that includes the centipedes, millipedes and insects (1). Hc appears to be found only in the scutigeromorph centipedes, also the only members of the uniramia in which the tracheal system ends blindly in the blood rather than proceeding all the way to the tissues.

The blood of Scutigera coleoptrata has the typical absorption spectrum of the arthropod Hcs, with an O_2 dependent band at 340 nm. Its O_2 affinity is low (19 mm Hg at pH 8.04 to 57 mm Hg at pH 7.46, 25 C) and its Bohr shift is normal (log P_{50}/ log pH = -0.87). Cooperativity appears at about 24-30 % HcO_2, rises to h = 6.8-10.7 and does not diminish appreciably at oxygenation states as great as 98 %. Although no physiological information is available, blood taken anaerobically from the pericardium may be either blue or colorless, indicating that in vivo changes in oxygenation are at least possible.

The native polymer is somewhat distinctive. Its molecular weight is about 2.81 x 10^6 d and electron micrographs appear to indicate more than 5 hexamers. We have suggested that it is a triantahexamer, with the 6 hexamers arranged as an octahedron (1). Its underlying structure, however, is quite typical of the arthropod Hcs. Two subunits with molecular weights of 72 and 78 x 10^3 d were resolved by SDS electrophoresis. The dimensions of the hexamers in electron micrographs are about the same as those of other arthropod Hcs and they have the same general appearance. Despite the absence of other examples, the existence of a triantahexamer is entirely consistent with current ideas concerning Hc assembly. The fundamental interest in the structure of the molecule, if any, lies in how polymerization is brought about.

In contrast, the evolutionary interest in the presence of a typical arthropod Hc in the uniramia as well as the crustaceans and chelicerates is considerable. As also pointed out recently, the distinctive quaternary structures of the molluscan and arthropod Hcs are not clearly related to respiratory function (1).

Invertebrate Oxygen Carriers
Ed. by Bernt Linzen
© Springer–Verlag Berlin Heidelberg 1986

In terms of a wide variety of respiratory properties an arthropod Hc may resemble a molluscan Hc, from which its quaternary structure differs fundamentally, more closely than another arthropod Hc, with which it shares many aspects of quaternary structure. Thus one cannot conclude that the presence of the same quaternary structure in the three different groups of arthropods is an example of convergent evolution resulting from strong selection to preserve a particular set of respiratory properties. The most viable alternative is that quaternary structure is a conservative feature which has been inherited by each because it does not influence respiratory function and therefore is not subject to strong selection.

In the past few decades very cogent arguments have been formulated to support the hypothesis of a polyphyletic origin of the arthropods (2,3). But if our interpretation of the relation between quaternary structure and respiratory function is correct, it means that either: 1) In the geological past arthropod-type Hcs occurred in at least three separate phyla, the separate ancestors of the three living arthropod groups, and in each case the phyla and/or their Hcs vanished without a trace. Or, 2) An arthropod Hc occurred in the common ancestor of the various arthropod groups, a hypothetical species that may or may not have been an arthropod itself. The second hypothesis, by far the most economical, implies a monophyletic Phylum Arthropoda.

In the present contribution we report the results of our attempts to separate additional subunits by electrophoresing them according to charge, and our examination of the recognition of centipede Hc by several antibodies of molluscan and arthropod Hcs.

METHODS

In the preliminary experiments whole blood was used without modification. After dissociation of the polymer by dialysis against 0.05 M EDTA at pH 8.9 (0.05 M Tris HCl), electrophoresis was carried out for 8-9 hr on 0.75 mm slab gels (10 % polyacrylamide)(4). Double diffusion immunoelectrophoresis was also performed (5). When a larger quantity (ca. 120 μl) of blood accumulated, the Hc was purified on a column of Sephacryl S-300 (1) and the immunoelectrophoresis was repeated.

RESULTS

Four attempts to demonstrate the presence in whole blood of more than two electrophoretically separable Hc subunits were unsuccessful, despite the use of material collected from a large number of individuals during two successive years

(Fig. 1A). Following gel filtration, only one band was observed (Fig. 1B).

Whole blood did not react with Hc antibodies prepared from the crustacean Bathynomus giganteus, the chelicerate Limulus polyphemus or the gastropod

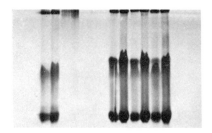

Fig. 1. PAGE of subunits of Scutigera coleoptrata Hc in the presence of 50 mM EDTA at pH 8.9. Whole blood, using two different pools (one applied to the left two and the other to the right six lanes) and several different concentrations.

molluscs Busycon contrarium and Murex fulvescens (Fig. 2A, B). Surprisingly, the centipede sample appeared to react with antibodies prepared from two cephalopod molluscs, Nautilus pompilius and Octopus dofleini (Fig. 2C). When a purified sample of centipede Hc was used, however, there was no sign of a reaction with any of the six antibodies tested (Fig. 2D).

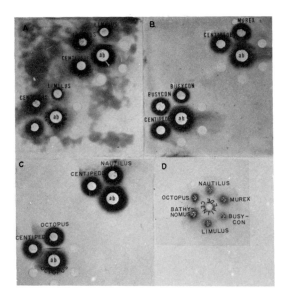

Fig. 2. Immunoelectrophoresis of Scutigera coleoptrata Hc. A-C. Whole blood. D. Purified Hc.

DISCUSSION

In spite of the degree of polymerization of centipedal Hc, we have been unable to demonstrate more than two different subunits. If assembly to the native triantahexamer requires more than two, we suggest that they cannot be separated by the electrophoretic techniques employed thus far.

The immunoelectrophoretic behavior of centipedal Hc may prove to be of interest when a truly wide variety of arthropod Hc antibodies is tested. Neither of the two available to us reacted with centipedal Hc and neither reacted with the antigen of the other. Thus the present evidence indicates only that centipedal Hc is no more distantly related to a chelicerate or a crustacean Hc than the latter two are to one another.

ACKNOWLEDGMENTS

Supported by NSF PCM 84-14856 (to C.P.M.) and MBC ESO 1908 (to C. and J. Bonaventura).

LITERATURE CITED

1. Mangum, C.P., Scott, J.L., Black, R.E.L., Miller, K.I. and Van Holde, K.E., Proc. Natl. Acad. Sci. USA 82, 3721-3725 (1985).
2. Manton, S.M., The Arthropoda, Clarendon, Oxford.
3. Anderson, D.F., Embryology and Phylogeny of Annelids and Arthropods, Pergamon, Oxford.
4. Davis, B.J., Ann. N.Y. Acad. Sci. 121, 404-427 (1964).
5. Margolis, J. and Kenrick, K.G., Nature (London) 214, 1334-1336 (1967).
6. Ouchterlony, A., Arkiv Kemi 1, 43-48 (1949).

IMMUNOLOGICAL CORRESPONDENCES BETWEEN THE HEMOCYANIN SUBUNITS
OF 86 ARTHROPODS: EVOLUTION OF A MULTIGENE PROTEIN FAMILY

Jürgen Markl, Walter Stöcker, Robert Runzler and Engelbert Precht
Zoologisches Institut der Universität München
Luisenstr. 14, 8000 München 2, F.R.G.

INTRODUCTION

One of the most striking features of arthropod hemocyanins is their remarkable subunit heterogeneity, which has been documented in a large variety of papers. Intensive studies on several hemocyanins have shown intraspecific differences between subunits in electrophoretic mobility, immunogenicity, oligomeric topology, primary structure, and oxygen binding function. Additionally, a marked interspecific diversity of the electrophoretic subunit patterns was observed. This, however, prohibits an easy comparison of data, because by no means it is obvious which subunit of a hemocyanin corresponds to a particular subunit of another hemocyanin. It has required much effort and the collaboration of several laboratories to analyse the interspecific correspondencies between the hemocyanin subunits of two xiphosura, two scorpions, and two spiders (1-4).
During the last years we have collected arachnids and crustaceans worldwide, and isolated their hemocyanin subunits, which then have been compared by immunochemistry. The results of this survey now enable us to classify subunit types according to differences in surface structures (which should, as far as we know, reflect differences in function), and to discuss the evolution of this multigene protein family. The aim of the present report is to refine and considerably enlarge our earlier data (5) on how the various subunits of 16S and 24S crustacean hemocyanins are correlated interspecifically, to show how crustacean subunits are related to those present in the larger hemocyanins (35S, 60S) from chelicerates, and finally, to reveal the origin of the 24S hemocyanins found in many modern spiders.

METHODS

Animal sources are enlisted in the figure legends. Collection of blood, isolation of hemocyanin, dissociation, and purification of subunits is described elsewhere (3,4,6). Subunit beta from <u>Callinectes</u> was provided by Bruce Johnson (Beaufort), and subunit gamma from <u>Panulirus</u> by Nell Soeter (Groningen). To preserve collected hemolymph under the special conditions of a tropic expedition, we forced it through a clean cotton towel to remove the clot, and then diluted it 1:1 with 0.1M Tris/HCl buffer, pH 7.5, containing 10mM calcium and 0.1% sodium azide, followed by a permanent storage on ice. The aggregation state of native hemocyanins was analysed by PAGE, and either ultracentrifugation or electron microscopy (EM was done in Groningen). Rabbit antisera were raised against dissociated hemocyanins as described (3,4,6). Comparing immune reactions was crucial especially with crustacean hemocyanins because of (i) the presence of non-respiratory proteins of similar size, (ii) incomplete dissociation, (iii) a subunit-specific tendency to denature, or to form hexameric reassemblies, under dissociation conditions. These difficulties caused mal-interpretations of results in the past, and convinced us that powerful but "blind" quantitative techniques like radioimmuno essay, ELISA, or immune complement fixation are inadequate to analyse and compare hemocyanin subunit mixtures, or purified subunits. Instead, we used various crossed and rocket immunoelectrophoresis techniques (3,4). Additionally, immuno blotting of alkaline PAGE patterns was applied; after immuno labeling, the blotted bands were stained by horseradish peroxidase/diamino benzidine (7). These methods allowed only semiquantitative estimations of the degree of immunological relationship, but enabled us to distinguish native hemocyanin subunits from denatured subunits, oligomeric assemblies, and other impurities.

Invertebrate Oxygen Carriers
Ed. by Bernt Linzen
© Springer-Verlag Berlin Heidelberg 1986

RESULTS and DISCUSSION

1. Subunit structure of hemocyanins from brachyuran crabs

In a broad survey including 32 crabs out of 11 distinct families, we found that their hemocyanin subunit composition follows one of the four possibilities shown in Fig. 1. Most of the species display a subunit pattern corresponding to that already published for Cancer, Carcinus and Hyas (6), and illustrated here for the swimming crab Portunus: it consists of three immunologically distinct subunit types, designated as alpha, beta and gamma. Gamma is immunologically related with alpha, and sometimes antigenically deficient compared to it; beta is immunologically unrelated with both. The different roles played by these subunits in the oligomeric assembly have been reported (8). Like in Portunus, frequently at least two immunologically identical alpha subunits were observed. In Cancer, one of these components was identified as inter-hexamer bridge (5,8), and designated as alpha'. Sometimes also a second beta subunit is present (e.g. in Callinectes, Fig. 3). As documented here for the rock crab Grapsus, some brachyuran hemocyanins completely lack a gamma subunit. The fiddler crab Uca lacks not only gamma, but also beta. A close relative of Uca, the ghost crab Ocypode, also has neither gamma nor beta subunits; remarkable was, however, the presence of a second, rather cathodic alpha subunit(Fig. 1).

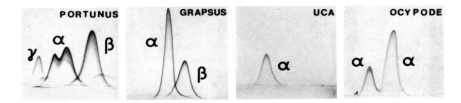

FIGURE 1:
Crossed immunoelectrophoresis to illustrate the four principle hemocyanin subunit patterns obtained from brachyuran crabs. Family-homologous antisera were applied. Immunologically corresponding subunits are identically designated. In the first dimension, the anode was on the left.

Our results from the Brachyura are summarized in Fig. 2. The species cover a broad environmental and activity range, from shallow water, intertidal, fresh water, and land, and from most agile to rather clumsy. As a confirmation of an earlier hypothesis (5,9), in all cases subunit alpha strongly cross-reacted interspecifically, whereas beta subunits behave much more variable. The widespread native hemocyanin aggregate is the 24S dodecamer, which was present throughout with only one exception: the fiddler crab Uca. Uca showed only 16S hexamers which corresponds to the absence of beta or gamma subunits. Much to our surprise, 24S hemocyanin was also found as main component in all three Ocypode species. The capability of Ocypode hemocyanin to form dodecameric aggregates without subunit beta may have something to do with the presence of the second alpha component. Electrophoretically, this subunit migrates in the range of beta, but doubtlessly carries no beta-typical antigenic determinants. Ocypode and Uca both belong to the same family; probably the situation within those Ocypodidae mirrors an ancient trend, leading from alpha hexamers via alpha/alpha dodecamers to alpha/beta dodecamers. However, since Ocypode is probably the most highly advanced crab at all, also other possibilities have to be considered. To clarify this point one has to answer the question whether or not beta is an invention of the comparatively modern Brachyura.

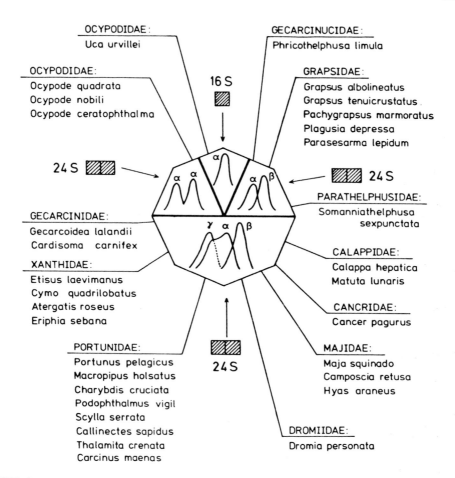

FIGURE 2:
Brachyuran crabs arranged according to the four principle hemocyanin subunit patterns introduced in Fig. 1 . The native aggregation states are indicated. Antisera were raised against hemocyanin from Cancer, Carcinus, Hyas, Callinectes, Grapsus, Matuta, Cardisoma, Gecarcoidea, Somanniathelphusa, Eriphia, and Ocypode. Moreover, a specific anti-Cancer beta antiserum was available. Cancer, Carcinus, Hyas, and Macropipus were purchased from the Biologische Anstalt Helgoland, FRG. Maja and Pachygrapsus stem from Ponza/Italy, and Callinectes from Beaufort/North Carolina, USA. Ocypode q. hemocyanin was provided by Bruce Johnson (Beaufort). Portunus, Charybdis, Podophthalmus, and Thalamita were collected and bled by Michael Walser (München) on the Philippines. Dromia was provided by Zvone Kralj (Aquarium of Piran/Yugoslavia) and Josef Breitenberger (München). In 1984 we collected Ocypode n., Ocypode c., Uca, Scylla, Phricothelphusa, Grapsus a., Parasesarma, Calappa, Camposcia, Etisus, Atergatis, Eriphia, Gecarcoidea, and Cardisoma on Phuket/Thailand, Matuta, Plagusia, Grapsus t., and Cymo on Tioman/Malaysia and Somanniathelphusa on Langkawi/Malaysia. We thank Prof. Phaibul Naiyanetr (Chulalongkorn University, Bangkok/Thailand) and Bamroongsak Chatananthawej (Marine Biological Center, Phuket/Thailand) for species determinations, and Prof. Leo Tan (Singapore Science Center) for useful information.

2. The subunits of 24S hemocyanins from Astacura and Anomura

Phylogenetically, the Astacura are doubtlessly older, and they also possess 24S hemocyanin. The freshwater crayfish Astacus leptodactylus has a hemocyanin composed of four distinct subunits: a disulphide bridged dimer and three monomers. In Homarus, only three monomers are present (Fig. 3). The most anodic of the Homarus subunits quantitatively denatures in our standard dissociation buffer of pH 9.6, and therefore in an earlier communication (6), was described as a "copper-free dimer". The use of a buffer system of pH 8.8, however, completely preserves the component in its native form. After having eliminated denaturation problems also in Astacus, we could cross-correlate the two sets of subunits. Then, a correlation of the Astacus system with crab hemocyanin subunits was attempted: after some mal-interpretations (5), the dimer and a related monomer were identified to form pendants to crab alpha. The second monomer was partially identical with Astacus alpha, and also cross-reacted with brachyuran alpha. With heterologous antisera, an antigenic deficiency towards Astacus alpha was observed. This closely resembles to the behavior of brachyuran gamma; therefore we used the designation gamma also for this particular astacuran subunit, although we have no proof whether or not there exists a true genetic correspondence. What remained was the question of a possible relatedness between the third Astacus monomer and brachyuran beta. By crossed immunoelectrophoresis, anti-Astacus antiserum was unable to precipitate Cancer beta, and vice versa. Therefore, we applied the immuno blotting technique, which is more efficient for a comparison of structurally rather distant antigens. This sensitive method revealed that anti-Astacus antibodies specific for the third monomeric subunit preferentially bind to brachyuran beta and vice versa, whereas alpha components are only weakly recognized. The results show that the beta subunit is not an invention of the Brachyura, but rather belongs to the common design of dodecameric Reptantia hemocyanins; consequently, the properties found in the Ocypodidae have no ancestral character (Fig. 3).
It was found in the present study, that 24S hemocyanin also occurs in a third decapodan group, namely the Anomura. Galathea, Pagurus, and Birgus 24S hemocyanins are composed of a number of electrophoretically distinct subunits; surprisingly, in each case all these subunits were immunologically identical. Thus, in terms of immunology, only one single subunit type is present. Two-dimensional immunoelectrophoresis yielded a clear cross-reaction with brachyuran and astacuran alpha, but also with astacuran (not with brachyuran!) beta. Thus, Anomura and Astacura must be closely related phylogenetically. The common presence of alpha- and beta-typical antigen determinants on a single subunit cannot be explained as the result of a gene fusion, because the size of the anomuran polypeptide chains is quite within the expected range (Mr = 76000-83000). It is much more likely to presume that the Anomura represent an ancestral feature, and that in later appearing species, alpha and beta antigen determinants have been separated by independent evolution of subunits (Fig. 3).

FIGURE 3:
Phylogeny of hemocyanin subunits of the crustacean supraorder Eucarida. From a morphologic-palaeontological viewpoint, a relatively late offspring of the Brachyura from palinuran ancestors, and a distant position of the Euphausiacea is well documented, whereas other relationships are still open for discussion. Caridea and Anomura are arranged according to our immunological data. The native aggregation states are indicated. Note that the occurence of 24S hemocyanin is correlated with the presence of beta subunits; several decapodan groups have independently abolished this subunit and, consequently, possess only 16S hemocyanin. In Ocypode, obviously a new version of dodecamer formation occurs. With anti-chelicerate (Eurypelma) antiserum in rocket immunoelectrophoresis, the Euphausia oligomer was recognized best. This documents the ancient character of hexameric alpha hemocyanins.

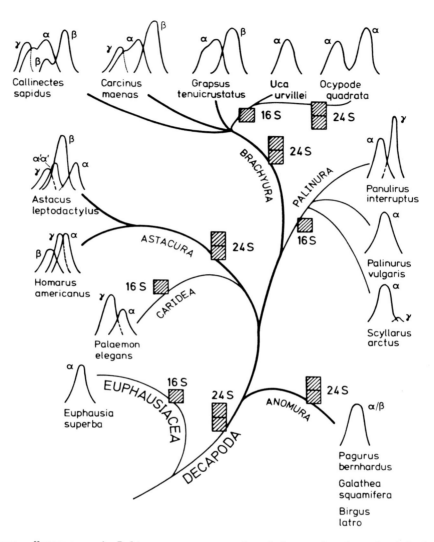

Astacus, Homarus and Palinurus were purchased from a local seafood dealer. Pagurus and Galathea were collected on Helgoland, FRG (we thank Tauchmeister Werner Schomburg for scuba diving opportunities). Scyllarus stems from Ponza/Italy. Palaemon and Euphausia blood was provided by Dr. Christopher Bridges (University of Düsseldorf, FRG). Panulirus was obtained from Prof. Joseph Bonaventura (Beaufort, USA), and Birgus was collected and bled on the Philippines by Michael Walser (München). Antisera were raised against hemocyanin of Astacus, Homarus, Panulirus, Palinurus, Palaemon, Galathea, Pagurus, and Euphausia, and against single subunits of Astacus and Homarus.
It should be noted that native Astacus hexamers consist of subunits beta and gamma. The disulphide bridged dimer M3' of the crayfish Cherax destructor (17) certainly is homologous to Astacus alpha'alpha', and the monomer M3 to alpha. In 1978, we compared Cherax monomers M1 and M2 (provided by Prof. Peter Jeffrey) to Astacus by SDS-PAGE. According to this, M1 (Mr = 84000) should correspond to gamma (Mr = 82000), and M2 (Mr = 78000) to beta (Mr = 76000).

3. The subunits of decapodan 16S hemocyanins (Palinura, Caridea)

According to convincing palaeontological records the brachyuran crabs, as the latest appearing decapodan group, evolved from ancestors related to spiny lobsters. Interestingly, spiny lobsters like Panulirus interruptus and Palinurus vulgaris possess 16S hemocyanin (10); we found this also in the related scyllarids. Panulirus, Palinurus and Scyllarus hemocyanins are composed of an immunologically homogeneous major subunit fraction, which corresponds to brachyuran and astacuran alpha (Fig. 3). Panulirus hemocyanin contains a second, cathodic subunit in addition, which has earlier been designated as c (11); in traces, this component was also detected in Scyllarus. With the homologous antiserum, Panulirus c appeared to be unrelated to Panulirus alpha; however, heterologous antisera showed common antigenic determinants. Moreover, weak cross-reactivities with astacuran alpha were observed, but no relationship with beta subunits. This is the typical behavior of a gamma subunit, and the component was designated accordingly (Fig. 3).
Also the 16S hemocyanin of the shrimp Palaemon elegans is constructed of two subunit types classified as alpha and gamma (Fig. 3). They show clear cross-reactivities towards both, astacuran and (weaker) palinuran hemocyanins. This is interesting, because little information exists about the phylogenetic position of the natantian superfamily Caridea within the Decapoda.
It should be pointed out that according to our results all alpha subunits most probably are coded on homologous genes. This is also supported by the fact that the primary structures of Astacus alpha (subunit "b") and Panulirus alpha are very similar (12). Also all beta subunits should be truly homologous. Gamma components, however, are defined as being antigenically deficient compared to alpha, if heterologous antisera are applied. Besides some alpha-typical antigen determinants, they behave rather distinct and species-specific. Only within, but not between, the systematic categories Brachyura, Astacura, Caridea, and Palinura we have enough evidence to classify the gamma subunits as homologous proteins. In earlier reports it was emphasized that dodecameric decapodan hemocyanins may have evolved from hexamers several times independently (5). However, their oligomeric subunit topology is very similar (8), and according to the present results, quite the contrary happened: an early invented beta subunit enabled the formation of dodecamers; this ability was repeatedly lost in later appearing species.

4. Krill hemocyanin: an ancient design

In order to view the situation from outside of the Decapoda, we have studied the 16S hemocyanin of the Antarctic Krill Euphausia superba. The orders Euphausiacea and Decapoda are combined as superorder Eucarida. Euphausia hemocyanin subunits exhibit a completely homogeneous peak in two-dimensional immunoelectrophoresis. In contrast to an earlier result (5), which later turned out to be a denaturation artifact, we could demonstrate that this subunit exclusively corresponds to component alpha of decapodan hemocyanins (Fig. 3). On the other hand, oligomeric Euphausia hemocyanin has preserved considerably more ancient antigen determinants compared to the Decapoda: in rocket immunoelectrophoresis, it was the only crustacean hemocyanin which could be precipitated by an anti-chelicerate (Eurypelma) antiserum. This means that indeed alpha-typical antigen determinants form the basic ancestral design of the subunit surface of decapod hemocyanin.

5. The subunits of larval crab hemocyanins

It was reported that larval crab hemocyanin is electrophoretically distinct from adult hemocyanin (13). The typical crab larvae are the planctic zoea and, after a series of molting cycles, the benthic megalops. The next stage is the first instar juvenile crab. We found that hemocyanin from zoea, megalops and juvenile of the sea spider Hyas araneus is composed of a single subunit type.

This component clearly cross-reacted with adult alpha subunits (Fig. 4), although electrophoretically it behaves somewhat different. Very similar results were obtained from larvae of the shore crab <u>Carcinus</u> <u>maenas</u>. The native aggregation state of <u>Hyas</u> and <u>Carcinus</u> larval hemocyanin is still unclear. It should also be investigated, at which ontogenetic stage hemocyanin synthesis switches over to the subunit pattern of the adults (Fig. 4). It is interesting that not only phylogenetically, but also ontogenically alpha subunits appear first. In several decapodan groups, we observe a loss of subunit beta (Fig. 3). As a possible genetic mechanism, some of those species might retain their larval hemocyanin through their whole lifetime.

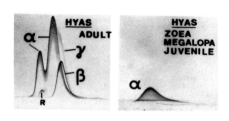

FIGURE 4:
Patterns of crossed immunoelectrophoresis of the hemocyanin subunits of <u>Hyas</u> <u>araneus</u> (Brachyura) versus anti-Hyas antiserum. In the first dimension, the anode was on the left. <u>Carcinus</u> <u>maenas</u> yielded comparable results. The three larval stages possessed the same kind of larval hemocyanin. Larvae were cultured, homogenized by ultra-sound, and the hemocyanin isolated by Beatrice Steiff, Biologische Anstalt Helgoland, FRG. R = reassembly product.

6. Immune precipitation of hemocyanin subunits from the other subphylum

We were particularly interested in detecting immunological relationships between crustacean hemocyanin subunits and those from the Chelicerata. By rocket immunoelectrophoresis, a precipitation of crustacean hemocyanins with an anti-chelicerate antiserum, or vice versa, was only successful if the hemocyanin was either denatured in 8M urea, or present in its oligomeric form. Therefore, we performed immuno blotting experiments with alkaline PAGE patterns of hemocyanin subunits from the horseshoe crab <u>Limulus</u>, the scorpion <u>Androctonus</u>, the tarantula <u>Eurypelma</u> and the hunting spider <u>Cupiennius</u> against anti-alpha (<u>Ocypode</u>) antiserum. In each case, the total set of subunits was recognized (Fig. 5; not visible for <u>Limulus</u>). Anti-beta and anti-gamma antisera give similar, but much weaker reactions. As judged semiquantitatively, cheliceratan subunits related to <u>Eurypelma</u> a, d and f are recognized best. This result is also illustrated in the right insert of Fig. 6. In the early Cambrium, from which we have the first fossil arthropodan records, both subphyla were already established. Thus, our results reflect a conservation of hemocyanin surface structures over at least 600 million years.

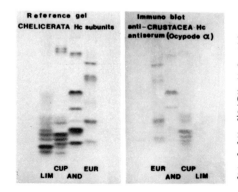

FIGURE 5:
Immuno blotting of alkaline PAGE patterns of the hemocyanin subunit mixtures of the xiphosur <u>Limulus</u> <u>polyphemus</u>, the scorpion <u>Androctonus</u> <u>australis</u>, and the spiders <u>Cupiennius</u> <u>salei</u> and <u>Eurypelma</u> <u>californicum</u> with anti-<u>Ocypode</u> antiserum. The recognition of <u>Limulus</u> subunits is not visible here, but is also documented. The reference was stained with Coomassie. The blot was stained with the horseradish peroxidase diamino benzidine system (7). The anode was at the bottom.

288

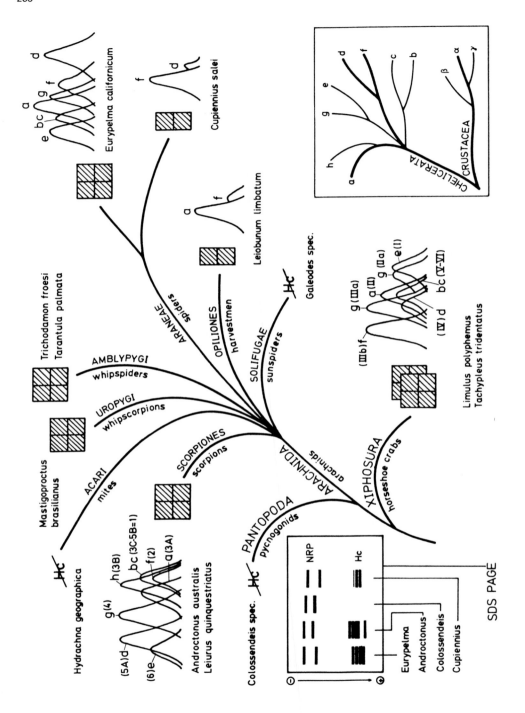

Eurypelma californicum

Cupiennius salei

Leiobunum limbatum

Galeodes spec.

Trichodamon froesi
Tarantula palmata

ARANEAE
spiders

OPILIONES
harvestmen

SOLIFUGAE
sunspiders

g (IIIa)
a (II)
g (IIa)
e (I)
bc (V-VI)
(IIIb)f
(IV)d

Limulus polyphemus
Tachypleus tridentatus

AMBLYPYGI
whipspiders

Mastigoproctus
brasilianus

UROPYGI
whipscorpions

ACARI
mites

SCORPIONES
scorpions

ARACHNIDA
arachnids

XIPHOSURA
horseshoe crabs

PANTOPODA
pycnogonids

Hydrachna geographica

g(4)
h(3B)
bc (3C-5B=1)
f(2)
a(3A)
(5A)d
(6)e

Androctonus australis
Leiurus quinquestriatus

Colossendeis spec.

NRP

Hc

Eurypelma
Androctonus
Colossendeis
Cupiennius

SDS PAGE

CHELICERATA

CRUSTACEA

6. Subunit correspondences of cheliceratan hemocyanins

In the subphylum Chelicerata we encounter three animal classes: the intertidal xiphosurs, the terrestrial arachnids, and the marine pycnogonids. Xiphosurs possess a highly complex 48-meric hemocyanin, composed of 8 immunologically distinct subunit types. 24-meric hemocyanins with similar structural features have been found in scorpions, whipspiders, whipscorpions, and in the tarantula Eurypelma (Fig. 6). Interspecific subunit correspondencies of xiphosuran, scorpion, and tarantula hemocyanins have been investigated (1-4); Fig. 6 summarizes the results. From these studies, also phylogenetic relationships between the 8 subunit types a - h could be derived (right insert in Fig. 6). Starting from such 24-mers, two phylogenetically distant species developed dodecameric hemocyanins composed of only two subunit types: the harvestman Leiobunum, and the hunting spider Cupiennius. Recent immuno blotting experiments have refined some earlier data (3,4); it is now clear that only the monomeric subunit type of Cupiennius hemocyanin corresponds to Eurypelma f, whereas the dimer, which acts as inter-hexamer linker, corresponds to a subunit phylogenetically closely related to f, namely Eurypelma d (Fig. 6).

7. Chelicerates which lack hemocyanin (Solifugae, Acari, and Pantopoda)

We found no hemocyanin in the sunspider Galeodes, and in the watermite Hydrachna. Both species possess a well developed tracheal system and therefore probably have no need of a respiratory pigment. The marine pycnogonids are a phylogenetic mystery – it is still debated whether or not they are arachnids, non-arachnidan chelicerates, or even mandibulates related to the Crustacea. Also remarkably, they completely lack respiratory organs. Most pycnogonids are far too small to be bled; however, we obtained extremely large Antarctic specimens (Colossendeis). Unfortunately, their hemolymph completely lacked hemocyanin. However, in SDS-PAGE we found two strong bands in a range typical for the second major blood protein of arachnids (left insert in Fig. 6), earlier designated as "non-respiratory protein" (14). With a specific anti-Eurypelma NRP antiserum, an immunologic relationship with the pycnogonidan protein could be confirmed. Since SDS bands of non-respiratory proteins from xiphosurs and crustaceans show much lower molecular masses, and moreover do not react with anti-Eurypelma NRP antiserum, the existence of this protein in Colossendeis blood is a good argument to put the pycnogonids close to the arachnids within the phylogenetic tree (Fig. 6).

FIGURE 6:
Subunit composition of the hemocyanins from various Chelicerata. The patterns of crossed immunoelectrophoresis, and also the native aggregation states are shown (60S, 35S, 24S; small rectangles correspond to 24S dodecamers). Most of the immunologic subunit correspondences have already been published (1-4). Homologous subunits are identically labeled using the designations a - h. Scorpion and xiphosuran subunits are additionally labeled according to their original designations (1,2). Amblypygi and Uropygi hemocyanin subunits were only electrophoretically studied (14). Leiobunum limbatum was collected in Bavaria/FRG. No hemocyanin (Hc) was found in Galeodes (collected in North Africa by Dr. Franz-Peter Fischer, Techn. Universität München), Hydrachna (collected in Sardinia/Italy) and Colossendeis (collected during the German Antarctis Expedition 1984/85, and provided by Heike Wägele, Universität Oldenburg, FRG). Left insert: SDS-PAGE of hemolymph proteins, showing that Colossendeis possesses the typical arachnidan non-respiratory protein (NRP). Right insert: Phylogenetic relations between the 8 cheliceratan subunits according to (4). Broader lines indicate that with anti-crustacean alpha antiserum, subunits a, d and f are recognized best.

9. Hemocyanin subunit structure, and the evolution of spiders

In a broad survey we have investigated the hemocyanin subunit composition of 40 spider species out of 25 families. Preliminary data are published elsewhere (5). The more ancient Orthognatha throughout possess 37S hemocyanin (24-mer), composed of 7 subunit types related to those of Eurypelma; also in a number of labidognath spiders, this structure occurs (Fig. 7). A large group of other labidognathan families, however, is characterized by a dodecameric Cupiennius-like hemocyanin, composed of only 2 subunit types. These dodecamers are of a striking immunological uniformity; thus, the whole group is undoubtedly monophyletic. It was very interesting for us to detect in two hemocyanins a stepwise transition from the 7-subunit to the 2-subunit particle: In the cribellate spider Filistata we found only hexameric hemocyanin; the patterns of crossed immunoelectrophoresis showed 5 subunit peaks (Fig. 7). Absent was heterodimer bc, which in native Eurypelma hemocyanin forms a central tetrameric ring structure, and in reassembly experiments is indispensable to exceed the hexamer level (15). The haplogyne spider Dysdera possesses a similar hexameric hemocyanin, but composed of only two different subunits in comparable proportions; immunologically they correspond to Eurypelma f and d (Fig. 7). The close relatedness towards Cupiennius hemocyanin is obvious; however, the specific structural role of subunit d as hexamer linker (16) is not (yet?) established here. Although it remains unclear whether or not Filistata and Dysdera hemocyanins represent true "missing links" (both hemocyanins quite as well could be secondarily developed forms), they can serve as models: It appears that an ancestral spider for whatever reason lost its ability to genetically express the heterodimer, and thus was restricted to hexameric hemocyanin with possibly negative consequences (drop of cooperativity, increase of colloidosmotic pressure and blood viscosity). Nevertheless, this spider became the ancestor of a large variety of highly advanced and most agile hunters and jumpers (Fig. 7).

FIGURE 7:
Phylogenetic tree of spiders as deduced from their hemocyanin subunit structure. Typical subunit patterns of crossed immunoelectrophoresis, and the native aggregation states are shown (35S, 24S, 16S; small squares represent 16S hexamers). The tree allows the conclusion that Entelegynae and Trionycha certainly are of polyphyletic origin. Haplogyne spiders are either direct progenitors, or secondary offsprings of spiders with Cupiennius-like hemocyanin. Since cribellate features have certainly been developed in evolution only once, the entire group must stem from cribellate ancestors.
Except of Argyope aurantia (collected at Beaufort, USA), Atrax formidabilis (provided by Struan K. Sutherland, Melbourne/Australia and transported by Dr. Bernhard Kempter, München), Cupiennius salei (cultured according to (14) in München), Dysdera spec. (from Egypt; provided by Ilse Tutter), and Eurypelma californicum (purchased from Carolina Biological Supply, USA), all spiders were collected in Europe:
In Germany: Araneus diadematus, Araneus umbriaticus, Meta segmentata, Linyphia marginata, Tetragnatha extensa, Pholcus phalangioides, Pardosa amentata, Tarentula fabrilis, Dolomedes fimbriatus, Micrommata rosea, Tegenaria atrica.
In Yugoslavia: Uroctea durandi (by Karin Heckel).
In Italy: Atypus affinis, Nemesia spec., Cyrtophora citricola, Argyope bruennichi, Theridion varians, Scotophaeus quadripunctatus, Haplodrassus signifer, Callilepis nocturna, Chiracanthium elegans, Clubiona terrestris, Liocranum rupicola, Salticus scenius, Philodromus collinus, Heriaeus hirtus, Synaema globosum, Thomisus onustus, Xysticus bifasciatus, Oxyopes lineatus, Menemerus taeniatus, Agelena labyrinthica, Pisaura mirabilis, Amaurobius fenestralis, Filistata insidiatrix, Dysdera crocata.
Antisera were raised against the hemocyanins of Eurypelma, Araneus, Argyope, Cupiennius, and Tegenaria, and against each subunit type of Eurypelma.

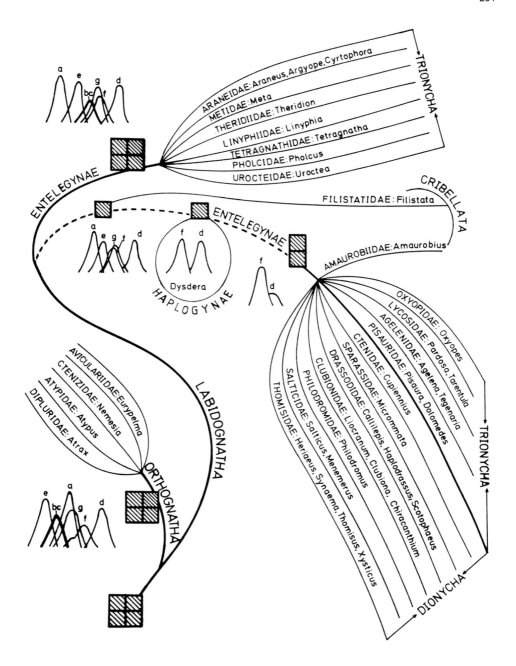

Acknowledgment: We thank the Biologische Anstalt Helgoland (FRG) and the Beaufort Marine Laboratory (USA) for research opportunities, Prof. Bernt Linzen and Prof. Ernst F.J. van Bruggen for their interest and support, Wilma G. Schutter and Dr. Trijntje Wichertjes (University of Groningen) for the EM experiments, and Dr. Bernhard Kempter for his important contributions to the already published parts of this report. Heide Storz performed the drawings. Supported by the Deutsche Forschungsgemeinschaft (Ma 843).

LITERATURE

(1) Lamy J., Lamy J., Weill J., Bonaventura J., Bonaventura C. & Brenowitz M. (1979): Arch.Biochem.Biophys. 196, 324 - 339.

(2) Lamy J., Compin S. & Lamy J. (1983): Arch.Biochem.Biophys. 223, 584 - 603.

(3) Markl J., Gebauer W., Runzler R. & Avissar I. (1984): Hoppe-Seyler's Z.Physiol.Chem. 365, 619 - 631.

(4) Kempter B., Markl J., Brenowitz M., Bonaventura C. & Bonaventura J. (1985): Biol.Chem. Hoppe-Seyler 366, 77 - 86.

(5) Markl J., Stöcker W., Runzler R., Kempter B., Bijlholt M.M.C. & van Bruggen E.F.J. (1983): In "Structure and function of invertebrate respiratory proteins" (Wood E.J., ed.), Life Chem. Reports 1, 39 - 42. Harwood Acad.Publ., London.

(6) Markl J. & Kempter B. (1981): J.Comp.Physiol. 140, 495 - 502.

(7) Towbin H., Staehelin T. & Gordon J. (1979): Proc.Nat.Acad.Sci.USA, 76, 4350 - 4354.

(8) Stöcker W., Raeder U., Bijlholt M.M.C., Schutter W.G., Wichertjes, T. & Markl J.: This volume, pp. 213-216.

(9) Markl J. & Kempter B. (1981): In "Structure, active site, and function of invertebrate oxygen binding proteins" (Lamy J., ed.). Marcel Dekker, New York, pp. 125 - 137.

(10) Markl J., Hofer A., Bauer G., Markl A., Kempter B., Brenzinger M. & Linzen B. (1979): J.Comp.Physiol. 133, 167 - 175.

(11) Gaykema W.P.J., Hol W.G.J., Vereijken J.M., Soeter N.M., Bak H.J., & Beintema J.J. (1984): Nature 309, 23 - 29.

(12) Schneider H.-J., Voll W., Lehmann L., Grisshammer R., Goettgens A. & Linzen B.: This volume, pp. 173-176.

(13) Terwilliger N. & Terwilliger R. (1982): J.Exp.Zool. 221, 181 - 191.

(14) Markl J., Markl A., Schartau W. & Linzen B. (1979): J.Comp.Physiol. 130, 283 - 295.

(15) Markl J., Kempter B., Linzen B., Bijlholt M.M.C. & van Bruggen E.F.J. (1981): Hoppe-Seyler's Z.Physiol.Chem. 363, 1631 - 1641.

(16) Markl J. (1980): J.Comp.Physiol. 140, 199 - 207.

(17) Murray A.C. & Jeffrey P.D. (1974): Biochemistry 13, 3667 - 3671.

DIVERGENCE PATTERN OF HORSESHOE CRABS: IMMUNOLOGICAL CONSTITUTION AND
IMMUNOLOGICAL DISTANCE OF THE ASIAN HORSESHOE CRAB'S HEMOCYANINS

H. Sugita
Institute of Biological Sciences
University of Tsukuba
Sakura-mura, Ibaraki 305, Japan

INTRODUCTION

Extant horseshoe crabs are assigned to two subfamilies, that
is, Limulinae and Tachypleinae. Limulinae includes only one American
species, Limulus polyphemus, and Tachypleinae includes three Asian
species, Tachypleus tridentatus, Tachypleus gigas, and Carcinoscorpius rotundicauda. These living species are morphologically
similar to fossil specimens of the genus Mesolimulus, so that one has
referred to them as "living fossils."

Hemocyanins of Asian horseshoe crabs, T. tridentatus, T. gigas,
C. rotundicauda showed, respectively, four, six, and six monomer
bands on polyacrylamide gel electrophoresis with Davis' buffer system
(1). However, Limulus hemocyanin showed eight or more monomers while
it was composed of eight antigenically different subunits (2).

In the course of immunological studies on the hemocyanins,
Sugita and Sekiguchi (3) revealed that gene duplication for hemocyanin monomer would happen in the lineage of T. gigas. Brenowitz
and Moore (4) suggested a possibility that Limulus hemocyanin was
led to heterogeneity of subunits by gene duplications for each
subunit type. Therefore, immunological comparison among hemocyanins
from four species is very interesting from phylogenic and evolutionary viewpoints of horseshoe crabs.

In the present paper, the systematic relationship of the horseshoe crabs is studied by examining the immunological constitution of
hemocyanin monomers and immunological distance of whole hemocyanins.

Invertebrate Oxygen Carriers
Ed. by Bernt Linzen
© Springer–Verlag Berlin Heidelberg 1986

MATERIALS AND METHODS

The Japanese horseshoe crab, Tachypleus tridentatus, was collected from the north coast of Kyushu, Japan, and the Southeast Asian horseshoe crabs, Tachypleus gigas and Carcinoscorpius rotundi-cauda, were kindly provided by Prof. Smarn Srithunya (Zoological Museum, Srinakharinwirot University, Bangsaen, Thailand). The American horseshoe crab, Limulus polyphemus, was supplied from the Marine Biological Laboratory, Department of Marine Resources, Woods Hole, Massachusetts, USA.

Acrylamide gel electrophoresis was done according to the method of Davis (5) and hemocyanin monomers were prepared from disc-type polyacrylamide gel. Purity of the monomers was monitored by vertical slab polyacrylamide gel electrophoresis. Antiserum to pure hemo-cyanin was prepared as described previously (6) and double immuno-diffusion tests were carried out in 1% agar plates composed of 52 mM Tris-glycine buffer (pH 8.9), 0.01% thimerosal, and 10 mM EDTA. Anti-T. tridentatus hemocyanin serum was tested for reactivity with hemocyanin in serum from each of four horseshoe crabs. Reactivity was measured by the quantitative microcomplement fixation method (7) and the results are given in immunological distance units.

RESULTS AND DISCUSSION

Heterogeneity of hemocyanin monomers is a common character in arthropods. The four monomer bands of T. tridentatus were named HT 1, HT 2, HT 3, and HT 4 in descending order of electrophoretic mobility. The six monomers of T. gigas and C. rotundicauda were designated, respectively, HG 1 to HG 6 and HR 1 to HR 6 in the descending order. Immunological comparison among these hemocyanin monomers was carried out using the gel slices containing respective monomer bands. Reaction of the antiserum to T. tridentatus hemo-cyanin with the four kinds of hemocyanins revealed that HT 2 consisted of three immunologically distinct monomers which were immunologically identical with HR 2, HR 3, and HR 4, respectively. Immunological similarity of hemocyanin monomers among the three Asian species is summarized in Table 1.

Table 1.   Immunological similarity of hemocyanin monomers

T. tridentatus	HT 4	HT 3	HT 2-1	HT 2-2	HT 2-3	HT 1
	‖	‖	‖	‖	‖	‖
C. rotundicauda	HR 6	HR 5	HR 4	HR 3	HR 2	HR 1
	‖	‖	⤢	‖	‖	‖
T. gigas	HG 6	HG 5	HG 4	HG 3	HG 1	HG 2

Hemocyanin monomers on both sides of "=" are immunologically identical. HT 2-1 to HT 2-3 mean immunologically distinct monomers which form only one band on polyacrylamide gel.

Hereafter, immunologically distinct components of C. rotundi-cauda hemocyanin are renamed 1, 2, 3, 4, 5, and 6 in descending order of electrophoretic mobility.   Therefore, T. gigas hemocyanin consists of immunological components of 2, 1, 3, 3, 5, and 6 in the descending order and T. tridentatus hemocyanin consists of subunits 1, 2, 3, 4, 5, and 6.   From these results of immunological comparison, two possible divergence patterns are proposed as shown in Figure 1.

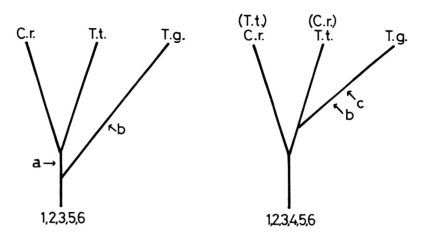

Figure 1.   Possible divergence patterns of Asian horseshoe crabs based on immunological constitution of hemocyanin monomers. C.r., C. rotundicauda; T.g., T. gigas; T.t., T. tridentatus; a, acquisition of 4; b, duplication of 3; c, deletion of 4.   The numbers at the root mean immunologically distinct monomers of the common ancestral species hemocyanin.   Antigenicity of each ancestral monomer is similar to that of each descendant monomer with the same number.

In order to determine the divergence pattern of the horseshoe crabs, immunological distance of whole hemocyanins between T. triden-tatus and other three species was estimated by microcomplement fixa-

tion (MC'F) method (7) using anti-T. tridentatus hemocyanin antiserum.   Preliminary immunological distances were 3.9 (T. triden-
tatus-C. rotundicauda), 8.7 (T. tridentatus-T. gigas), and 64 (T.
tridentatus-L. polyphemus).

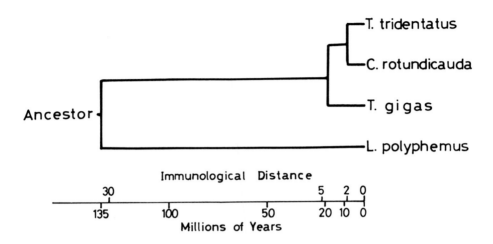

Figure 2.   Phylogenetic relationships of horseshoe crabs.   It is
estimated that American and Asian horseshoe crabs diverged 135
million years ago (8).

Divergence pattern of the horseshoe crabs was inferred as shown
in Figure 2 from the results of MC'F measurements.

REFERENCES

1. Sugita, H., and Sekiguchi, K., in Invertebrate Oxygen-Binding
   Proteins (Lamy, J., and Lamy, J., eds.), pp. 247-255, Marcel
   Dekker, New York (1981).
2. Brenowitz, M., Bonaventura, C., Bonaventura, J., and Gianazza,
   E., Arch.Biochem.Biophys. 210, 748-761 (1981).
3. Sugita, H., and Sekiguchi, K., Life.Chem.Rep.(Suppl.1) 361-364
   (1983).
4. Brenowitz, M., and Moore, M., in Physiology and Biology of
   Horseshoe Crabs (Bonaventura, J., Bonaventura, C., and Tesh,
   S., eds.), pp. 257-267, Alan R. Liss, New York (1982).
5. Davis, B.J., Ann.N.Y.Acad.Sci. 121, 404-427 (1964).
6. Sugita, H., and Sekiguchi, K., J.Biochem.(Tokyo) 78, 713-718
   (1975).
7. Champion, A.B., Prager, E.M., Wachter, D., and Wilson, A.C., in
   Biochemical and Immunological Taxonomy of Animals (Wright,
   C.A., ed.), pp. 397-416, Academic Press, London (1974).
8. Shishikura, F., Nakamura, S., Takahashi, K., and Sekiguchi, K.,
   J.Exp.Zool. 223, 89-91 (1982).

Physiological role and modulation
of hemocyanin function

OXYGEN BINDING BY HEMOCYANIN : COMPENSATION DURING ACTIVITY
AND ENVIRONMENTAL CHANGE

Brian McMahon
Department of Biology
University of Calgary
Calgary, Alberta, Canada   T2N 1N4

In the majority of decapod crustaceans oxygen consumption can be controlled to fit the animals' current metabolic demand. The mechanisms are complex, involving control over diverse processes including gill ventilation, perfusion of both gills and tissues and change in the oxygen-binding characteristics of the animals respiratory carrier molecule hemocyanin (Hcy). Although it is apparent that these are all inter-related and integrated functions which act together to control gaseous exchange at both gills and tissues, only the latter mechanism can be treated in this account.

The functioning of oxygen carrier molecules can be adjusted at several levels. Briefly, functional decapod crustacean hemocyanins are large copper containing proteins which are aggregate molecules built from subunits (monomers) of approx. 70,000 Daltons.   Both the structure of the subunits, the size of aggregate, and the subunit composition within the aggregate are variable and each variant may have differing oxygen (and perhaps carbon dioxide, H^+ and other ion binding properties). Thus at the genetic level a species may be able to manufacture one or more unique hemocyanins which allow adaptation to their own particular ecological, behavioural and physiological matrix. Within an animals lifetime (acclimation) hemocyanin performance could be varied in the long term, i.e., with developmental stage or with season either by the manufacture of one or other variant hemocyanins as above or by modulation of the binding properties of the existing molecule. Such affects could occur by change in the non-protein constituents of the molecule (i.e., see Mangum 1983a, 1983c) or by changes in microstructure or microhabitat of the molecule induced by changes in composition of the hemolymph. Several blood borne factors (termed modulators) have been shown to influence hemocyanin oxygen binding characteristics in 'in vitro' experiments. However rather less is known of their actions 'in vivo'. The present paper thus

Invertebrate Oxygen Carriers
Ed. by Bernt Linzen
© Springer–Verlag Berlin Heidelberg 1986

concentrates on analysis of measurements of both hemolymph oxygen and CO_2 levels and the levels of several known modulator substances from a series of samples taken from animals undergoing compensation for change in either oxygen demand or oxygen supply. For the sake of brevity I will focus largely on changes in oxygen binding occurring during activity and during the transition from water to air breathing. These are chosen as common natural occurring events in the lives of the animals concerned. In some cases consideration of the time course of changes in modulator concentration allows us to attempt to elucidate the relative roles of these putative modulator substances as they vary 'in vivo'.

Modulation of hemocyanin oxygen binding during activity

During maintained strenuous activity the energy demand of, particularly muscular, tissue increases, and, if aerobic work is to increase in proportion, oxygen supply, oxygen uptake and oxygen delivery all must increase concomitantly. During strenuous activity oxygen supply to the gills is increased by increase in ventilation volume. Increased perfusion, however, can allow increased oxygen transport and hence increased oxygen supply to tissues only to the extent that oxygen uptake can be maintained. Since the residence time for blood in the gills is reduced at high flow rates this can be ensured only by an increase in the efficiency of oxygen transfer across the gill barrier, i.e., an increase in the transfer factor (TO_2; the amount of O_2 passing per torr gradient). Since oxygen uptake is mediated by diffusion alone then the mechanisms by which this can occur are limited to: -

a) increase in the oxygen partial pressure difference (gradient) across the gills

b) decrease in the thickness of the gill epithelium

c) increase in gill surface area

d) increased participation of hemocyanin in facilitation of O_2 transfer.

While there is no clear evidence for the existence of either b) or c) in crustacean gill systems, a) and d) are commonly occurring related processes pertinent to this account. They may be illustrated by reference to work carried out over the last several years in this laboratory on oxygen transport during activity. Several species of decapod crustaceans including air and water breathers, runners and swimmers are involved.

In Cancer magister resting in well aerated water but equipped with a mask and probes to allow measurement of ventilation volume, oxygen consumption, branchial pressures, etc. the frequency of ventilatory pumping is enhanced by disturbance and oxygen partial pressures elevated slightly over that occurring in undisturbed animals (McMahon 1984). Under these conditions, post-branchial (arterial, PaO_2) and pre-branchial (venous, PvO_2) oxygen partial pressures and contents (CaO_2 and CvO_2) are 75 and 15 torr and 0.48 and 0.22 mM, respectively. If we fit these values on an oxygen equilibrium curve plotted under conditions (temp. 10°C, pH = 7.88) similar to those found 'in vivo' (Fig. 1A; taken from McMahon et. al. 1979) we see that although a reasonable amount of oxygen (per unit blood flow) is delivered to tissues, hemocyanin plays a minor role with the majority (60%) of O_2 released from the dissolved fraction. In unmasked animals, lower circulating O_2 levels occur due to differences in respiratory pumping pattern (details in McMahon 1984). Under these conditions ($\bar{x}PaO_2$ = 28±9 torr, $\bar{x}PvO_2$ = 7±2 torr) a similar amount of oxygen is delivered to tissues, but hemocyanin now plays a greatly increased role carrying more than 80% of the O_2 delivered to tissues (Fig. 1).

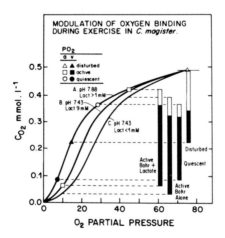

Figure 1. Modulation of hemocyanin oxygen binding 'in vivo', prior to, in and following 30 min. exhausting exercise in Cancer magister CO_2 = Oxygen content of hemolymph in $mmol.L^{-1}$. Histograms show amount of oxygen delivered to tissues. Solid area is the amount delivered from Hcy bound O_2, clear area is amount delivered from solution. 'In vitro' oxygen binding characteristics are from McMahon et. al. (1979). 'In vivo' hemolymph acid-base and oxygenation status data and levels of hemolymph lactate (Lact) from McDonald et. al. (1979); McMahon (1984).

If we use these latter (undisturbed) values as a basis for comparison 20-30 minutes of strenuous enforced activity apparently increases both arterial and venous oxygen levels since immediately following activity, PaO_2 and PvO_2 have risen to 45±12 and 10±4 torr respectively. The increase in PaO_2 increases the amount of dissolved oxygen released to tissues, but this is slight and counterbalanced by the increase in PvO_2 which decreases the Hcy bound O_2 release. Since the a-vO_2 difference has increased from 21-31 torr, the overall effect of these changes would be little change in O_2 delivery to tissues but a reduction in the contribution of Hcy. The above however does not take into account the effect of changes in concentration of two or more modulators known to be present in activity.

During activity, marked acidosis occurs in hemolymph, resulting partially from build up of PCO_2 and partially from build up of lactic acid (measured as lactate) in hemolymph. For the sake of clarity, let us consider the effect of the protons, assumed to be released in stoichiometrically equal amounts to the lactate measured, separately. C. magister has only a moderate Bohr effect ($\Delta \log P_{50}/\Delta pH = -.079$, McDonald et. al. 1979) but nonetheless the increase in protons (pH falls to 7.43) would be sufficient to reduce oxygen affinity substantially (P_{50} rises from 15 to 26 torr; Fig. 1A-C). This positive Bohr shift would be adaptive, facilitating oxygen delivery to tissues allowing substantially greater release of Hcy bound oxygen to the tissues despite the increased hemolymph to tissue O_2 difference (Fig. 1). It is, however, interesting to note that if one plots the actual values for PO_2 and CO_2 measured immediately after exercise in C. magister (McMahon et. al. 1979) they did not fit upon the oxygen equilibrium curve characteristic of blood at pH 7.43 (Fig. 1C; calculated from curve A using the Bohr factor above) but are substantially left shifted. A curve of similar form drawn through these points thus more accurately reflects the 'in vivo' oxygen status of the blood at this time. Clearly one or more factors existing in hemolymph are responsible for this increase in O_2 affinity.

Several possibilities occur. The most likely candidate is lactate which increases >10 fold in the blood during strenuous activity. Originally proposed as a factor effecting Hcy O_2 binding by Truchot (1980) lactate has been shown to effect O_2 binding of Hcy from many (but not all species) of Crustacea but not those of other groups (Mangum 1983a, b). In fact, the magnitude of the effect observed 'in vivo' is similar to that predicted from the relationship between lactate and O_2 affinity in this species 'in vitro' by Mangum (1983b) and Grahame et. al. (1983). Ca^{++}, a second modulator known to increase O_2 affinity may increase in the hemolymph of crustaceans during acidotic conditions (deFur et. al. 1980). Possible modulation by Ca^{++} and lactate in concert is described below but the interactions between these two ions and H^+ are complex and are discussed in detail in a separate paper in this volume (Bridges and Morris 1985). The combined potentiating effects on O_2 affinity seem not to be of great significance here at this maximal exercise level, where they seem simply to act as a brake on the affects of the Bohr shift so that depletion of oxygenation at the gills is not extreme. It is possible that more dramatic effects occur in more controlled voluntary activity or during the development of the activity period or under more stressful situations, i.e., elevated temperature.

Pertinent data are not available for C. magister but are, however, available for lactate and proton effects on oxygen affinity during sustained swimming in the Blue Crab Callinectes sapidus (Booth 1982; Booth et. al. 1982). Blue crabs lifted by a hook from the substratum will swim 'voluntarily' for periods of up to 60 minutes at 20°C. This length of exercise is non-exhausting and, after an initial brief spurt involving anaerobic activity, is apparently fueled largely aerobically. Nonetheless, lactate and proton competition for modulation of the oxygen carrier function occurs as illustrated in Fig. 2. Even at the onset of exercise, lactate is produced in muscle and released into the blood resulting in a marked acidosis (to pH 7.35) occurring within two minutes of the start of swimming at 20°C.

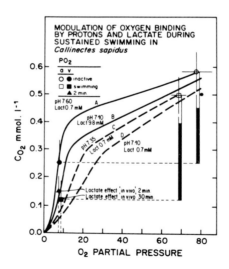

Figure 2. Modulation of hemocyanin O_2 binding 'in vivo' at 2 and 30 minutes during a period of sustained swimming in Callinectes sapidus at 20°C. Curves A and B drawn through PO_2 and CO_2 (small symbols) data points measured 'in vivo'. Curves C and D predicted from 'in vitro' data to fit 'in vivo' levels of pH and lactate concentration. Data from Booth et. al. (1982).

At this time, lactate concentration in blood is only 2.4 mM but this represents a 3.5 fold increase over quiescent levels, and since lactate effects are greatest at low concentrations (Mangum 1983b) this release of lactate is able to balance out the majority of the Bohr effect (Fig. 2). Over the remainder of a 30 minute exercise period, lactate increased to approximately 10 mM and pH fell to 7.1 but little change occurred in the location of the O_2 equilibria 'in vivo' (Fig. 2). The effect of lactate modulation here again is thus to stabilise the oxygen equilibrium curve preventing too large a decrease in O_2 affinity as a result of acidosis during exercise.

If C. sapidus are exercised to exhaustion (30 min) at higher temperatures (30°C, well within the natural range) the effects are more dramatic. We now see the cumulative effects of increase in temperature which decreases oxygen affinity both directly by change in the specific heat (ΔH) of the reaction and indirectly via the associated increase in protons (Bohr effect) as well as additional energy expenditure As above the results can be best visualised on oxygen equilibrium curves (Fig. 3). Curve A was constructed from data generated 'in vitro' for both proton and lactate effects to represent the conditions existent in hemolymph of animals resting at 30°C (pH 7.63 lactate 0.4 mM; Booth 1982). These animals were not masked and thus had lower circulating PO_2 levels than the unmasked animals described above (see also Mangum et. al. 1985). Mean levels of post-, and particularly prebranchial oxygen partial pressures measured 'in vivo' from quiescent animals at 30°C, fit closely to this predicted 'in vitro' curve.

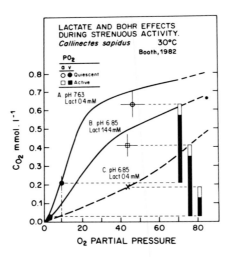

Figure 3. Modulation of hemocyanin O_2 binding 'in vivo' in Callinectes sapidus in exhausting exercise at 30°C. Reworked from data in Booth (1982).

In exercise the combined effects of increased temperature and activity result in severe acidosis ($\bar{x}$ venous pH falls to 6.85, Booth 1982). Under these conditions P_{50} predicted from the 'in vitro' relationships used for curve A is above 75 torr. Perhaps due to a combination of high oxygen demand and limitation of O_2 uptake at the gills, the animals are not able to elevate the oxygen partial pressures of hemolymph leaving the gills. Under these conditions, and the pronounced Bohr shift (Fig. 3A-C) hemolymph leaving the gills would be able only to reach 25% saturation.

In fact mean postbranchial oxygen content measured 'in vivo' at this time is elevated considerably above this level. The reason is almost certainly lactate accumulation in hemolymph since if one plots a curve for hemolymph adjusted to similar conditions to those of curve C but with 14 mM lactate (equivalent to 'in vivo' levels) added, this curve now closely approximates the 'in vivo' oxygen status. The adaptive value of the positive lactate effect is now much more easily seen. With lactate present at this level oxygen affinity is elevated sufficiently that postbranchial hemolymph can be over 50% saturated and an amount of oxygen equivalent to that delivered at rest can occur even during exhausting activity. The data suggest that limitation of oxygen delivery to tissues in high temperature activity occurs at the gill rather than at the level of Hcy function. Adequate performance of Hcy under these conditions of maximum effort is, however, dependent on elevation of O_2 affinity by lactate.

Regulation of oxygen uptake in different activity states thus clearly involves a balance between proton and lactate influences on Hcy oxygen binding. The balance is actually more complex than that described since both lactate and protons influence cooperativity as well as affinity (Booth 1982; Bridges et. al. 1984) and since other factors such as Ca^{++}, Mg^{++} and urate may influence either directly, or via the lactate/proton balance above (see Bridges and Morris, this volume). The magnitude of the lactate effect is clearly related to the Bohr effect (Mangum 1983b). Thus in animals such as Coenobita compressus (McMahon 1984; Wheatly and McMahon 1985) where the Bohr effect is very small, although hemolymph becomes acidotic during sustained voluntary running activity, the resulting Bohr shift does not have the severe implications seen in Callinectes at similar temperature. The absence of an appreciable lactate effect in this species (M.G. Wheatly pers. comm.) is thus not surprising. Fig. 4 demonstrates the role played by the Bohr effect in promoting increased tissue oxygenation similar to that described for Cancer magister above.

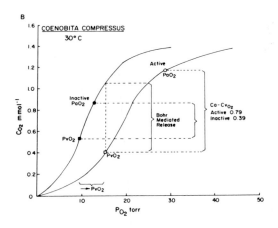

Figure 4. Proton modulation of hemocyanin O_2 binding during sustained walking in Coenobita compressus in the absence of a lactate effect. Figure from McMahon (1984).

Functioning of Hcy in the transition between aquatic and aerial respiration

Cancer productus, the Japanese or Red Rock Crab, is common on gravel and stony substrates in temperate West Pacific coastal waters. Like the Green Shore Crab, Carcinus maenas from Atlantic waters, smaller individuals of C. productus are often found exposed to air in the intertidal. An animal stranded by the tide may protect itself from the worst of the rigors of the aerial environment in several ways. In rocky areas, it may take shelter between rocks or under plant material and thus be exposed to air for the length of the intertidal period. When on softer substrates, however, the animals may be able to burrow into the substratum which may remain more or less moist. The animals' responses to these conditions differ (deFur et. al. 1983) and thus we will examine the effects of air exposure separately.

Despite their common occurrence in the littoral, circulating hemolymph PO_2 levels of C. productus fall sharply during air exposure. Postbranchial (PaO_2) and prebranchial (PvO_2) fall from 58-16 and 19-8 torr respectively during 4h emersion. Hemolymph PCO_2 levels also increase, and acidosis results from this as well as from an increase in hemolymph lactate, presumably resulting from increased anaerobic metabolism. Compensation for this acidosis may involve dissolution of internal calcium carbonate stores and hence an increase in hemolymph [Ca^{++}]. All of these substances, H^+, Ca^{++}, lactate and CO_2, have

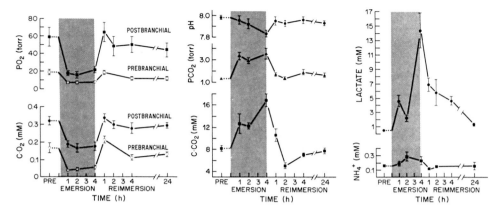

Figure 5. Time course of changes in circulating levels of O_2, CO_2, pH, lactate and NH_4^+ during and after 4 h air exposure in C. productus. Data from deFur et. al. (1980), deFur and McMahon (1984a, b).

been implicated in control of hemocyanin (Hcy) O_2 binding. The combined effects of these modulators on Hcy oxygen binding during air exposure can be seen on the oxygen dissociation curves plotted in Fig. 6. Curves shown are for C. productus at 10°C and, like those used previously, are located based upon 'in vivo' rather than 'in vitro' data. In this case the general form has been taken from 'in vitro' curves produced for this species by the mixing method (details in deFur 1980) but these curves have been fitted to the mean levels of pH and PO_2 and $[O_2]$ measured from samples taken 'in vivo' at different times during the experimental regime (Fig. 5A, B, and see deFur and McMahon 1984). P_{50} determined 'in vitro' for a pH equivalent to that of resting postbranchial hemolymph shows little discrepancy under resting normoxic conditions.

Immersed animals

The 'in vivo' situation for animals immersed in normoxic water (pH 7.97, PCO_2 1.4 torr, levels typical of small individuals of this species at 10°C) is represented by curve B in Fig. 6. Under these somewhat stressful experimental conditions, post- and prebranchial oxygen partial pressures are relatively high, post- and prebranchial oxygen contents are correspondingly high and thus, although oxygen delivery to tissues is normal (adequate) for this species at this temperature, a large proportion (63%) of oxygen delivered comes from the dissolved fraction while most of the bound oxygen remains in venous reserve. This is typical of slightly disturbed animals in experimental conditions as pointed out for C. magister above (and see McMahon and Wilkens 1983).

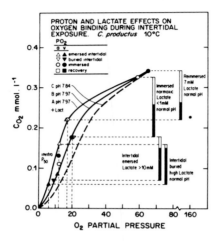

Figure 6. Modulation of hemocyanin O_2 binding before, during, and following air exposure in an intertidal crab Cancer productus. 'In vitro' oxygen binding characteristics from deFur and McMahon (1984a). 'In vivo' oxygenation, acid-base and lactate data from deFur and McMahon (1984a, b) and deFur et. al. (1983). Curve A represents lactate effects with pH compensation. Curve B represents conditions of immersed animals prior to air exposure. Curve predicts situation occurring during air exposure but without increase in lactate (lact).

Emersed animals

Following 4h air exposure in the laboratory, hemolymph had become acidotic (Fig. 5) and due to the moderate Bohr effect seen in this species ($\Delta \log P_{50}/\Delta pH = -1.0$, deFur and McMahon 1984a) we should expect oxygen affinity to decrease. Curve C (Fig. 6) represents a curve adjusted from Curve B using the relationship above to represent this new 'in vivo' pH (7.84). If we were to use this 'in vitro' curve together with the observed decrease in circulating oxygen partial pressures (Fig. 5A) to predict the situation for O_2 delivery 'in vivo' we would conclude that the carrier was almost completely depleted and the amount of O_2 delivered to tissues seriously reduced, a situation which would seem maladaptive as suggested generally for the Bohr shift in decapod crustaceans by Mangum (1983). Plotting the PO_2 and CO_2

levels measured 'in vivo', however, show this prediction to be false i.e., there is no correspondence between the measured levels and the predicted relationship (Fig. 6C). In fact the 'in vivo' points still lie very close to the original curve (6B) showing again the involvement of modulator substances other than protons acting to increase oxygen affinity, limit the expression of the Bohr effect and thus allow increased delivery of oxygen to the tissues, despite the evident oxygen depletion (Fig. 6B). Clearly, under these circumstances, hemocyanin plays a much greater role in tissue oxygen delivery.

As noted above, levels of at least 4 substances known to affect Hcy O_2 binding change during air exposure (Fig. 5). All are metabolites, except Ca^{++} which is often an indirect resultant of anaerobic metabolism in crustaceans. As in activity above lactate is the most likely potential modulator. Increases in hemolymph lactate content of 10-14 mM have commonly been reported for this species during air exposure (Fig. 5, from deFur and McMahon 1984; McMahon et. al. 1984). While the magnitude of the lactate effect has not been described for C. productus Hcy, if a sensitivity equivalent to that of other cancerid species (Truchot 1980; Mangum 1983) is assumed, the reported 13 mM increase would be more than sufficient to allow the adjustment observed (Fig. 6C-B).

Ca^{++}, another factor known to augment O_2 affinity in 'in vitro' experiments, also increases significantly in the hemolymph during air exposure. The increase in concentration in this particular experimental series reported (1.7 mM) is sufficiently small that a substantial direct effect may be discounted here. Much larger increases (11.7 mM) however, have been reported for this species during air exposure (deFur et. al. 1980) and thus this cannot be discounted in all cases. Additionally, it is now known (Bridges and Morris, this volume) that complex correlations exist between Ca^{++} and lactate effects on Hcy oxygen binding. These have not been investigated in this species. PCO_2 increases four-fold in emersion, however no CO_2 specific effect (see below) has been reported for Hcy of this or any other cancerid crab. In any case, since CO_2 was the buffer involved in the construction of the pH/P_{50} relationship used to position curve C, this factor would have been already taken into account as part of the normal complete Bohr shift.

Following air exposure upon reimmersion, oxygen levels rise rapidly reaching or exceeding pre-emersion values within 1h (Fig. 5). The rates of clearance of each of the effector substances above differ markedly (Fig. 5) and may play different roles as oxygen affinity changes during recovery. At 1h $[H^+]$ and PCO_2 are almost completely restored while [lactate] and $[Ca^{++}]$ remain elevated for 8-24h. Under these circumstances we would expect to see an increase in oxygen affinity (left shift) as the positive effects of lactate and Ca^{++} are freed from the opposition of protons. The location of 'in vivo' O_2 values sampled early in recovery (Fig. 6A) confirm an oxygen equilibrium located considerably to the left of that shown in the original curve (B) demonstrating the extent of this modulation. The resulting high O_2 affinity in combination with high perfusion and ventilation levels results in rapid recovery of hemolymph O_2 levels. Once this has been achieved, Hcy plays little role on O_2 transport, the amount of O_2 carried (per unit blood-flow) is reduced and the adaptive significance of this extended positive modulation is unclear. Incidentally, this is another reason for the very high O_2 levels found in decapod crustaceans following handling or disturbance (McMahon and Wilkens 1983).

Circumstances in which positive modulation of Hcy O_2 binding by lactate has a clear adaptive significance, however, do occur in the littoral. Not all individuals in the intertidal are equally air exposed. deFur et. al. (1983) examined animals stranded naturally on the beach and showed that often animals were able to bury into the substratum and, where the substrate was moist, some water was visible exiting the branchial cavities. Hemolymph samples taken from these naturally exposed animals showed an equivalent or greater oxygen depletion to that seen in air exposure, but no acidosis. Measurement of interstitial water PO_2 'in situ' confirmed that the limited water available to these buried animals was severely hypoxic but nonetheless allowed the animals to maintain acid-base balance. Fig. 6, curve A clearly shows the adaptive role of the positive modulators (lactate and perhaps Ca^{++}) produced in hypoxic conditions. Hemolymph sampled from these naturally buried animals had lower oxygen levels than those of animals air exposed in the laboratory experiments. Mean PvO_2 was 6 and mean PaO_2 was as low as 12-17 torr depending on the amount of water in the substratum (deFur et. al. 1983). When these data are plotted on oxygen dissociation curves produced above for an equivalent condition i.e., 10°C, high pH, high lactate (Fig. 6A), the enhanced O_2 affinity can be seen to increase O_2 uptake and delivery of a reasonable oxygen supply to the tissues despite the severe hypoxia. Here lactate and perhaps Ca^{++} are not acting simply to ameliorate the negative effects of H^+ released concomitantly. The H^+ is buffered separately, perhaps by ionic changes with the remaining branchial water but the end products of hypoxic metabolism allow adjustment of oxygen affinity and thus increased effectiveness of O_2 transport from the hypoxic interstitial water. An interesting parallel is seen with the situation described for three subtidal burrowing crabs described by Taylor et. al. (1985) all of which have potentially hypoxic habitats and show pronounced lactate effects.

The respiratory variables of these animals which had been trapped at Friday Harbor or at the Bamfield Marine Station, flown to Calgary and maintained for up to several months did not vary significantly in terms of any respiratory variable from animals trapped and tested within hours (deFur et. al. 1983). This contrasts with the work presented for <u>Carcinus maenas</u> by Houlihan and others at Leiden SEB conference several years ago, where a correlation between several indices of performance, such as endurance and the scope for activity with length of holding time were observed.

The situation seen during emersion in the green shore crab <u>Carcinus maenas</u> (Truchot 1975a; Taylor and Butler 1978; deFur 1980) is dissimilar to that described for the air-exposed animals above in that oxygen consumption can be maintained or even increased in air. Nonetheless the animals show depletion of circulating O_2 levels and acidosis of a magnitude similar to those observed in <u>C. productus</u> above. Oxygen affinity seems similar in both cases when compared at equivalent pH and temperature and thus one might reasonably expect the existence of a positive modulator system similar to that described above.

Possible events occurring at the level of Hcy O_2 binding are illustrated in Fig. 7. Curve A is plotted from the 'in vitro' O_2 binding data given by Truchot (1975a) for conditions typical of immersed animals (pH 7.82, PCO_2

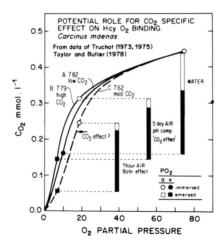

Figure 7. A potential role for a CO_2 specific effect on hemocyanin O_2 binding in <u>Carcinus maenas</u>. Curve A calculated for low CO_2 level and pH of immersed animals. Curve C predicted for initial air exposure conditions of respiratory acidosis. Curve B predicted for conditions of chronic air exposure after compensation for acidosis. 'In vitro' data from Truchot (1973, 1975a, b). 'In vivo' data for initial air exposure from Taylor and Butler (1978).

approx. 1 torr, 15°C; Truchot 1975b). Oxygen partial pressures reported for animals under similar conditions by Taylor and Butler (1978) plotted on this curve show high PaO_2 and a high delivery of oxygen from the soluble fraction, conditions typical of disturbed animals above. Emersion is initially accompanied by a sharp acidosis peaking at 1-4h (Truchot 1975b; deFur 1980) and marked reduction of both circulating O_2 levels (Taylor and Butler 1978) and of the a-v PO_2 difference. This combination, as we have seen before, allows greater participation of Hcy in oxygen uptake and transport but also involves substantial depletion of the venous reserve (Fig. 7C). Reasoning from the situation described for C. productus above, we could postulate modulators other than H^+ may be involved here. No significant increase in lactate occurs during air exposure in C. maenas (Butler and Taylor 1978) and, especially as the hemolymph of this species has reduced lactate sensitivity compared with that of the cancerid crabs (Truchot 1978; Mangum 1983) a significant lactate effect seems unlikely in air exposure. Several proposed factors may substitute for the absence of lactate. The first factor which allows restoration of O_2 affinity is active compensation for the acidosis which is evident early in this species and becomes virtually complete in long term air exposure (Truchot 1975b). A further factor however becomes evident here, PCO_2 which rises quickly upon emersion is not affected by the pH compensation above and remains highly increased over immersed levels. Earlier, Truchot (1973) reported a marked pH independent augmentation of O_2 affinity occurring by increase in PCO_2 in C. maenas. If we include such an effect on the oxygen equilibria of (Fig. 7) we see the resulting further increase in O_2 affinity in long term air exposure would have important effects allowing restoration of some venous reserve capacity and hence some scope for activity.

In fact, viewed in the present sequence, the occurrence of a CO_2 specific effect here in an intertidal species, which is modified to allow limited aerial oxygen uptake and which suffers no anaerobic metabolism in air is extremely attractive. With the acquisition of aerial O_2 uptake capability, emersion results in respiratory acidosis rather than metabolic (lact-) acidosis. Under these circumstances, a CO_2 specific effect could simply replace the lactate effect in offsetting the maladaptive features. As before, a product of emersion is utilised to improve O_2 uptake during emersion.

Reports of such CO_2 specific effects on O_2 binding are established for hemoglobin, although the effect is of opposite direction to that above. The existence of CO_2 specific effects in hemocyanin containing systems is controversial (reviewed by Burnett and Infantino 1983) but they have been implicated for other crab species (Young 1972; Arp and Childress 1981) and described for shrimp and prawn species (Weber and Hagerman 1981; Bridges et. al. 1984) as well as for non-crustacean Hcy systems (Mangum and Lykkeboe

1979). To my knowledge, this is the first tentative attempt to ascribe a
function for these effects. Hopefully, it may act to stimulate additional
work in this area.

If we continue to progress in series upwards through the littoral, we
should end up with some discussion of Hcy performance in the land-crabs, of
which much has been said, but very little decided in recent years. There is a
controversy as to whether the air breathing crabs have a hemocyanin which has
lower oxygen affinity than the standard range continues (i.e., see discussion
in Mangum 1980, 1983a, b). One of the problems here is that considerable
variability occurs in O_2 affinity within the air-breathing crabs. I would
like to illustrate, and perhaps explain, some reasons for this apparent
variance in affinity by reference to a number of examples of animals from
apparently similar habitats but with functionally different hemocyanins.

The Australian land crab Holthuisana transversa is interesting in that, in
contrast with most air-breathing crabs studied, it has entered the terrestrial
habitat from freshwater rather than the sea, and also is a truly bimodal
breather, equally at home for extended periods in air or in water, as when its
burrow is flooded (MacMillen and Greenaway 1978). Recently, Greenaway,
Burnett and myself (unpublished) examined the maintenance of hemolymph
oxygenation and acid-base status with variation in temperature in animals
acclimated to both water and air. For today's purpose, we need consider only
hemolymph oxygenation and acid-base status in animals confined either in air
or in water at 25°C.

Oxygen equilibrium curves constructed 'in vitro' for Holthuisana transversa
were published by Greenaway et. al. (1983a, b). Since H. transversa probably
spends the majority of its time in air, we will consider air breathing first.
As previously, we will attempt to fit data determined from measurements on 'in
vivo' samples on the existing 'in vitro' curves. The relevant set of curves
(A, Fig. 8) were constructed on hemolymph pooled from several animals in air
at pH 7.6 and 7.3 (Greenaway et. al. 1983b). The solid line represents an
adjustment of these curves to pH = 7.43, the mean pH level measured 'in vivo',
using the relationship $\Delta \log P_{50}/\Delta pH$ = -0.33 published by Greenaway et. al.
(1983a, b). If we plot 'in vivo' PaO_2 and PvO_2 data for animals acclimated to
air on this curve we see a situation, commonly observed in 'air breathing'
crabs, where circulating oxygen levels at rest are low and located on the
steep portion of the O_2 equilibrium curve (McMahon and Burggren 1979; Wheatly
et. al. 1984; Wheatly et. al. 1985). The amount of oxygen delivered to
tissues per unit blood flow is actually less in this species than in most of
the 'air-breathing' crabs since O_2 capacity is considerably lower in H.
transversa than in the 'marine' land crabs (see McMahon and Wilkens 1983 for
review), but is in line with the very low level of oxygen consumption seen in

this species (Greenaway et. al. 1983a, b). These authors recorded somewhat higher levels of oxygenation in air breathing H. transversa. Possibly the animals in the earlier study were more disturbed by sampling and other procedures or were becoming dehydrated, which causes increased circulating PO_2 in this species (Fig. 8).

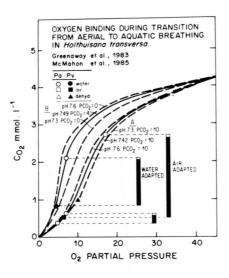

Figure 8. Change in oxygen binding characteristics of Hcy in an Australian Land Crab Holthuisana transversa during transition between aerial and aquatic breathing. Curve A, dashed lines redrawn from, solid line calculated from 'in vitro' data for air breathing group from Greenaway et. al. (1983b). Curve B, as above from water breathing group (Greenaway et. al. 1983a). Symbols, data from McMahon, Burnett and Greenaway (unpublished).

In animals confined underwater, oxygen partial pressures were further (significantly) depressed. In aerated water at 25°C, pH is similar (pH = 7.49) to that of air equilibrated animals above but PaO_2 falls to 7±2 and PvO_2 to 4±2.5 torr. If we plot these levels on the curve utilised above, O_2 delivery to tissues would be clearly insufficient. Since O_2 consumption in this species is elevated in water (Greenaway et. al. 1983a) some modulation of Hcy O_2 binding must be implicated here.

Greenaway et. al. (1983a) also produced curves for a separate group of animals breathing only water. These curves were produced from pooled hemolymph by the same method and at the same pH range but clearly are of markedly higher affinity. The 'in vivo' PO_2 levels plotted on these curves (Fig. 8B) show reduced but reasonable oxygen delivery to tissues. At first sight the only difference is PCO_2 which was 10 torr in the construction of curves A but <1 for curves B. Now this could be another interesting example of CO_2 specific activity, which again would fit nicely into the story developed to this point, except that the relationship here is inverted, i.e., in all other cases reported for crustaceans CO_2 augments rather than diminishes oxygen affinity! Also the two sets of curves in Greenaway et. al. study were produced from different groups of animals and levels of other modulator substances were not assessed.

Change in the hemolymph concentrations of several other modulator substances discussed above were measured in the later study but unfortunately offer no clear alternate solution. A small decrease in Ca^{++} and larger decrease in Mg^{++} accompanying 4d immersion in water would have only a small effect on O_2 binding and would cause effects of opposite direction than that observed. Hemolymph lactate levels were doubled in water breathing animals as compared with the airbreathing group (McMahon, Burnett and Greenaway unpub.) at 25°C, but the actual concentration difference is small (<1 mM). Mangum (1983) stresses the potential importance of small changes in lactate concentration. Unfortunately, the magnitude of the lactate effect is not known for this species. The Bohr effect however is relatively small (see above) and since large lactate effects are usually found in conjunction with large Bohr shifts (Mangum 1983) it seems unlikely that a shift of this magnitude could result entirely from the observed increase in lactate. Urate levels and possible effect on O_2 binding (Morris and Bridges, this volume) are unknown in this species. Thus we are left with either another 'unknown' factor or an inverted CO_2 specific effect to explain the change. Although the concept of an inverted CO_2 specific effect occurring at high CO_2 levels in air breathing forms is attractive, and is possible, clearly much work remains to be done and many other avenues tested before it can be accepted. It is interesting that all the modulating factors except protons so far discussed increase O_2 affinity. To achieve fully independent (from H^+) control over O_2 affinity, a modulator acting to decrease O_2 affinity would seem desirable.

A somewhat similar situation is seen in another animals with freshwater origins, the Red Swamp Crayfish *Procambarus clarki*. The air breathing credentials of this animal may come as a surprise to some but it is another burrowing form which survives long periods of drought. In an artificial burrow with water available it spends approx. 33% of its time voluntarily in air (Hankinson, Lorenzetti, and McMahon, unpublished) and may survive for a month in humid air at 25°C (McMahon and Stuart 1985), longer at lower temperatures (J. Huner, pers. comm.) with no access to water. The oxygen equilibrium curve in Fig. 9 was generated 'in vitro' using whole blood pooled from animals adapted to water breathing alone, circumstances which are typical

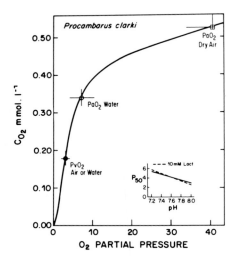

Figure 9. Oxygen binding by hemocyanin of the crayfish <u>Procambarus clarki</u>. Curve produced 'in vitro' using mixing method. Symbols are data for oxygen pressures and contents of post- and prebranchial hemolymph 'in vivo'.

for this animal when its habitat is flooded. Oxygen affinity is (P_{50} = 2.5 torr, pH 7.7, 20°C; McMahon 1984) higher than that for <u>H</u>. <u>transversa</u>. Additionally, the Bohr effect is small and this Hcy is apparently also without a lactate effect (Mangum 1983b). Blood samples taken from quiescent animals adapted to purely aquatic breathing again show very low circulating O_2 pressures (McMahon 1984; Stuart and McMahon 1984). Due to the high affinity Hcy these low 'in vivo' levels again fit astride the steep portion of the oxygen equilibrium curve. The high O_2 affinity is particularly appropriate for this species which often finds only hypoxic water either in the burrow or in their stagnant swamp environment. Similarly either water may become very hypercapnic, thus explaining the reduction in Bohr effect. It is of interest to note that when the animals are air exposed both oxygen and CO_2 partial pressures rise substantially. There are obvious parallels with the situation seen for <u>Holthuisana</u> above but in this species also more work is needed.

An essentially similar situation is also seen in a 'marine' species, the Land Hermit Crab <u>Coenobita brevimanus</u>. These animals carry a water store in their 'mobile homes' (McMahon and Burggren 1979) and are also bimodal breathers. The gills are reduced in area and circulating PO_2 levels are very low as in <u>Procambarus</u> and <u>Holthuisana</u> above, but again span the sharp portion of the O_2 equilibrium curve yielding an adequate amount of oxygen to tissues (Fig. 10). O_2 delivery is enhanced by high CO_2 capacity characteristic of

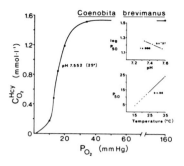 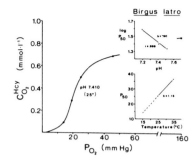

Figure 10. Oxygen binding characteristics of the hemocyanin of two closely related anomuran crabs <u>Coenobita brevimanus</u> and <u>Birgus latro</u> showing different oxygen affinities related to extent of dependence on aerial respiration.

this genus. Similar patterns are seen in 2 other species of <u>Coenobita</u> (<u>C</u>. <u>clypeatus</u>, McMahon and Burggren 1979; and <u>C</u>. <u>compressus</u> (Fig. 4), McMahon 1984; Wheatly et. al. 1984). In all cases the Bohr effect is small and in the only case investigated (<u>C</u>. <u>compressus</u>, Wheatly, pers. comm.) the lactate effect is negligible. There is no information on effects of other modulators in <u>Coenobita</u>.

The last species is a true air breathing crab, the Coconut or Robber Crab <u>Birgus latro</u>. Although very closely related to <u>Coenobita</u> this animal is perhaps as terrestrial a crab as can be found, with an exceptionally well developed 'lung' (Harms 1932) and very reduced gills (Cameron 1981). As an adult, <u>Birgus</u> cannot survive extended submergence in water and thus is not a bimodal breather. Associated with this dependence on aerial gas exchange it has an oxygen carrier of different O_2 binding properties from the other decapods listed above, showing a relatively low O_2 affinity (Fig. 10). Circulating blood PO_2 levels are higher than in the species above but, again because of the lower O_2 affinity, can be seen to sit easily on the shoulder of the curve in quiescent animals. Oxygen capacity is lower than that of <u>Coenobita</u> but relatively high for a decapod crustacean. Since this animal has rather less exposure to variation in environmental oxygen and CO_2 levels in its entirely aerial environment it also exhibits a larger Bohr effect ($\Delta\log P_{50}/\Delta pH = -0.9$) than the other 'air-breathing' crustaceans discussed above. Lactate effects have not been investigated. McMahon and Burggren (unpublished) showed differences between 'in vivo' and 'in vitro' hemocyanin oxygen binding characteristics for <u>Birgus</u> suggesting the occurrence of modulator effects, but made no attempt to assess the possible factors involved.

318

We have discussed three examples of animals, each with considerable air breathing potential but each having to rely on oxygen uptake from depleted water for some part of their life cycle. Hcy O_2 binding characteristics of these animals show high O_2 affinity in water. _Birgus latro_, a species very closely related to _Coenobita_ has a low O_2 affinity pigment, clearly well adapted to function in air. To this extent, it seems that _Birgus_ and _Coenobita_ differ in their respiratory habitat and thus have developed Hcy's of different O_2 binding properties. However, in the bimodal breathers described above, circulating O_2 levels rise in especially dry air. In _H. transversa_ O_2 affinity falls to maintain efficient Hcy function in air. This almost certainly results from the action of specific modulator substances, unfortunately of unproven identity at the time of writing. A question raised here is whether the O_2 affinity of all such animals is extremely variable depending on the circumstances. This would explain some of the immense variability seen in the literature. The larger question raised is whether the differences in Hcy performance between animals like _Coenobita_ and _Birgus_ has (or needs) a genetic origin or whether the action of modulator substances is sufficient for long term acclimation. A pointer is given in the work of Bridges et. al. (1984) who show that change in lactate concentration is sufficient to allow for the observed changes seen in seasonal acclimatisation in an intertidal prawn _Palaemon elegans_.

This account has concentrated on changes in the O_2 binding and delivery properties of Hcy. It is however obvious that the changes in O_2 affinity described here must also affect CO_2 binding and release by the Haldane effects now demonstrated for many crustaceans (Truchot 1976; Booth 1982; McMahon 1984; Morris et. al. 1984; Taylor et. al. 1985). Thus, where modulation affects the amount of O_2 bound and released from Hcy, similar effects on CO_2 transport and release occur and may be important in the animal's overall compensatory responses to increase in activity or to environmental change.

Literature Cited

Arp, A.J., and Childress, J.J., Science 214, 559-561 (1981).
Booth, C.E., Ph.D. Thesis, University of Calgary. Calgary, Alberta, Canada (1982).
Booth, C.E., McMahon, B.R., and Pinder, A.W., J. Comp. Physiol. 148, 111-121 (1982).
Bridges,C.R., and Morris, S., this volume, pp. 341-352 (1986).
Bridges, C.R., Morris, S., and Grieshaber, M.K., Resp. Physiol. 57, 189-200 (1984).
Burggren, W.W., and McMahon, B.R., J. Exp. Zool. 218, 53-64 (1981).
deFur, P.L., Ph.D. Thesis, University of Calgary (1980).
deFur, P.L., and McMahon, B.R., Physiol. Zool. 57(1), 137-150 (1984a).
deFur, P.L., and McMahon, B.R., Physiol. Zool. 57(1), 151-160 (1984b).

deFur, P.L., Wilkes, P.R.H., and McMahon, B.R., Respir. Physiol. 112, 247-161 (1980).

deFur, P.L., McMahon, B.R., and Booth, C.E., Biol. Bull. 165, 582-590 (1983).

Graham, R.A., Mangum, C.P., Terwilliger, R.C., and Terwilliger, N., Comp. Biochem, Physiol. 74A, 45-50 (1983).

Greenaway, P., Bonaventura, J., and Taylor, H.H., J. exp. Biol. 103, 225-236 (1983a).

Greenaway, P., Taylor, H.H., and Bonaventura, J., J. exp. Biol. 103, 237-251 (1983b).

Harms, J.W., Z. Wiss. Zool. 140, 167-290 (1932).

Mangum, C.P., in Structure and Function of Invertebrate Respiratory Proteins (Wood, E.J., ed), Life Chemistry Rep., Suppl. 1, pp. 333-352, Chur - London - New York, Harwood Academic Publishers (1983a).

Mangum, C.P., Mar. Biol. Lett. 4, 139-149 (1983b).

Mangum, C.P., in The Biology of Crustacea (Bliss, D.E., and Mantel, L.H., eds.), vol. 5, pp. 373-429, Academic Press (1983c).

Mangum, C.P., and Lykkeboe, G., J. Exp. Zool. 207, 417-430 (1979).

Mangum, C.P., McMahon, B.R., deFur, P.L., and Wheatly, M.G., J. Crust. Biol. 5, 188-206 (1985).

McDonald, D.G., McMahon, B.R., and Wood, C.M., J. exp. Biol. 79, 47-58 (1979).

McMahon, B.R., in Structure and Functions of Respiratory Pigments (Lamy, J., and Truchot, J.P., eds.), Springer-Verlag (1985).

McMahon, B.R., and Burggren, W.W., J. exp. Biol. 79, 265-281 (1979).

McMahon, B.R., and Wilkens, J.L., in The Biology of Crustacea (Bliss, D.E., and Mantel, L.H., eds.), vol. 5, Academic Press (1983).

McMahon, B.R., McDonald, D.G., and Wood, C.M., J. exp. Biol. 80, 271-285 (1979).

McMahon, B.R., Burnett, L.E., and deFur, P., J. Comp. Physiol. 145B, 371-383 (1984).

MacMillen, R.E., and Greenaway, P., Physiol. Zool. 51, 231-240 (1978).

Morris, S., Taylor, A.C., Bridges, C.R., and Grieshaber, M.K., J. Exp. Zool. 233, 175-186 (1984).

Morris, S., and Bridges, C.R., this volume, pp. 353-356 (1986).

Stuart, S., and McMahon, B.R., Ist International Congress of Comparative Physiology, Liège, August 1984, Abstract A 71.

Taylor, E.W., and Butler, P.J., J. Comp. Physiol. 127, 315-323 (1978).

Taylor, A.C., Morris, S., and Bridges, C.R., submitted, J. Comp. Physiol. (1985).

Truchot, J.P., C.R. Acad. Sci. Paris 276, 2965-2968 (1973).

Truchot, J.P., Resp. Physiol. 24, 173-189 (1975a).

Truchot, J.P., Respir. Physiol. 23, 351-360 (1975b).

Truchot, J.P., J. Comp. Physiol. 112, 282-293 (1976).

Truchot, J.P., J. exp. Zool. 214, 205-208 (1980).

Weber, R., and Hagerman, L., J. Comp. Physiol. 145, 21-27 (181).

Wheatly, M., McMahon, B.R., Burggren, W.W., and Pinder, A.W., submitted, J. exp. Biol. (1985).

Young, R.E., J. Exp. Mar. Biol. Ecol. 10, 183-192 (1972).

GAS EXCHANGE AND GAS TRANSPORT IN THE TARANTULA EURYPELMA CALIFORNICUM - AN OVERVIEW

R. Paul

Zoologisches Institut, Universität München, Luisenstr. 14, 8000 München 2, F.R.G.

Gas transport in spiders is influenced by two peculiar features: (i) The unique structure of the so-called book-lungs which are essentially a stack of very flat invaginations - the sacculi - from the air space (Fig. 1,a). The matter of debate is whether these are ventilated or whether gas exchange is simply by diffusion. (ii) The blood does not only serve as a medium of convection for dissolved gases etc., but also as a hydraulic fluid for leg extension. So, hydraulic function interferes with transport function.

We have studied these two aspects of spider physiology in the tarantula, Eurypelma californicum. By combining our results on the respiration and circulation of this animal with data on blood P_{O_2} and pH obtained by Angersbach (1) and by Loewe and Brauer de Eggert (2), we are beginning to understand oxygen transport in this species .

For more than fifty years it has been discussed whether spiders ventilate their book-lungs or not. Several observations have been made which indicate ventilatory movements but their amplitude has not been determined, not even in the order of magnitude. We have measured 'ventilatory' volume changes in the lungs by a micro-barometric method (3,4,5), but their amplitude is very small. Several experiments (5) indicate that they do not represent effective ventilation. I conclude that oxygen transport into the book-lungs occurs essentially by diffusion. Using anatomical data on Eurypelma book-lungs (Focke 1981) it can also be demonstrated theoretically that diffusion suffices for oxygen transport. By putting into Fick's diffusion equation a) the measured oxygen consumption rates, b) the known anatomical dimensions (and an assumed 0.3 μ for cuticle thickness, cf. (6,7)), c) the hemolymph P_{O_2} values determined by Angersbach (1), and d) the diffusion constants for oxygen in air and in chitin, one can calculate the P_{O_2} values along the pathway of oxygen from outside air through the lung slits, atrium and sacculi to the hemolymph, at rest and during recovery (Fig. 1).

Invertebrate Oxygen Carriers
Ed. by Bernt Linzen
© Springer–Verlag Berlin Heidelberg 1986

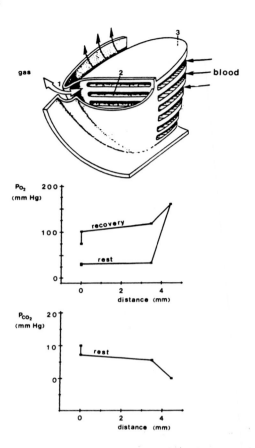

Fig. 1, a. Schematic representation of spider lung with stigma (lung slit, left), atrium and sacculi with alternating air and blood spaces. b. Calculated P_{O_2} values along the diffusion path from lung slit (Point 1) to the tips of the sacculi (Point 3). P_{O_2} at Point 1 is assumed to be 160 Torr for our calculation. In the blood, Angersbach (1) has measured a P_{O_2} of ca. 28 Torr at rest, and of 74 Torr during recovery. Assuming a diffusion barrier of 0.3 μ between sacculi space and blood, the P_{O_2}'s at Point 3 are 31 and 102 Torr, respectively. At Point 2, one arrives at 32 Torr at rest, and 119 Torr during recovery.
c. For P_{CO_2}, an analogous calculation can be made only for the resting state, using the values measured by Loewe and Brauer de Eggert (2): 10 Torr in the arterial blood and practically 0 Torr in the atmosphere.

From these calculations, three conclusions can be drawn: (i) Diffusion is sufficient to account for the measured oxygen consumption rates at rest and even during recovery. At least at rest this also holds true for carbondioxide. (ii) The main diffusion barrier is the cuticula of the sacculi. In Tegenaria its thickness has been estimated at 0.2 μ (7). Diffusion in air is no problem over a small distance (up to 10 mm). (iii) At rest the stigmata can be closed almost totally without impairment of oxygen supply. After onset of activity the opening of the stigmata raises the P_{O_2} in the atrium and subsequently in the blood. The important point is that the stigma and the atrium function to control the gas exchange by changing the geometry of the path of diffusion.

Next we have to consider the hydraulic mechanisms of the tarantula. A flattening of the prosoma produces blood pressures up to 300 Torr, and with the hemolymph serving as hydraulic fluid, leg extension is performed. Fig. 2 shows parallel recordings of heart amplitude and blood pressure in the prosoma. At low pressures cardiac activity is very regular. During activity, heart amplitude is reduced, and sometimes the heartbeat stops completely. At a blood pressure of 90 Torr heart activity and consequently, oxygen supply are strongly diminished. This is also indicated by the venous P_{O2} which sometimes drops to 0 Torr (1).

rel. heart amplitude

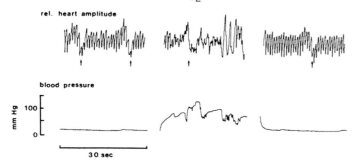

blood pressure

mm Hg

100

0

30 sec

Fig. 2. Relative heart amplitude (upper trace) and blood pressure in the prosoma of the tarantula, Eurypelma californicum. Deflections of the upper trace and gaps marked by arrows are artifacts caused by data processing.

It was found that the increase of heart frequency from resting to recovery level follows a typical pattern. The increase begins already during activity but the maximum is reached only after 2-5 minutes. Heart frequency remains high for more than 1 hour. From these findings and additional experiments it can be concluded: (i) During strong activity the high blood pressures in the prosoma interfere with or even interrupt perfusion and hence oxygen supply to the prosoma and its appendages. (ii) Since the demand for oxygen is high with the onset of activity, the relatively slow increase of heart frequency must lead to an oxygen deficit. (By measuring directly $\dot{V}_{O2}$, Herreid (1982) and Anderson and Prestwich (1985) reported also an oxygen deficit in tarantulas.)

According to our data and to data of Angersbach three different metabolic states of Eurypelma can be distinguished: (i) The **resting state** which is characterized by low oxygen consumption (minimum 0.3 ml O_2 h^{-1} for a 15 g Eurypelma), low perfusion rate (minimum heart frequency: 8 min^{-1}), low blood P_{O2} values and a blood pH of about 7.5. The stigmata are nearly closed. (ii) During **activity** the heart frequency begins to rise, however, at higher blood pressures in the prosoma, perfusion and therefore, oxygen supply of the prosoma comes to a halt. In addition, gas exchange at the stigmata may sometimes be interrupted at the onset of activity because the stigmata are being closed (5). All these factors must lead to an oxygen debt. During maximal activity glycolysis is probably the main energy source (see also (10)). In Brachypelma, Anderson and Prestwich (1985) have found a strong increase of blood lactate after onset of activity which is probably the reason for the drop of blood pH. In Cupiennius, Linzen and Gallowitz (1975) found that the leg muscles have only few mitochondria, and the enzyme pattern in both leg and prosoma muscles is strongly glycolytic. (iii) During **recovery** the blood lactate must be removed and the glykogen reservoir restored. Therefore the metabolism must be high. The recovery state is characterized by high oxygen consumption (maximum 4 ml O_2 h^{-1}), high perfusion rate (maximum heart frequency: 90 min^{-1}), high blood P_{O_2} and very low blood pH values (1). The stigmata are open. Arterial P_{O_2} and heart frequency have a typical course when passing from resting to recovery state: a fast rise in the first minute followed by a slower increase reaching a maximum in 2-5 minutes. The opening of the stigmata reduces the diffusion resistance at that point and the rise of arterial P_{O_2} ensues from the increase of P_{O_2} in the atrium and in the sacculi. Because of the higher P_{O_2} gradient from the lung air space to the blood, the diffusional flux is increased. In parallel the perfusion rate increases and also the transport capacity of the hemolymph due to the higher saturation of hemocyanin.

If another activity bout occurs within the recovery state, both venous and arterial P_{O_2}'s decrease (1). The drop in arterial P_{O_2} can be explained by the fact that P_{O_2} in the sacculi is already maximum and the drop of venous P_{O_2} cannot be compensated by (higher) ventilation. But the return to the previous venous P_{O_2} values is faster in the recovery state because diffusional oxygen flux is higher. The arterial pH shows a short drop after activity but soon reaches again, and even passes the starting value (1). The venous pH drops more sharply and during a time-span of about 20 to 40 minutes a high arteriovenous pH difference is maintained. Because the time course of arterial pH return mirrors the time course of expired CO_2 it is probable that the drop of pH is compensated by the release of CO_2 (Fig. 3). In this context it is interesting to note that CO_2 has a strong influence on the width of the stigmata.

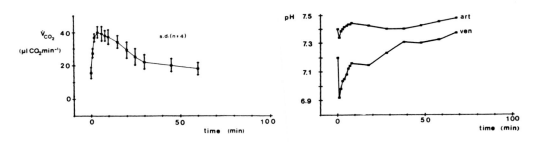

Fig. 3. Time course of CO_2 expired (left diagram: data from 4 animals) and hemolymph pH (right diagram; data of Angersbach (1)) after an activity bout.

Finally, the role of hemocyanin in gas transport in the tarantula can be assessed by comparing the aforementioned values with hemocyanin-oxygen equilibrium curves obtained by Loewe(12) and A. Savel (unpublished). As already shown by Angersbach (1) only 50 percent of the blood oxygen transport capacity is utilized during

rest. After activity, arterial P_{O_2} rises quickly to a level beyond 60 Torr, and
the hemocyanin becomes almost completely saturated. At the beginning of recovery
the venous P_{O_2} values are very low (1). That means that the transport capacity of
the hemocyanin is almost totally used. The following rise of venous P_{O_2} seems to
indicate a lower utilization of the blood oxygen transport capacity, but this is
not so because the very low venous pH (7 to 7.2) will shift the oxygen dissociation curve of the hemocyanin to the right. So it can be concluded that the capacity of the hemocyanin is utilized to a high extent even at higher venous P_{O_2}
values. In accordance the time-span of high arterial P_{O_2} (and also high oxygen
consumption) is more or less identical (about 20 min) to the time-span of high ar-
terio-venous pH difference.

The data gathered for the tarantula by several works allow to compare the respiratory and circulatory functions of this animal to those of an equally-sized mammal
(Table 1).

Table 1. Physiological comparison

	15 g <u>Eurypelma</u> (at 22 °C)		15 g eutherian mammal (37 °C)	
	rest	maximum	rest	maximum
$\dot{V}_{O_2}$ (ml h^{-1})	0.4	4	29[13]	190[14]
Heartrate (min^{-1})	20	90	500[15]	1200[13]
Cardiac output (ml blood min^{-1})	0.7[x]	4.3[x]	6.2[13]	21[x]
a-v difference of O_2 (μl O_2/ml blood)	9.3[1]	15.6[1]	77[x]	150[xx]
Respiratory pigment (mg/ml blood)	53[1] (hemocyanin)		150[13] (hemoglobin)	
Molecular weight of O_2-carrying protein per binding site (g)	70,000		16,000	

[x] calculated using Fick's principle
[xx] at a venous P_{O_2} of about 40 Torr

First of all, tarantula blood has a markedly lower oxygen transport capacity, which is related to the fourfold higher molecular weight of the O_2-carrying protein (70,000 vs. 16,000 per subunit) and to the much lower concentration of the respiratory protein. Secondly, whereas the ratio of oxygen consumption (eutherian mammal : _Eurypelma_) is about 73 at rest and 48 at maximum $\dot{V}_{O_2}$, the ratio of cardiac output is about 9 at rest and 5 at maximum $\dot{V}_{O_2}$. The ratio of cardiac output is about 9 times smaller than the ratio of $\dot{V}_{O_2}$. From the cardiac output and heart frequency the stroke volume of the _Eurypelma_ heart can be calculated to be 35 to 48 µl. This volume is in agreement with an experimental estimate which is about 40 µl. This in turn confirms that the calculated cardiac output is in the right order of magnitude.

So it is evident that the low oxygen transport capacity of the tarantula blood (ratio eutherian mammal : _Eurypelma_: about 9) is compensated by a relatively high perfusion rate, which is mainly achieved by a relatively large stroke volume and not by an extraordinarily high heart rate.

Acknowledgement. This work was supported by the Deutsche Forschungsgemeinschaft (Pa 308/1-1, Li 107/24-6).

1. Angersbach, D., J. Comp. Physiol. _123_, 113-125 (1978).
2. Loewe, R., and Brauer de Eggert, H., J. Comp. Physiol. _134_, 331-338 (1979).
3. Paul, R., Fincke, T., Linzen, B., First Congress of Comparative Physiology and Biochemistry, Liège (Belgium), (1984).
4. Fincke, T., Diplomarbeit, Zoologisches Institut der Universität München (1985).
5. Fincke, T., Tiling, K., Paul, R., Linzen, B., this volume, p.327 - 331 (1986).
6. Focke, P., Diplomarbeit, Zoologisches Institut der Universität München (1981).
7. Strazny, F., and Perry, S.F., J. of Morphology _182_, 339-354 (1984).
8. Herreid, C.F., Energetics of pedestrian arthropods. In: Herreid, C.F., Fourtner C.R. (eds), Locomotion and energetics in arthropods. Plenum Press, New York (1982).
9. Anderson, J.F., and Prestwich, K.N., J. Comp. Physiol. B _155_, 529-539 (1985).
10. Prestwich, K.N., Physiol. Zool. _56_, 122-132 (1983).
11. Linzen, B., and Gallowitz, P., J. Comp. Physiol. _96_, 101-109 (1975).
12. Loewe, R., J. Comp. Physiol. _128_, 161-168 (1978).
13. Schmidt-Nielsen, K., Why is animal size so important? Cambridge University Press,, Cambridge, New York, Melbourne (1984).
14. Dejours, P., Principles of comparative respiratory physiology. Elsevier/North-Holland Biomedical Press, Amsterdam (1981).
15. Prosser, C.L., Comparative animal physiology. W.B. Saunders Company, Philadelphia, London, Toronto (1973).

RELATIONS BETWEEN RESPIRATION AND CIRCULATION IN THE TARANTULA, EURYPELMA CALIFORNICUM

T. Fincke, K. Tiling, R. Paul, and B. Linzen
Zoologisches Institut, Universität München, Luisenstr. 14, 8000 München 2, F.R.G.

Although numerous investigations have dealt with respiration and circulation in spiders, there are still many aspects which are poorly or not at all understood. The present paper - in conjunction with the accompanying paper by Paul (1) - focuses on (i) the function of the book-lungs and, (ii) the relation between hemolymph circulation as a means of gas convection and hemolymph serving as a hydraulic fluid for leg extension. Both organs - the lungs and the hemolymph - are working in series to supply the spider's tissues with oxygen, and their performance must be matched to each other and to the requirements of the different tissues. Only if the convective functions in gas exchange are understood will it be possible to fully appreciate the molecular functions of the respiratory pigment in spiders, hemocyanin.

1. Respiration

In spite of Willem's observation of heart-synchronous movements of the book-lungs (2), both Kaestner (3) and Krogh (4) regarded them as diffusion lungs. More recently, Hill (5) repeated Willem's visual inspection and proposed a "hemolymph bellows theory" for the mechanism of ventilation of the sacculi (the leaf-like air spaces). We have tested this hypothesis using the large tarantula, Eurypelma californicum, by attaching small (1 ml) manometers equipped with pressure transducers, (i) over the lung opening (a special preparation guarantees that only volume changes in the lungs were measured) and, (ii) to the cuticle overlying the heart to record heart activity. The spider was held in place by a clamp but was free to move its legs on a styrofoam ball (6). The transducers allowed to record both rapid pressure changes (ventilation) and the gradual fall in pressure due to oxygen consumption. Fig. 1 gives examples of three types of ventilatory pattern: Type I are heart-synchronous fluctuations with an amplitude of not more than 0.5 % of total lung volume (the latter estimated to be 10 µl per lung in a 15 g animal). These are typical of the resting state. Type II are non-heart-synchronous fluctuations with lower frequency (ca. 6-10 min^{-1}), occurring mainly in the recovery state. Type III are non-heart-synchronous fluctuations with normally much lower frequency and maximum amplitude of 5 % of total lung volume. All these fluctuations are probably due to movements of the posterior atrium wall, caused by cardiac activity (type I) or by the atrium retractor muscle (type III). One can observe directly heart-synchronous variations of the width of the stigmata, and heart-synchronous pulsations of the atrium wall. It is also possible to observe that the non-heart-synchronous ventilation runs parallel with movements of the posterior border of the stigma.

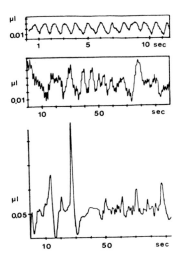

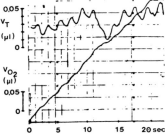

Fig. 2. Dual recording of pressure ($\hat{=}$ volume) changes in a tarantula lung. Lower trace, unfiltered signal showing (in upward direction) decrease of gas volume with shallow "ventilation ripples" superimposed. Clearly, O_2 consumption is much greater than ventilatory volume changes. Upper trace, signal filtered through a high pass ($f_C = 0.2$ Hz), and further amplified 2.5x to show ventilation (essentially Type I).

Fig. 1. Three recordings of ventilatory pressure ($\hat{=}$ volume) changes in tarantulas. A micromanometer equipped with a pressure transducer was attached over one of the four lung slits. Top: Type I ventilation, very small (ca. 0.02 µl) breaths synchronous with the heartbeat. Centre: Type II, non-heart-synchronous movements superimposed on Type I. This is typical for the recovery state. Note different time scale. Bottom: Type III, very strong excursions, in most cases occurring immediately after activity.

Since a likely mechanism by which the sacculi could be ventilated, is a dilation caused by Bernoulli forces of the streaming hemolymph, we have calculated these forces from the dimensions of the lung hemolymph space and an estimate of cardiac output, and have found that such forces would be much too low to detach the upper sacculus integument from the cuticular struts, since the integument is pressed against the struts by hydraulic pressure in the opisthosoma (about 10 Torr at rest and higher during activity). Ventilatory volume changes at the level of the sacculi can thus be discarded. Fig. 2 shows a dual recording of the pressure change in one lung (resting state): The high pass filtered signal reflects tidal volume, the unfiltered signal, O_2 uptake with ventilation superimposed. It is evident that tidal volume is much smaller than the volume of oxygen consumed during one breath, clearly indicative of a diffusion lung. Typically, the tidal volume is 1/3 to 1/5 of the O_2 volume consumed per breath (in man the relation is 25 : 1!). There is also no correlation between tidal volume and O_2 consumed during each particular breath. The foregoing considerations do not hold always for the Type II and Type III ventilations, where sometimes the fluctuations are higher than the rate of pressure decrease (Fig. 3a, recovery period). But not in all experiments the higher V_{O_2} after activity is correlated with Type II or Type III ventilation (Fig. 3b, recovery period).

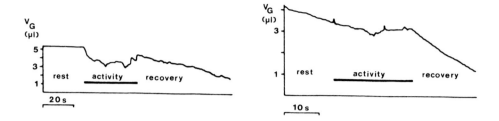

Fig. 3. Two recordings of pressure ($\hat{=}$ volume) changes from a tarantula lung. Note steeper slope during recovery (repayment of oxygen debt) and irregular changes during activity, perhaps due to circulatory perturbation. Net O_2 uptake appears to be zero during this period.

Fig. 3 shows two examples of the pressure changes at rest, during activity and recovery. $\dot{V}_{O_2}$ is clearly higher during recovery than at rest as shown by the steeper slope, indicating repayment of an oxygen debt. During activity there are irregular pressure changes, but overall no net uptake of oxygen. In this phase circulation is perturbed and also transient closure of the stigmata is observed especially at the onset of activity.

2. Circulation

The hemolymph serves a dual role in spiders: transport, and hydraulic transmission of force during leg extension. The latter is bound to interfere with the former; if tissue perfusion is thus interrupted the provision of energy must switch to anaerobic mechanisms. Wilson and Bullock (7) have put forward the hypothesis that the rapid fatigue observed in many spiders is caused by a circulatory collapse: during activity the prosoma would loose blood to the opisthosoma due to the high pressures generated in the prosoma, and pumping activity of the heart would not suffice to restore the prosomal blood supply in order to sustain the mechanism of hydraulic leg extension. We have measured heart activity, blood pressure in the anterior and posterior parts of the body and two other parameters which are related to the gross displacement of blood: carapace depression, which causes the high pressures in the prosoma during activity, and volume and diameter changes of the abdomen. Heart (pericardium) contractions were observed by illuminating the heart with cold laser light and recording the movements with a video camera. Evaluation from the monitor screen was by means of several solar cells attached

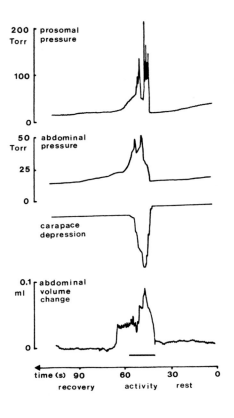

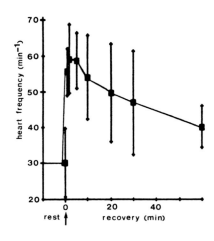

Fig. 4. Time course of heart frequency in tarantulas after 15-60 sec of locomotory activity. Means + S.D., data from 13 animals. Start of external stimulation is indicated by arrow.

Fig. 5. Parallel recording of prosomal and abdominal pressure (top) and of carapace depression and abdominal diameter change (bottom). The measured diameter change corresponds to a volume change of 0.1 ml. The time axis runs from right to the left.

to the video screen on the image contours, and an automatic recording set-up. For the blood pressure, syringes were introduced and connected to pressure transducers. Carapace depression was measured using an inductive displacement transducer positioned on the carapace, and the volume changes of the abdomen by a plethysmographic method or diameter changes by a stethographic method. Heart rate rises almost immediately after onset of activity and reaches a maximum after 2-5 min. The subsequent decline to resting values takes an hour (Fig. 4). In some cases, the heart stops beating completely during activity. This occurs at about 80 Torr of prosomal pressure (1), but if the animal fatigues, the critical value drops to 40 Torr. It should be remembered that normally the heart pumps blood in both directions, anterior and posterior. We have observed that the proportion of anteriorly and posteriorly directed blood is variable. After the onset of activity, more blood flows backward than before. After the end of an activity period, the anteriorly-directed blood flow increases again strongly.

In the open circulatory system of spiders, pressure in the different body regions is the resultant of four components: activity of the heart, the musculi laterales of the prosoma, the sub-cuticular muscle sheet and the dorso-ventral muscles of the abdomen (8), and the elasticity of the integument. Keeping this in mind it is easy to understand the parallel recordings of Fig. 5. With onset of activity the carapace is pulled down, leading to high prosomal pressure. Indeed a small quantity of blood (less than 0.1 ml, lowest trace) flows back from the prosoma into the opisthosoma. This alone would not suffice to raise also the abdominal pressure, since the opisthosoma is very distensible, but at the same time the abdominal muscles contract to raise the pressure. These observations show that a shift of hemolymph from the prosoma to the opisthosoma is negligible, and "circulatory failure" is unlikely as the cause of fatigue (see also (8)). But it is clear that the regular pumping of blood from the opisthosoma to the prosoma is interrupted for a certain period with consequence of impaired supply with oxygen and other solutes. This would not only explain the venous P_{O_2} falling to almost zero, but also the extremely steep fall in venous pH after activity (6).

In conclusion, the present work has shown that the book-lungs of Eurypelma function as diffusion lungs, and although volume changes in the lungs are measurable, these have virtually no effect on O_2 uptake. This shifts the regulation of O_2 supply (i) to the control of the lung slits and (ii) to the rate of perfusion. During strong locomotory activity which is dependent on hydraulic force transmission, blood flow to the prosoma comes to a temporary halt, which makes it likely that energy supply becomes dependent on the remaining oxygen stores (hemocyanin-O_2) and subsequently on anaerobic mechanisms (9).

Acknowledgement. This work was supported by the Deutsche Forschungsgemeinschaft (Li 107/24-6, Pa 308/1-1).

1. Paul, R., this volume, p. 321-326 (1986).
2. Willem, V., Arch. Neerl. Physiol. Homme 1, 226-256 (1917).
3. Kaestner, A., Z. Morphol. Ökol. Tiere 13, 463-558 (1929).
4. Krogh, A., The comparative physiology of respiratory mechanisms, p. 55-61, Dover, New York (1941).
5. Hill, D.E., Peckhamia 1, 41-44 (1977).
6. Angersbach, D., J. Comp. Physiol. 123, 113-125 (1978).
7. Wilson, R.S., and Bullock, J., Z. Morphol. Tiere 74, 221-230 (1973).
8. Anderson, J.F., and Prestwich, K.N., Z. Morph. Tiere 81, 257-277 (1975).
9. Prestwich, K.N., Physiol. Zool. 56(1), 122-132 (1983).

CRAB HEMOCYANIN FUNCTION CHANGES DURING DEVELOPMENT

Nora B. Terwilliger, Robert C. Terwilliger and Robert Graham
Department of Biology, University of Oregon
Eugene, OR 97403, USA
Oregon Institute of Marine Biology
Charleston, OR 97420, USA

The megalops and first instar juvenile stages of the Dungeness crab, Cancer magister, contain hemocyanin whose subunit composition differs from that of the adult crab (1). The larval and early post-larval hemocyanins resemble adult hemocyanin in molecular weight, appearance in electron micrographs and in copper content. However, adult 25S hemocyanin contains one subunit not seen in either megalops or juvenile hemocyanin; in addition, the relative amounts of the other subunits are not the same between adult and larval stages. The 16S larval hemocyanin also differs in subunit composition from that of the adult. Thus, the hemocyanin of C. magister undergoes an ontogenic change in structure as the crab develops from megalops to adult.

An important question is whether differences can also be found in the functional properties of these molecules. In this paper, we compare the O_2 binding properties of purified hemocyanin from megalops, first instar juvenile and second instar juvenile C. magister with those of the adult crab hemocyanin.

Hemolymph was collected from adult, megalops and juvenile crabs as described previously (1). The blood was fractionated on a BioGel A-5m (200–400 mesh) column (1.8 x 110 cm), equilibrated with 0.05 ionic strength Tris-HCl (pH 7.5), made 0.1 M in NaCl, 10 mM in $MgCl_2$ and 10 mM in $CaCl_2$. The 25S hemocyanin fraction was pooled and used for O_2 binding studies. Second instar juvenile hemocyanin was assayed as in (1) and found to have the same subunit composition as megalops and first instar juvenile hemocyanin. Samples were dialyzed against either the A-5m column buffer or a saline consisting of 455 mM NaCl, 11 mM KCl, 13 mM $CaCl_2$, 18 mM $MgCl_2$ and 22 mM Na_2SO_4 buffered with 0.05 I Tris-HCl (2). Adult and juvenile hemocyanin samples were treated identically for purposes of direct comparison.

O_2 equilibrium curves of purified hemocyanin were measured at $20^{\circ}C$ using a tonometric method (3) with a Zeiss PMQ II spectrophotometer equipped with a temperature-controlled cell holder. It was necessary to carefully monitor pH because of the hemocyanin's striking Bohr effect. Experiments in the presence of lactate were carried out as described in Graham et al (4).

The O_2 affinities of purified 25S hemocyanin from megalops, first instar and second instar juvenile crabs appear to be identical. The O_2 affinities of megalops and early juvenile crab hemocyanins are about 50% lower than that of adult C. magister under the same conditions of purification, buffer and oxygen binding pro-

Invertebrate Oxygen Carriers
Ed. by Bernt Linzen
© Springer-Verlag Berlin Heidelberg 1986

CRAB HEMOCYANIN FUNCTION CHANGES DURING DEVELOPMENT

cedures. A comparison of the O_2 binding curves of first instar juvenile and adult hemocyanins at pH 7.85 is shown in Fig. 1. The O_2 affinities of hemocyanin from

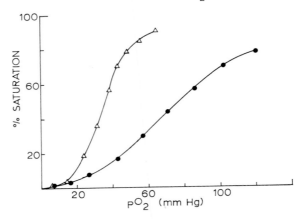

Fig. 1. Oxygen binding curves of C. magister adult and juvenile hemocyanins, pH 7.85, 20°C. △, adult; ●, juvenile.

all stages studied were measured over a pH range of 7.4-8.2, which includes the physiological range measured in the adult crab, and in the two different buffer systems mentioned above. Hemocyanins from all of these stages of development show a strong Bohr effect with ∅ = -1.0 to -1.2; the Bohr curves of the megalops and juvenile hemocyanins are approximately parallel to those of the adult hemocyanin.

Adult C. magister hemocyanin shows an increase in O_2 affinity in the presence of lactate (4). Megalops and first instar juvenile hemocyanin also show a lactate effect. Fig. 2 illustrates the increase in O_2 affinity as a function of lactate concentration in first instar juvenile hemocyanin. The O_2 affinities of megalops and first instar hemocyanin, like adult hemocyanin, are most sensitive to changes in concentrations of lactate at lactate levels found in the adult crab, approx. 0.05 - 5 mM lactate (4). The blood lactate concentration in individual megalopa was not measured due to the small size of the animal and the small volume of the blood sample. The magnitude of the lactate effect can be calculated from a log P_{50} vs log (lactate) plot. The slopes of the lines, -.291 for juvenile and -.287 for adult hemocyanin (20°C) are very similar, indicating that both hemocyanins respond comparably to L-lactate.

This study is the first to show a functional change in hemocyanin at different stages of crab development. The O_2 affinities of hemocyanins of megalops, juvenile and adult C. magister share similar sensitivities to H^+ ion and L-lactate. Yet the O_2 affinity of megalops and juvenile hemocyanin is markedly lower than that of adult

CRAB HEMOCYANIN FUNCTION CHANGES DURING DEVELOPMENT

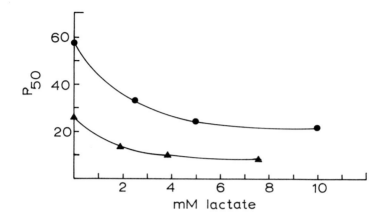

Fig. 2. Graph showing the effect of lactate concentration on the oxygen affinity
of C. magister hemocyanin. ▲ , adult; ● , juvenile.

crab hemocyanin. This same difference in O_2 affinities is present in C. magister
physiological saline as well as in a purification buffer. Thus the changes in C.
magister hemocyanin which occur during development are functionally as well as
structurally significant. The questions we now ask are when during development
does adult hemocyanin first appear, and how does the lower O_2 affinity of larval
and juvenile hemocyanins affect O_2 transport.

Acknowledgment

This work was supported by NSF Grant PCM 82-67548.

References

1. Terwilliger, N.B. and Terwilliger, R.C., J. Exp. Zool. 221, 181-191 (1982).
2. Holliday, C.W., Mykles, D.L., Terwilliger, R.C. and Dangott, L.J., Comp. Biochem. Physiol. 67A, 259-263 (1980).
3. Benesch, R., MacDuff, G. and Benesch, R.E., Analyt. Biochem 11, 81-87 (1965).
4. Graham, R., Mangum, C., Terwilliger, R.C. and Terwilliger, N.B., Comp. Biochem.
Physiol. 74A, 45-50 (1983).

SEXUAL AND SEASONAL CHANGES OF Hc FROM PALINURUS ELEPHAS

B. GIARDINA, M. CORDA, M.G. PELLEGRINI, A. BELLELLI, A. CAU, S.G. CONDO[1] and M. BRUNORI.

Dept. of Exp. Med. and Biochem. Sciences, II^ University of Rome; Inst. of Chemistry, Fac. of Medicine, University "La Sapienza", Rome; Inst. of Biochem. and Inst. of Zool., University of Cagliari, Italy.

INTRODUCTION

The structural and functional properties of hemocyanin from the mediterranean lobster Palinurus elephas have been recently published (1). The results obtained have shown that:
a) the protein exists in only two aggregation states (monomeric and hexameric);
b) the monomeric subunits are not homogeneous, but belong to at least 3 or 4 different classes; c) the O_2 binding of the hexamer is cooperative, and both pH and $[Ca^{++}]$ dependent; d) the Bohr effect of the hexamer in the presence of Ca^{++} is negative and similar to that of the previously characterized Panulirus interruptus hemocyanin; e) the monomers are non cooperative and have relatively low O_2 affinity (somewhat similar to that of the T state of the hexamer).

During the course of this previous characterization we observed slight variations of both electrophoretic patterns and functional properties of hemocyanin from different preparations. In order to elucidate the nature of this phenomenon we have analyzed all over the year hemocyanin samples from single individuals both male and female. The results of such an analysis are the object the present note.

RESULTS AND DISCUSSION

Alkaline electrophoresis under conditions of complete dissociation of the hexamers has been performed on hemocyanin samples extracted from single animals (male and female), which were collected monthly.

Two main facts are evident from the results obtained and reported in fig. 1:
a) in all the specimens examined (up to 30) the subunit composition of the hemocianin molecule is sex dependent since an additional subunit is found in female's

Invertebrate Oxygen Carriers
Ed. by Bernt Linzen
© Springer-Verlag Berlin Heidelberg 1986

hemolymph (see the lower part of the gels towards the positive electrode); b) for both male and female individuals the electrophoretic pattern is strikingly variable during the year as far as the relative proportions of the subunits constituting the native hexamer is concerned.

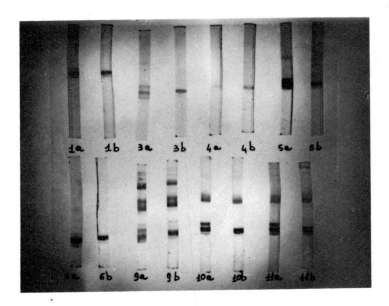

FIG. 1: Alkaline electrophoresis of dissociated subunits of hemocyanin from single lobsters. The number of each couple of gels refers to the month and the letter to the sex of the animal (a = female, b = male). Conditions: acrylamide 5%, buffer Tris-glycine EDTA pH 9,6, samples dialyzed overnight against the same buffer.

The presence of a sex-linked subunit, although surprising, is not novel since it has already been reported for another arthropodal hemocyanin (Callinectes sapidus, ref. 2), for which however a functional comparison between male and female hemocyanin is lacking.

In order to ascertain the possible functional significance of the additional subunit found in female individuals of Palinurus elephas we have performed a set of O_2 equilibria as a function of pH on hemocyanin from single individuals of bot sexes, collected in the same period of the year.

The data (fig. 2) have unequivocally established that, as far as the reversible reaction with O_2 is concerned, male and female hemocyanins of Palinurus elephas do posses binding parameters (such as O_2 affinity, cooperativity and

Bohr effect) almost identical. In conclusion the O_2 transport properties of the molecule are indipendent of the above mentioned structural difference indicating that the functional role of the sex-linked subunit, if any, has to be searched elsewhere, and possibly in some other carrying property(e. g. hormone binding).

O_2 binding curves were also determined on hemocyanin samples from animals (both male and female) collected in different period of the year. It was found that the seasonal changes observed in the relative proportions of the subunits constituting the hexameric molecule are paralleled by functional changes, which find their expression mainly in the cooperative character of O_2 binding that reaches its maximum value during the summer (see fig. 2).

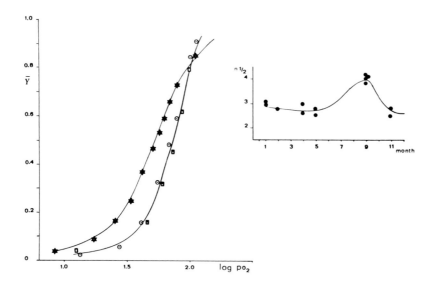

FIG. 2: Oxygen equilibrium curves for P. elephas hemocyanin. Conditions: 100 mM Tris-HCl buffer pH 7. 0 plus 10 mM $CaCl_2$; T = 20° C. September (°) male Hc and (□) female Hc; March (*) male and female Hc. The inset shows the variations of $n_{1/2}$ with the yearly cycle of the hemocyanin composition.

In particular from May to September we observed a marked increase of the Hill coefficient ($n_{1/2}$) measured at pH 7. 0, 20° C and in the presence of 10 mM Ca^{++}; the value of n = 2. 5 observed in May increases to n = 4 in September.

This increase in cooperativity is accompanied by a slight decrease of O_2 affinity (p $_{1/2}$ 55 mmHg vs 75 mmHg). Although it remains to be seen if these season-dependent functional changes could be related to periodic migration of the animals towards higher depths, it seems reasonable to correlate these structural and functional variations with the physiological process of molt, which is known to be accompanied by a large decrease of the hemocyanin concentration in the hemo-lymph. In conclusion, even if it is not possible, at the moment, to indicate une-quivocally the physiological needs which have to be met by such structural and functional variations of hemocyanin, they seem to be of particular interest at least because they may be taken as biochemical clues of yet poorly understood biological processes. Finally the whole body of the data indicates the existence, among these giant respiratory proteins, of a sophisticated mechanism of function modulation which may be achieved not only through the interaction with small ions but also through changes in the ratio of the various subunits. Moreover this phenomenon should be well kept in mind in considering the structural and functio-nal variability which has been sometimes reported in the literature (3,4).

REFERENCES

1. Bellelli, A., Zolla, L., Giardina, B., Costantini, S., Cau, A. and Bruno-ri, M. Biochim. Biophys. Acta, 1985 in press.
2. Horn, E.C., Kerr, M.S. Comp. Biochem. Physiol. 29, 493-508 (1969).
3. Durliat, M., Vranckex, R., Herberts, C., Lachaise, F. CR Seances Soc. Biol. Ses. Fil. 169 (4), 862-867 (1975).
4. Markl, J., Hofer, A., Bauer, G., Markl, A., Kempter, B., Brenzinger, M. and Linzen, B.-J. Comp. Physiol. 133, 167-175 (1979).

MODULATION OF HAEMOCYANIN OXYGEN AFFINITY BY L-LACTATE
- A ROLE FOR OTHER COFACTORS

C.R.Bridges & S.Morris

Institut für Zoologie IV, Lehrstuhl für Stoffwechselphysiologie
Universität Düsseldorf, D-4000 Düsseldorf, FRG

ABSTRACT: Evidence for the modulation of crustacean haemocyanin oxygen affinity by L-lactate is reviewed. The specificity of the lactate effect, its magnitude and correlation with pH, Bohr effect, cooperativity and calcium concentration is shown. The magnitude and the identity of dialysable non-lactate effectors of haemocyanin oxygen affinity are discussed.

INTRODUCTION

Since the finding that organic phosphates decrease the oxygen affinity of haemoglobins (1, 2) large advances have been made in understanding the role of organic molecules in oxygen affinity modulation (3). In haemocyanins only inorganic molecules were thought to modulate oxygen affinity (4, 5).

The presence of a dialysable factor which increased oxygen affinity was evident from the early work on Carcinus maenas and Cancer pagurus (6). In a later study a temperature dependent factor which increased oxygen affinity of the blood of animals acclimated to higher temperature was found, which was removed to some extent on dialysis (5). At the same time, in a separate study, it was also shown that dilution in an isotonic solution caused a decrease in C. maenas haemocyanin oxygen affinity despite the presence of calcium in the solution (7). In the same study recombination experiments using plasma and pelleted haemocyanin showed that the original oxygen affinity could be restored. Gel-filtration experiments also suggested the presence of a dialysable factor with a M.W < 5,000 (7). In these early studies the influence of lactate and Mg^{++} ions were not, however, fully appreciated.

The finding that L-lactate increased haemocyanin oxygen affinity (8) in C. maenas and in C. pagurus was a new milestone in the understanding of haemocyanin oxygen affinity modulation. The effect of L-lactate on oxygen affinity has also been demonstrated in Cancer magister (9) and in vivo in Callinectes sapidus (10). The presence of the lactate effect in other crustaceans has been reviewed (11,12) and also quantitative details of the effect in a number of species is available (13). Further evidence on the molecular basis of the lactate effect together with data on the binding of L-lactate to haemocyanin has also recently been made available (14). The present

Invertebrate Oxygen Carriers
Ed. by Bernt Linzen
© Springer–Verlag Berlin Heidelberg 1986

review attempts to synthesize the available information on the effect of lactate and other effectors on haemocyanin oxygen affinity and provide suggestions for further areas of investigation.

LACTATE FORMATION

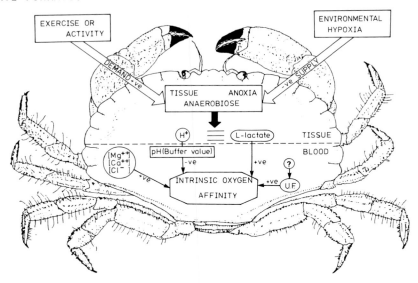

Figure 1. Formation of L-lactate and modulation of intrinsic oxygen affinity by inorganic and organic effectors.

The formation of L-lactate is known to occur in crustaceans under exposure to environmental stress such as hypoxia (15) and also during exercise (10,16,17) as shown in Figure 1. Concomitant with lactate production H^+ ions are produced decreasing oxygen affinity through the Bohr effect. The efflux rates of each ion will therefore be important in controlling the response of the pigment. Most known modulators such as the inorganic ions Mg^{++}, Ca^{++} and Cl^- act positively in that they increase oxygen affinity as does lactate unlike the addition of 2,3—DPG to haemoglobin which decreases oxygen affinity. Protons and temperature changes are the only know negative effectors of intrinsic oxygen affinity for haemocyanin.

SPECIFICITY OF THE LACTATE EFFECT

A number of workers (9, 11, 14) have tested the specificity of the lactate effect in C. magister and C. sapidus by using analogs of L-lactate and their data are summarised in Figure 2. From this work a model has been proposed to describe the allosteric interaction of L-lactate with haemocyanin which suggest that the lactate binding site interacts with the 4 positions of the chiral carbon of L-lactate (9). This theory explains the ability of the binding site to distinguish the difference between D- and L- lactate. The difference in the magnitude of the effect on log P_{50} can be seen in Figure 2.

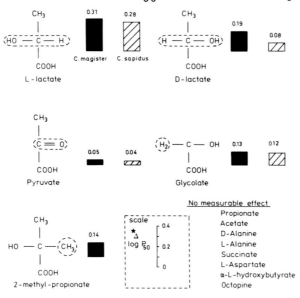

Adapted from Mangum (1983a), Graham et al. (1983), Johnson et al. (1984)

★ Calculated Difference in log P_{50} of dialysed blood in the presence and absence of 10 mM of the indicated compound

Figure 2. Specificity of response of oxygen affinity of C. magister (pH=7.75, 13.5C) and C. sapidus (pH=7.50, 20C) haemocyanin to small organic compounds.

Removal of the hydroxyl group, which appears to play a central role in the effect, as in propionate or replacement with a keto group as in pyruvate reduces the effect on oxygen affinity. In glycolate the hydroxyl group is free to rotate and therefore oxygen affinity is increased in both species. This rotation is hindered by the methyl group in D-lactate. The methyl group and the hydrogen are not distinguished by the binding site as substitution of a methyl group for the hydrogen, as in 2-methyl propionate, increases the oxygen affinity to the same extent as D-lactate in C. magister.

QUANTIFICATION OF LACTATE EFFECT
The lactate effect has been quantified by a number of workers

(8,10,13). Figure 3 describes one of the methods used (13). Briefly whole blood is collected and divided into two aliquots. One is dialysed against a Ringer of similar ionic composition to that of the original blood for 24 hrs. The untreated aliquot is used directly to determine oxygen dissociation curves for whole blood at various pH values. From the dialysed fraction a 100 μl sample is centrifuged to pellet the haemocyanin, 10μl of ringer is removed and replaced with 10μl of Ringer containing varying concentrations of L-lactate. The sample is remixed and oxygen dissociation curves at different pH values again determined (Fig.3A). At a given pH the log P_{50} can be determined for a known lactate concentration and plotted as in Figure 3B. A plot of log P_{50} against log L - lactate concentration provides a lactate effect coefficient. The difference in oxygen affinity between non-dialysed blood and dialysed blood at the same lactate

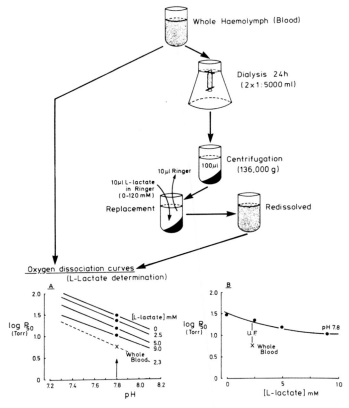

Figure 3. Schematic description of methodology used to quantify the lactate effect and determine the magnitude of the unidentified factor (13).

concentration and pH is defined as the effect of the unidentified factor(s) (U.F.). Table I details values for the lactate effect co-efficient from the literature and our own work. Values range from -0.04 in Ocypode saratan to -0.55 in the Palaemon serratus(pH=7.8). In all studies the effect of lactate on P_{50} was exponential with the largest effect reached between 0-5 mM lactate. Comparisons between different methods show good agreement (cf.Homarus vulgaris). Interestingly whole blood values were higher than the corresponding value for dialysed blood (cf C. callinectes and O. saratan). The Caridea appear to show higher values than the Brachyura or Astacura.

Table I. Magnitude of Bohr and lactate effects in dialysed haemolymph

Species	Temp °C	Bohr Effect	pH	[1.] $\dfrac{\Delta \log P_{50}}{\Delta \log [\text{L-lactate}]}$		Source
1. Homarus vulgaris	15	-1.00	7.8	-0.107	*	Bridges et al. (1984)
			7.4	-0.189		
1a. " "	15	-1.17	7.9	-0.158		Bouchet & Truchot (1985)
2 Austropotamobius pallipes	15	-0.45	7.9	-0.194	*	Morris et al. (1986a)
			7.4	-0.175		
3. Crangon crangon	10	-0.52	7.8	-0.175	*	Bridges et al. (1984)
			7.4	-0.286		
4. Palaemon elegans	10	-1.21	7.8	-0.560	*	Bridges et al. (1984)
			7.4	-0.629		
5. Palaemon serratus	10	-0.66	7.8	-0.550	*	Morris et al. (1985a)
			7.4	-0.655		
6. Carcinus maenas	15	-0.62	7.8	-0.096		Truchot (1980)
			7.2	-0.094		
7. Cancer pagurus	15	-1.00	7.9	-0.211		Truchot (1980)
			7.4	-0.237		
8. Cancer magister	13.5	-1.22	7.8	-0.247		Graham et al. (1983)
			7.4	-0.237		
9. Callinectes sapidus	20	-1.10	7.4	-0.319 W.B.		Booth et al. (1982)
9a. " "	20	-0.97	7.8	-0.213	*	This study
			7.4	-0.197		
9b. " "	25	-1.19	7.5	-0.257		Johnson et al. (1984)
10. Atelecyclus rotundatus	10	-0.92	7.8	-0.333	*	Taylor et al. (1985)
			7.4	-0.444		
11. Goneplax rhomboides	10	-0.62	7.8	-0.181	*	" " "
			7.4	-0.202		
12. Liocarcinus depurator	10	-1.40	7.8	-0.294	*	" " "
			7.4	-0.392		
13. Maia squinado	15	-0.38	7.8	-0.134 W.B.	*	Bridges et al. (1984)
			7.4	-0.175	"	
14. Ocypode saratan	30	-0.67	7.8	-0.162 W.B.	*	Morris & Bridges (1985)
			7.4	-0.240	"	
14a. " "			7.8	-0.044	*	Morris & Bridges (1985)

1.Calculated coefficient W.B.=Whole Blood * = Replacement Technique

A lactate effect is absent in the marine gastropod Busycon contrarium (11) and also in the cephalopods Eledone cirrhosa and Sepia officinalis (Bridges, in prep.). Interestingly metabolites such as octopine, alanopine and strombine also have no effect on oxygen

affinity (11, Bridges, in prep). No lactate effect was observed in
<u>Limulus polyphemus</u> (18), <u>Procambarus clarkii</u> (12) or in <u>Coenobita</u>
<u>compressus</u> (19) and <u>Coenobita clypeatus</u> (Morris & Bridges, in prep).

CORRELATION OF LACTATE EFFECT WITH OTHER FACTORS
a) pH - Bohr Effects
From Table I and Figure 4 it is evident that in some species the
lactate effect coefficients at pH 7.4 are higher than at pH 7.8.
This would be physiologically more beneficial to the animal as
protons produced at the same time as lactate and effluxing into the
blood would lower pH. A direct correlation between the Bohr effect
and the lactate effect is not possible since even if lactate and H^+
ions are released stoichiometrically H^+ ions are buffered and
equality of the values is not necessary for the counteraction of the
two effects. Perhaps a useful correlation could be obtained when the
buffer value of the blood was taken into account. A general trend may
be discerned that increased Bohr effects are associated with larger
lactate effects (Fig.4), but much will depend upon the relative
efflux rates of lactate and H^+ ions. Situations will also occur where
only the Bohr effect is operative.

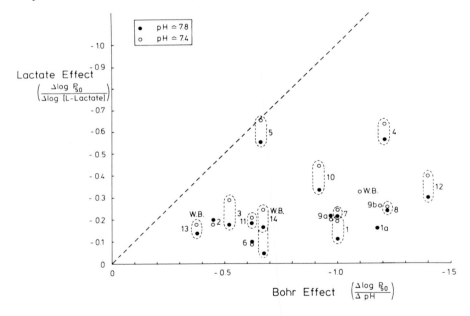

Figure 4. Plot of lactate effect agsinst the Bohr effect for different species
(denoted by the numbers in table I). WB = whole blood. Dotted line
encircles measurements on the same species.

In a separate study (20) a correlation between the lactate concentration and the Bohr effect in O. saratan was found. The Bohr effect decreasing with increasing lactate concentrations in whole blood. This was correlated with an exponential decrease in the buffering value of the blood. An allosteric change in the structure of the haemocyanin may occur such that the number of proton buffering sites are decreased when lactate is added to the haemocyanin of this species. In other studies (8, 10, 14) no pH dependency was found in C. maenas, C. magister and C. sapidus. Studies on C. sapidus however using the replacement technique (Fig.3) also found no pH dependency indicating that the pH effect is species specific and not due to a methodological difference.

b) Cooperativity

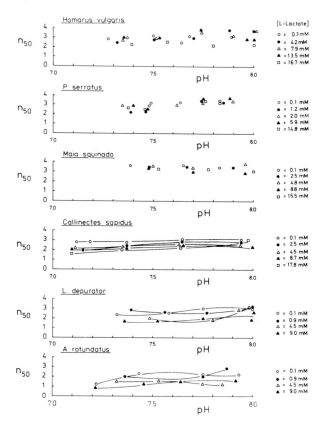

Figure 5. Interaction of lactate concentration and cooperativity (n50) in H. vulgaris (13), P. serratus (21), M. squinado (13), C. sapidus, present study, and L. depurator and A. rotundatus (22).

In initial studies on C. magister it was shown that L-lactate slightly decreased cooperativity. In later studies in other species, however, no effect of lactate on cooperativity could be detected (12, 13, 20, 21). This is shown in figure 5 for H. vulgaris, P. serratus and Maia squinado. Although initial in vivo studies on C. sapidus (10) haemocyanin showed no effect of lactate on haemocyanin cooperativity a more detailed study (14) showed a decrease in cooperativity with increasing lactate concentrations. This was confirmed in the present study (Fig.5). Further evidence for lactate dependent cooperativity was also found in Liocarcinus depurator and Atelecyclus rotundatus (22) (Fig. 5). From theoretical considerations and our actual measurements it is evident that in haemocyanins where cooperativity is lactate dependent a decrease in cooperativity accompanied by an increase in oxygen affinity must involve a lowering of the energy of the T state, facilitating oxygen binding. Physiologically this will increase the venous reserve raising venous saturation at a given oxygen tension. At the same time arterial saturation will be slightly depressed. This lactate dependent change in cooperativity agrees with the theory (14) that lactate binds between subunits therefore affecting cooperativity.

c) Calcium concentration

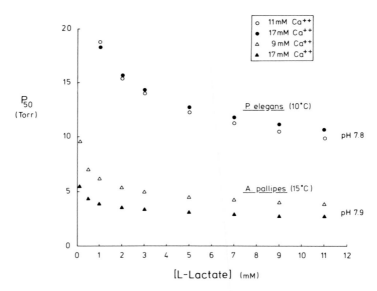

Figure 6. The effect of calcium concentration on the lactate effect in A. pallipes (24) and in Palaemon elegans (Taylor et al, in prep.)

In a previous study (23) it was suggested that lactate interacts directly with divalent cations at high concentrations of lactate. Investigations in C. magister were unable to show any effect of lactate on the level of free calcium and magnesium in the blood (9). It is well known that calcium can exert a specific effect on haemocyanin increasing oxygen affinity (4). Recent experiments with A. pallipes (24) have indicated a marked dependence of the lactate effect on calcium concentration in dialysed blood (Figure 6). At low calcium concentrations (9mM) the effect of lactate on oxygen affinity was larger than at high calcium concentrations (17mM). It may also be noted that a large specific effect of calcium on oxygen affinity is present in this species (24). In P. elegans a similar change in Ca^{2+} had no effect on the lactate effect (Fig.6) and no specific action of calcium on haemocyanin oxygen affinity could be shown (Taylor et al.in prep.). Again it would appear that lactate interactions are species specific although both these species may show changes in blood calcium levels (25, Taylor et al. in prep.).

DIALYSABLE NON-LACTATE EFFECTORS OF HAEMOCYANIN OXYGEN AFFINITY

When the oxygen affinity of dialysed and non-dialysed haemolymph is

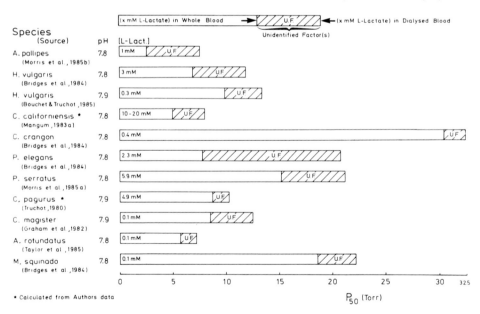

Figure 7. Magnitude of non-lactate effectors of haemocyanin oxygen affinity in different species.

compared, as in Figure 3B, for a number of species at the same pH and lactate concentration, oxygen affinity is greater in non-dialysed blood. This effect originally shown in C. pagurus (8) and later quantified in other species (9, 11, 12, 13) has been defined as the effect of unidentified factor(s), U.F.(13). The magnitude of this effect is shown in a number of species in figure 7. The largest reported values are 13 Torr in P. elegans (13) and 6 Torr in P. serratus (21). Other values range between 1 and 5 Torr. In H. vulgaris similar values for the U.F. have been obtained independently by different workers (13, 26). A difference between the oxygen affinity of non-dialysed and dialysed blood, at the same pH and lactate concentration is absent in C. sapidus, Squilla empusa, Sicyonia ingentis (12), L. depurator and Goneplax rhomboides (22) which all show a lactate effect. In C. clypeatus (Morris & Bridges in prep.) and P. clarkii (12) both effects are absent. L. polyphemus haemocyanin does not exhibit a U.F. effect (12) and in the molluscs so far studied no U.F. effect has been observed (11, In prep).

No changes in the magnitude of the U.F. effect could be found in P. elegans haemolymph collected in summer and winter (13). Other studies on P. elegans blood from a different region, however, showed a U.F. effect of only 10 Torr (21). Recent work on H. vulgaris exposed to hypoxia (35 Torr, 40 h) indicated that whole haemolymph oxygen affinity increased to an extent that could not be accounted for by L-lactate alone (26). It was suggested that this effect was due to the increase in non-lactate dialysable factors in the haemolymph, although other effectors such as calcium were not monitored. In a study on aerial exposure of A. pallipes , however, no significant change in the magnitude of the U.F. effect could be detected (27). The modulating role of the U.F. therefore remains to be verified. Elegant plasma recombination experiments between P. serratus and P. elegans blood (21) have shown that the effect due to the U.F. is present within the plasma and is not species specific. The magnitude of the effect is, however dependent upon the haemocyanin sensitivity.

IDENTIFICATION OF U.F.

It was initially thought that the U.F. could be another anaerobic metabolite, but no change in oxygen affinity was observed when a number of known anaerobic metabolites were added to dialysed blood (Fig.2) (9, 11, 14). As part of a wider study of cofactor modulation of haemocyanin oxygen affinity the effects of the products from

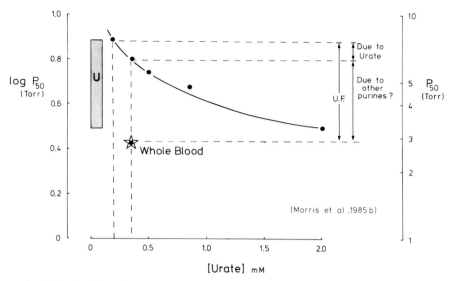

Figure 8. Effect of urate on oxygen affinity in A. pallipes haemocyanin at 15C, 9mM Calcium, 1mM Lactate and pH 7.8. Maximum effect of urate indicated by the bar (U). Whole blood oxygen affinity and urate concentration are shown by the star symbol (27).

purine catabolism have been investigated (27). The results shown in Figure 8 indicate that the log P_{50} decreases exponentially with increasing urate concentration. Using a similar log transformation as with L-lactate a urate coefficient of -0.394 could be calculated. This compares with a lower value of -0.194 for L-lactate in the same species (Table I). The effect of urate appears to be both independent of pH and cooperativity , as is the lactate effect in this species. Urate is present in A. pallipes haemolymph (0.35mM) but at this concentration it can only account for approximately 20% of the measured U.F. effect (Fig. 8). If urate concentrations were saturating then urate could account for 80% of the U.F. effect. The identity of the effectors which account for the other 80% in whole blood remain unknown but other purines may be present in haemolymph and these may influence oxygen affinity. Further studies on these purines and the interaction of known effectors of haemocyanin oxygen affinity are now in progress. It is evident, however, that in our knowledge of both the lactate and U.F. effects a number of questions still remain to be answered.

Financial support for this study was provided by the DeutscheForschungsgemeinschaft to C.R.B (Gr 456/10 - 1) and by the Royal Society London to S.M.

REFERENCES

1. Benesch, R.E. and Benesch, R., Biochem.Biophys.Res.Comm. 26, 162-167 (1967).

2. Chanutin, A. and Churnish, R., Arch. Biochem. Biophys. 121, 96-101 (1967).

3. Isaacks, R.E. and Harkness, D.R., Amer. Zool. 20, 115-129 (1980).

4. Truchot, J.-P., Respir. Physiol. 24, 173-189 (1975).

5. Brouwer, M., Bonaventura, C., Bonaventura, J., Biochemistry 17, 2148-2154 (1978).

6. Truchot, J.-P., C. R. Acad. Sci. Paris. 272, 2706-2709 (1971).

7. Harris, R.R., Chantler, E.N. and Bannister, W.H., Comp. Biochem. Physiol. 52A, 189-191 (1975).

8. Truchot, J.-P., J. Exp. Zool. 214, 205-208 (1980).

9. Graham, R.A., Mangum, C.P., Terwilliger, R.C. and Terwilliger, N., Comp. Biochem. Physiol. 74A, 45-50 (1983).

10. Booth, C.E., McMahon, B.R. and Pinder, A., J.Comp.Physiol. 148, 111-121 (1982).

11. Mangum C.P., in Structure and Function of Invertebrate Respiratory Proteins (E.J.Wood,ed) Life Chem. Reports Suppl.1. pp. 333-352, Harwood Academic Press, Chur -London - New York (1983a).

12. Mangum, C.P., Mar. Biol. Lett. 4, 139-149 (1983b).

13. Bridges, C.R., Morris, S. and Grieshaber, M.K., Respir. Physiol. 57, 189-200 (1984).

14. Johnson, B.A., Bonaventura, C. and Bonaventura, J., Biochemistry 23, 872-878 (1984).

15. Bridges, C.R. and Brand, A.R., Comp. Biochem. Physiol. 65A, 399-409 (1980).

16. Phillips, J.W., McKinney, R.J.W., Hird, F.J.R., and MacMillan, D.L., Comp. Biochem. Physiol. 56B, 427-433 (1977).

17. McDonald, D.G., McMahon, B.R. and Wood, C.M., J.Exp. Biol. 79, 45-58 (1979).

18. Mangum, C.P., Mol. Physiol. 3, 217-224 (1983c).

19. Wheatly, M.G., McMahon, B.R., Burggren, W.W. and Pinder, A.W. In press.

20. Morris, S. and Bridges, C.R. , J. Exp. Biol. 117, in press (1985).

21. Morris, S., Bridges, C.R. and Grieshaber, M.K., J.Exp.Zool. 234, 151-155 (1985a).

22. Taylor, A.C., Morris, S. and Bridges, C.R., J. Comp. Physiol. 155, in press (1985).

23. Jackson, D.C. and Heisler, N., Respir. Physiol. 49, 159-174 (1982).

24. Morris, S., Tyler-Jones, R. and Taylor, E.W., J.Exp. Biol. in press (1986).

25. Morris, S., Tyler-Jones, R., Bridges, C.R. and Taylor, E.W., J. Exp. Biol. In press (1986).

26. Bouchet, J.Y. and Truchot, J.-P. Comp. Biochem. Physiol. 80A, 69-73 (1985).

27. Morris, S., Bridges, C.R. and Grieshaber, M.K., J.Exp.Zool. 235, 135-139 (1985b).

NOVEL NON-LACTATE COFACTORS OF HAEMOCYANIN OXYGEN AFFINITY IN CRUSTACEANS

S. Morris and C.R. Bridges

Institut für Zoologie IV, Lehrstuhl für Stoffwechselphysiologie, Universität Düsseldorf, D - 4000 Düsseldorf 1, F.R.G.

ABSTRACT: Some purine bases and their derivatives have been demonstrated to increase the oxygen affinity of a number of crustacean haemocyanins. Urate has been shown to interact with L-lactate and Calcium ions which have similar potentiating effects on oxygen affinity.

INTRODUCTION

The effect of L-lactate increasing the oxygen affinity of crustacean haemocyanin (1) is, alone, insufficient to account for the difference in the oxygen affinity, expressed as P_{50}, between dialysed and non-dialysed haemolymph (pH and L-lactate concentration constant). Although there is now some evidence that this effect of L-lactate may be influenced by the concentration of Ca^{2+} in the haemolymph (2,3) this interaction cannot account for this difference in the oxygen affinity. The greater affinity of whole haemolymph is due to unidentified factors (U.F.) (1,3,4).

REPORT

Recent studies on the haemolymph of _Austropotamobius pallipes_ (5,6) have demonstrated that a second organic molecule, urate, also markedly increases haemocyanin oxygen affinity (Fig. 1). The potentiating effect of urate (pH and L-lactate constant) has been demonstrated for the haemocyanin of several species (Table 1). The presence or absence of a U.F. effect cannot always be correlated with the concentration of urate in the haemolymph, cf. _C. maenas_ and _C. clypeatus_ (Table 1) the latter showing no sensitivity to urate. Throughout this report concentrations of test substances are given as those in the dialysis Ringer because urate concentrations in dialysed haemolymph were consistently higher than in the Ringer and often exceeded the physical solubility for urate in water (6). This has been interpreted as a possible indication of urate binding by the Hc. This may not be unreasonable in the light of the similarity between Hc and the cuproflavoprotein urate oxidase. It is apparent that the extent of the U.F. and urate effects is not dependent on the concentration of factors in the haemolymph (8) but to a large extent on the haemocyanin sensitivity and the absence of a simple correlation between native urate concentration and the magnitude of the two

Invertebrate Oxygen Carriers
Ed. by Bernt Linzen
© Springer–Verlag Berlin Heidelberg 1986

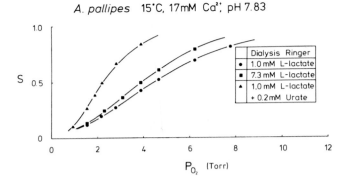

Figure 1. Oxygen equilibrium curves constructed for A. pallipes haemolymph dialysed against Ringer with different concentrations of L-lactate and urate.

Species	Temp. °C	U.F. Effect	Urate Effect Δlog P_{50}	ΔP_{50} (Torr)	Native [Urate] mM
Carcinus maenas	15	−	0.41	5.6	0.05
Palaemon serratus	15	+	0.36	19.5	0.44
Homarus vulgaris	15	+	0.57	8.7	0.31
Austropotamobius pallipes	15	+	0.43	2.0	0.24
Callinectes sapidus	20	−	0.19	2.7	0.46
Coenobita clypeatus	25	−	None	None	0.97
Limulus polyphemus	20	−	None	None	0.28

Table 1. Species investigated for the presence of a U.F. (+ = effect present) effect and a urate effect. All data are shown for pH 7.8 and native inorganic ion and L-lactate concentration.

effects is, therefore, not suprising. Urate alone cannot account for all of U.F. (Fig. 2) and may in vivo account for a small part of the total.

In an attempt to identify remaining contributions to U.F. other purine bases and derivatives were tested. Adenine and hypoxanthine, together with IMP, were the most effective but less effective than urate (Fig. 2). Introducing adenine and hypoxanthine together with urate to the dialysis Ringer did not result in an affinity increase beyond that achieved by urate alone. A similar absence of an additive effect was observed when L-lactate was added to a solution already containing urate (Fig. 2). A maximal urate effect greatly reduces or abolishes the lactate effect. Although the presence of large amounts of urate may possibly account for the absence of a L-lactate effect

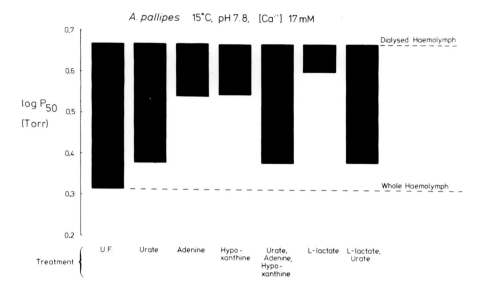

Figure 2. Schematic representation of the increase in haemocyanin oxygen affinity (Δlog P50) due to the presence of different test substances in comparison to the U.F. effect which is responsible for the difference in the affinity of non-dialysed and dialysed haemolymph. The result of various combinations of these substances is also shown.   Ringer concentrations: 0.2 mM Urate, 0.5 mM Adenine, 0.5 mM Hypoxanthine, 7.3 mM L-lactate.

in the whole haemolymph of some species (4) recent investigations using dialysed haemolymph (8) indicate haemocyanin insensitivity to be more important. Urate and purine structures may influence the lactate effect in whole haemolymph so that it differs from that determined in dialysed solutions (9). Work with A. pallipes haemolymph demonstrated the affinity of Hc for oxygen to be markedly Ca^{2+} dependent (6). Additionally both the U.F. effect and that of urate were reduced by increasing the concentration of Ca^{2+} in the dialysis Ringer (Fig. 3). There is little evidence at present to implicate a direct interaction between Ca^{2+} and organic ions. Instead it would seem likely that the allosteric change resulting in increased oxygen affinity is finite in extent and that this limit can be approached by increasing the concentration of various effector substances which influence the Hc to differing extents. The affinity of Hc for each effector will almost certainly be reflected by the relative efficacy of the different structures. Using competitive inhibition as the distinguishing criterion it is possible, in the case of A. pallipes, to rank the various effectors in order of increasing efficacy: L-lactate < urate < Ca^{2+}. It is clear that the effect of each of these

356

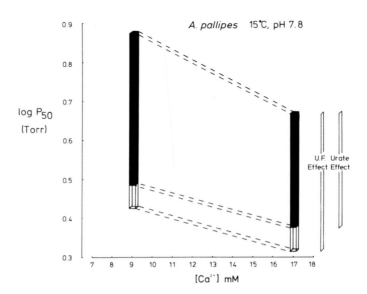

Figure 3. Schematically demonstrating the potentiating effect of Ca-ions on the oxygen affinity of A. pallipes haemocyanin and also the apparent suppression by Ca-ions of both the U.F. and urate effect in this species.

ions on Hc oxygen affinity will be species dependent.

The role in vivo of purine bases is a subject for further investigation and at present is open to speculation. The involvement, however, of urate and related structures in the catabolism of ATP and AMP suggests one possible mechanism whereby the circulating level of such effectors may respond to physiological stress and impart a true modulating role to these substances.

Financial support for this work was provided by the Royal Society London (S.M.) and the DFG (Gr 456/10-1, CRB).

REFERENCES

1.Truchot, J.-P., J. Exp. Zool. 214, 205-208 (1980).

2.Morris, S., Tyler-Jones, R. and Taylor, E.W., J. exp. Biol. (In Press).

3.Morris, S., Tyler-Jones, R., Bridges, C.R. and Taylor, E.W., J.exp.Biol. (In press).

4.Bridges, C.R., Morris, S. and Grieshaber, M.K., Resp.Physiol.57, 189-200 (1984).

5.Mangum, C.P. Mar. Biol. Lett. 4, 139-149 (1983).

6.Morris, S., Bridges, C.R. and Grieshaber, M.K., J.Exp. Zool.235, 135-139 (1985).

7.Morris, S., Bridges, C.R. and Grieshaber, M.K., J. Comp. Physiol. (In Press).

8.Morris, S., Bridges, C.R. and Grieshaber, M.K., J.Exp.Zool.234, 151-155 (1985).

9.Morris, S. and Bridges, C.R. J. exp. Biol. 117 (In Press).

THE EFFECTS OF DIFFERENT CATIONS ON THE ELECTRICAL PROPERTIES OF THE CHANNEL FORMED BY MEGATHURA CRENULATA HEMOCYANIN

G. Menestrina, C. Porcelluzzi and R. Antolini
Dipartimento di Fisica
Universita' di Trento
38050 Povo (TN) Italy

Megathura crenulata hemocyanin incorporates into lipid bilayers forming ionic channels through them which are mainly cation selective. The channels have a fixed orientation with respect to the side of protein addition, which is called the cis side (1).
The hemocyanin channel has several conductance states and the probability of occupancy of each state and the transition rates between the different states are regulated by the applied voltage (2). Using an appropriate holding voltage, cis side positive, and very short voltage pulses it is possible to study the electrical properties of the pore in the most conductive state, that we call open state, as a function of the applied voltage and the ionic composition of the medium.
Under these conditions the conductance voltage characteristic of the channel has in general a sigmoidal shape, the conductance being lower when the trans compartment is made more positive than the cis, as is shown in Fig. 1 for the case of a KCl solution.

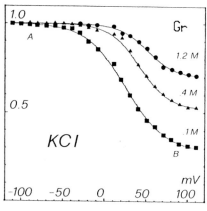

applied voltage

Fig. 1 Voltage dependence of the relative conductance Gr, given by the ratio G/Ga, for the hemocyanin channel in a KCl solution of varying concentration.
Solid lines are drawn according to eq. 1. The voltage regions where the two states of limiting conductances Ga and Gb are attained, are indicated.

This shape is strongly reminiscent of the conductance voltage characteristic of natural excitable cells and can be explained using a classical two state voltage gating model:

$$G = Gb + (Ga - Gb)/(1 + \exp q(V - Vo)/kT)$$

1.

where G is the mean pore conductance, Ga and Gb the conductance in the two limiting states, q is the gating charge and V the applied voltage.

Invertebrate Oxygen Carriers
Ed. by Bernt Linzen
© Springer–Verlag Berlin Heidelberg 1986

It is apparent from Fig. 1 that the shape of the conductance voltage
curve depends on the ionic strength of the solution. This is the
result of the fact that both the limiting conductances, Ga and Gb,
saturate with the ionic concentration either with monovalent cations
(Li,Na,K,Cs) and with divalent cations (Mg,Ca,Ba,Sr) (3), an example
with KCl is shown in Fig. 2.

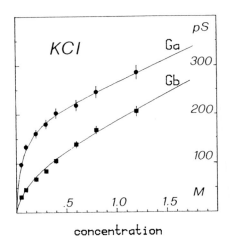

Fig. 2 Concentration dependence of the hemocyanin channel
conductance in its two possible conformational states,
Ga and Gb, using a KCl
solution.
Solid lines are best fit to
the experimental points using
eqs. 1-4.

concentration

This saturation can be explained assuming that channel conductance
is linearly proportional to a local ion concentration at the pore
entrance, which in turn is determined by a discrete negative charge
fixed on the channel in proximity of its mouth. We can write:

$$Ga,b = \pi r^2 / l \; w \cdot z \cdot e \cdot Ca,b \qquad\qquad 2.$$

$$Ca,b = Co * \exp(- z \cdot e \cdot \Psi a,b \, / \, kT) \qquad\qquad 3.$$

$$\Psi a,b = Qa,b \, / \, 4 \pi \varepsilon \; * \exp(- x d) \, / \, d \qquad\qquad 4.$$

where r,l are radius and length of the pore, Co,w,z are
concentration, mobility and valence of the cation in the bulk
solution, Ca,b, $\Psi a,b$,Qa,b are local cation concentration, local
potential and fixed charge on the channel in state a or b
respectively, d is the effective distance between the fixed charge
and the pore entrance, e,k,T have their usual meaning and ε,x are
dielectric constant and Debye-Huckel coefficient of the solution.
Though this model is quite simplistic we have found that eqs. 1
through 4 can describe the pore conductance as a function of voltage
and ionic concentration with a wide number of cations just using a
different fixed charge for each one. The major finding of this
fitting is that the charge of the channel is maximum when K+ is the
main cation present in the solution, decreases with Li+ and Na+ and
is a minimum when divalent cations are used, see Fig. 3. The
explanation of this result lies most probably in the fact that
divalent cations and small monovalent cations can bind to hemocyanin
thereby neutralizing a part of its charge (4).

Two salt experiments in which small amounts of Ca2+ or Tb3+ were added to a KCl solution allowed the determination of the constants

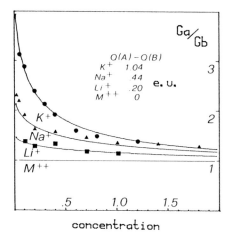

Fig. 3 Ratio between the channel conductance in the two states a and b as a function of the salt concentration with several different cations.
Solid lines are best fit to the points using eqs.1-4, the difference in fixed charge between the two states is indicated.

of binding of these cations to the hemocyanin channel. They were found to be in good agreement with published data on the binding of the same cations to other hemocyanins, as determined by NMR under physiological conditions (4,5).

Finally, since protons are also known to be effective in neutralizing the hemocyanin negative charge, we have examined the effects of varying the pH on the properties of the pore. We have found that indeed lowering the pH decreases the conductance of the open state both at high and at low ionic strength and irrespectively of the membrane surface charge, an example for the case of a KCl solution is shown in Fig.4.

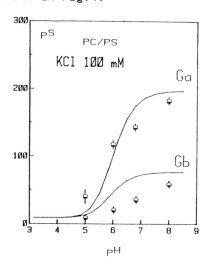

Fig. 4 pH dependence of the hemocyanin channel conductance in the two conformational states, Ga and Gb, in a KCl solution.
Solid lines are drawn according to eqs. 1-4.

In addition we have also observed that the selectivity between K+ and Cl-, which originates in the model from their being counterion

and coion respectively, also decreases lowering the pH and is lost
near the isoelectric point, pH 4.4, as expected (6).
The observed pKs of the effects of both Ca2+ and protons are shown
in Tab. 1 and compared to the corresponding values for some chemical
groups usually present on proteins. They indicate that the negative

	Group	pK H+	Ca++
	-O⁻	7.2	1.5
	-S⁻	8.4	-
	$\bigcup_{N}^{N-}$	6.0	-
	-COO⁻	2.4	1.3
	(-COO⁻)₂	4.3	3.0
M. Crenulata hcy	Cond.	5.4	2.4 ⎱
	Shift	5.4	2.7 ⎰
J. Edwardsii hcy	-COO⁻	3.6	-
	$\bigcup_{N}^{N-}$	6.3	-
L. Hierosolima hcy		-	1.9
P. Interruptus hcy		-	2.0

Table 1 Comparison between
the pKs for the binding of
protons and calcium ions
either to various functional
groups commonly found on
proteins, upper part, or to
hemocyanins, as determined by
electrophysiological properties, middle part, or
spectroscopic techniques,
lower part.

charge influencing the ionic pathway of the channel is mostly
composed of carboxyl groups, with the probable involvement of
histidines.

REFERENCES

1. Latorre, R., and Miller, C., J.Membrane Biol. 71, 11-30 (1983).
2. Menestrina, G., Maniacco, D., and Antolini, R., J.Membrane Biol.
   71, 173-182 (1983).
3. Menestrina, G., and Antolini, R., Biochim.Biophys.Acta 688, 673-
   684 (1982).
4. Andersson, T., Chiancone, E., and Forsen, S., Eur.J.Biochem. 125,
   103-108 (1982).
5. Menestrina, G., Biophys.Struct.Mech. 10, 143-178 (1983).
6. Menestrina, G., and Porcelluzzi, C., Biochim.Biophys.Acta in
   press

ELECTRICAL PROPERTIES OF IONIC CHANNELS FORMED BY HELIX POMATIA HEMOCYANIN IN PLANAR LIPID BILAYERS

F. Pasquali, G.Menestrina and R.Antolini
Dipartimento di Fisica
Universita' di Trento
38050 Povo, (TN) Italy

In 1972 Pant and Conran (1) could show that the hemocyanin extracted from the blood of the giant Keyhole Limpet Megathura crenulata was able to increase the conductance of black lipid membranes by several orders. We have then shown that two other hemocyanins, those extracted from Aplysia and Paludina vivipara are able to form channels in BLM (2,3). We have now consistently observed the formation of single ionic channels in planar bilayers after the interaction with Helix pomatia hemocyanin as well.
Planar lipid bilayers were formed by the apposition of two monolayers following the Montal technique (4). Monolayers were prepared using a lipid mixture comprised of phosphatidylcholine (PC) and phosphatidylserine (PS) dissolved in n-pentane, PC:PS ratio was 9:1 and total lipid concentration 12.5 mg/ml. n-Pentane was allowed to evaporate before apposing the monolayers. Helix pomatia

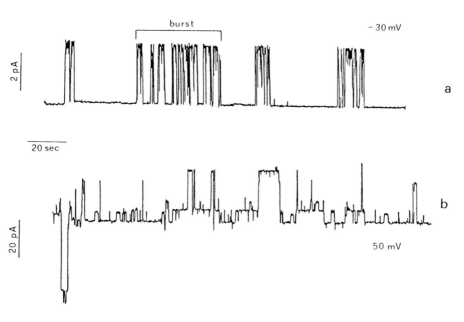

Fig.1 a) Current fluctuations due to the opening and closing of one single Helix pomatia hemocyanin channel of about 100 pS conductance. Fluctuations are grouped in bursts as indicated.
b) Current fluctuations due to the opening and closing of a few H. pomatia channels of different conductances all present in the same membrane.

Invertebrate Oxygen Carriers
Ed. by Bernt Linzen
© Springer–Verlag Berlin Heidelberg 1986

hemocyanin was a kind gift of prof. B. Salvato, Dipartimento di Biologia (PD) I, it was added in small amounts to only one of the two membrane bathing solutions (cis-side). Current was detected through Ag-AgCl electrodes and amplified by means of an operational amplifier (AD 515 K) with virtual ground in the cis compartment. After the addition, followed by intense stirring, of <u>Helix</u> <u>pomatia</u> hemocyanin to a voltage clamped planar lipid membrane, current pulses sometimes appear from the bare membrane current level to a well defined value whose magnitude suggests that an ionic channel is formed, see Fig.1a. Pulses follow one another in close succession indicating fast open-close fluctuations;sometimes the current remains at the bare membrane level for a long period, then fluctuations start again. This fluctuation pattern could last even more than three hours during which the conductance of the open channel did not change. Because of the similarity of this behaviour to that of the acetylcholine-receptor channel, early observed by Neher and Sakmann with the patch recording technique, we have decided to call each group of square pulses between long intervals of apparent inactivation a "burst".

The amplitude of the square pulses is not the same from channel to channel in fact even applying the same voltage individual channels can display different values of conductance in different membranes. Also on the same membrane, when it contains more than one channel at a time, current jumps of different amplitudes can be observed, see Fig. 1b. The frequency distribution of the conductance jumps in different experiments is shown in Fig 2.

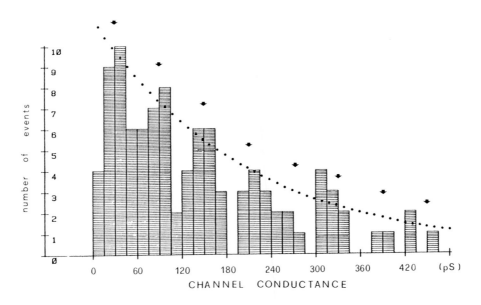

Fig.2 Frequency distribution of the appearance of a channel of a given amplitude, 100 events have been used. A constant spacing of 60 pS has been indicated by arrows. Dotted line is drawn according to eq. 1 using p = 0.75.

Conductance values collect into equally spaced broad peaks whose height decreases as conductance increases. The spacing between the peaks is roughly 60 pS which means that each peak can be calculated from another just by adding or subtracting a multiple of this discrete conductance unit.

We introduce here the hypothesis that the channels originate from the aggregation of a discrete number of equal subunits each of conductance 60 pS. This hypothesis is consistent with the shape of the histogram of Fig. 2, in fact if we assume that the aggregation probability, i.e. the probability to go from an n-mer to an (n+1)-mer, is a constant of value p, we can write:

$$N(n) = No * p^{(n-1)} \qquad \qquad 1.$$

where N(n) is the number of n-mers observed and No the number of monomers observed. As shown in Fig. 2 this exponential dependence fits reasonably well to our data.

Since all the channels, irrespectively of their conductance, fluctuate most of the time between one single high conductance value and one single low conductance value, which last corresponds to the bare membrane level, we are lead to conclude that all the subunits within an aggregate are forced to fluctuate simultaneously between their open state, conductance 60 pS, and their closed state, conductance 0 pS. Hence they form the structure which has been named the "gatling gun" (5) and which has been observed to occur in the Na-channel reconstituted in lipid bilayers (6), i.e. an aggregate of parallel and equal channels gated all together by the applied voltage.

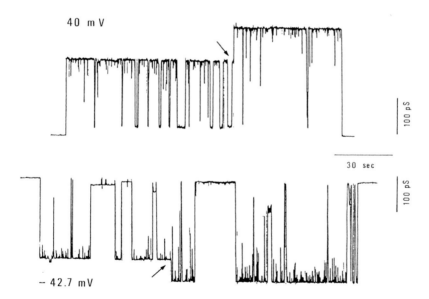

Fig.3 Examples of channels which increased their conductance during the experiment, at the moment indicated by arrows, without changing their gating properties.

The best evidence for this hypothesis is probably the seldom observation we have done of one channel which increased abruptly its conductance by a multiple of 60 pS during the experiment, without changing its fluctuating properties, two examples are shown in Fig.3. The fact that the channel fluctuates only between the highest and the lowest conductance level either before or after the voltage jump indicates that this is not due to the opening of a new independent channel but by the addition of a new unit to a preexisting structure, which are then forced to fluctuate simultaneously.

Helix pomatia hemocyanin channels have also many other interesting features which can be derived from a kinetic analysis of their gating behaviour (7). As shown in Fig. 1a fluctuations can be grouped in intervals, called bursts, separated by relatively large iterburst periods during which the channel is silent. The time distribution between burst and interburst intervals is also voltage dependent. Open to closed fluctuations within one burst show a complex kinetic behaviour which is compatible with a multistate model which employs one single open state and six different closed states. All transitions, which may be grouped in two classes, fast and slow, are voltage dependent in the sense that both high positive and high negative potentials tend to close the channels. A much similar behaviour has been recently described in detail for the channels of the gap junctions between amphibian blastomers (8), indicating that this hemocyanin pore is at all an interesting model for the understanding of the molecular mechanisms underlying the physiological functioning of the naturally occurring ionic channels of the excitable tissues.

REFERENCES

1. Pant, H.C., and Conran, P., J.Membrane Biol., 8, 357-362 (1972).
2. Antolini, R., and Menestrina, G., Biochim.Biophys.Acta, 649, 121-124 (1981).
3. Menestrina, G., and Antolini, R., Biochem.Biophys.Res.Commun., 88, 433-439 (1979).
4. Montal,M. and Mueller,P., Proc.Natl.Acad.Sci USA, 69, 3561-3566 (1972).
5. Miller,C., Philos.Trans.R.Soc.Lond.(Biol.), 299, 401-411 (1982).
6. Hanke,W., Boheim,G., Barhanin,J., Pauron,D., and Lazdunski,M., EMBO J., 3,509-515 (1984).
7. Menestrina, G., Pasquali, F., and Antolini, R., Biophys.Struct. Mech., 10, 169-184 (1984).
8. Spray,D.C., Harris,A.L. and Bennet,M.V.L., J.Gen.Physiol. 77, 77-93 (1981).

Allosteric interaction -
theories and experiments

INFERENCE OF ALLOSTERIC UNIT IN CHLOROCRUORIN, ERYTHROCRUORIN, AND HEMERYTHRIN ON THE BASIS OF THE MONOD-WYMAN-CHANGEUX MODEL

K. Imai[a], S. Yoshikawa[b], K. Fushitani[a,c], H. Takizawa[d], T. Handa[d], and H. Kihara[e]

[a]Department of Physicochemical Physiology, Osaka University Medical School, Osaka 530, Japan; [b]Department of Biology, Faculty of Science, Konan University, Kobe 658, Japan; [c]Department of Zoology, University of Texas, Austin, Texas 78712, U.S.A.; [d]Faculty of Science, Science University of Tokyo, Tokyo 162, Japan; [e]Department of Physics, Jichi Medical College, Tochigi 329-04, Japan.

SUMMARY

The allosteric unit of oxygen binding macromolecules, i.e. the minimum unit which substantially preserves the cooperativity of the whole molecule, was inferred from oxygen equilibrium data on the basis of the Monod-Wyman-Changeux (MWC) allosteric model. Two methods were used: one, a graphic method yielding a linear plot (Decker et al., 1983), and the other, a numerical method assuming symmetry of the oxygenation curve (Colosimo et al., 1974).

The analysis by means of the linear MWC plot indicated the following. In *Potamilla* chlorocruorin the number of sites for oxygen binding involved in heme-heme interactions is six, that is, the allosteric unit for this protein is the one containing six heme groups. The allosteric unit of *Lumbricus* erythrocruorin under moderate pH conditions is the one containing 12 heme groups, but the size of the unit tends to become smaller in extreme oxygen saturation ranges and/or under extreme pH conditions. The allosteric unit of *Lingula* hemerythrin contains eight oxygen binding sites and, therefore, the unit is the whole molecule (octamer). The numerical method gave similar results under certain experimental conditions but diverse numbers of the binding sites contained in the allosteric unit under other conditions, indicating that it is difficult to infer the allosteric unit from a given oxygenation curve determined under a single set of conditions.

The relation between the inferred allosteric units and the molecular structures is discussed.

INTRODUCTION

Annelid chlorocruorin (Chl) and erythrocruorin (Ec) have similar molecular architecture, i.e. bilayered hexagonal rings composed of 12 subunits (submultiple). They are characterized by a large molecular

Invertebrate Oxygen Carriers
Ed. by Bernt Linzen
© Springer–Verlag Berlin Heidelberg 1986

mass ranging from 2.7 and 3.8 MDa and a large number of heme groups
(72 to 192 per molecule), compared to those of the vertebrate hemoglobin. The constituent polypeptide chains are heterogeneous and not
all the chains are reported to be associated with heme groups, yielding lack of one-to-one correlation between polypeptide chains and heme
groups (see review articles, e.g. (1,2)). Because of such a complex
structure, it is not self-evident what is the minimal unit of Chl or
Ec which preserves the native cooperativity of the whole molecule.

Recently, we determined accurate oxygen equilibrium curves of
Chl from *Potamilla leptochaeta* (3), Ec from *Lumbricus terrestris*
(Fushitani, K., Imai, K., and Riggs, A.F., in preparation; also see
their paper in this book), and hemerythrin (Hr) from *Lingula unguis*
(4) under a variety of experimental conditions by using an automatic
oxygenation apparatus (5-7). In the present study, we attempted to
infer the allosteric units for these macromolecular oxygen carriers
by analyzing the oxygenation data in the framework of the Monod-
Wyman-Changeux (MWC) allosteric model (8).

METHODS

The MWC model

By analogy with the vertebrate tetrameric hemoglobin we assume
that a unit of the macromolecular oxygen carriers described above can
take two alternative quaternary structures designated as T (tense)
and R (relaxed). The unit is defined by the minimal one that preserves
the native cooperative properties of the whole molecule (allosteric
unit). It is implicitly assumed that the whole molecule is a homo-
oligomer composed of the allosteric units. Interactions among the
units in the molecule are absent or very weak compared to those among
the subunits within each unit. The T and R states differ in the number and energy of intersubunit bonds and are in equilibrium. Let L_0
be equilibrium constant for $R \rightleftarrows T$ in the absence of oxygen. Let K_T and
K_R be intrinsic association equilibrium constants of oxygen for the T
and R states, respectively ($K_T < K_R$). Then, the binding polynomial (9,
10) for oxygen binding is given by

$$P = (1 + c\alpha)^N + L_0(1 + \alpha)^N \tag{1}$$

where N is number of oxygen binding sites per allosteric unit, $\alpha = K_R x$, $c = K_T/K_R$, and x is partial pressure of oxygen. Fractional saturation of the oxygen carrier with oxygen is given by

$$Y = \frac{1}{N}\frac{d\ln P}{d\ln x} = \frac{1}{N}\frac{P'}{P} = \frac{\alpha(1 + \alpha)^{N-1} + L_0 c\alpha(1 + c\alpha)^{N-1}}{(1 + \alpha)^N + L_0(1 + c\alpha)^N} \tag{2}$$

where $P' = dP/d\ln x$.

Linear MWC plot

From Eq.(2) it follows that

$$\log \frac{1 + H/\alpha}{H/\alpha - c} = (N - 1)\log \frac{1 + c\alpha}{1 + \alpha} + \log L_0. \tag{3}$$

Here, $H = Y/(1 - Y)$. Eq.(3) indicates that, if the oxygenation accords to the MWC model and the values of K_R and c (or K_T and K_R) are already known, plotting of the two logarithmic terms against each other should yield a straight line with a slope of $N - 1$ and an intercept of $\log L_0$ on the ordinate (11).

Eq.(3) can be expressed in a slightly different way to show clearer physical meanings of each term (3). By making a substitution: $Y/(1 - Y)/x = H/x = Q$, Eq.(3) is written as

$$\log \frac{K_R - Q}{Q - K_T} = (N - 1)\log \frac{1 + K_T x}{1 + K_R x} + \log L_0 \tag{4}$$

and, further, as

$$\log Z = (N - 1)\log X + \log L_0, \tag{5}$$

where $Z = (K_R - Q)/(Q - K_T)$ and $X = (1 + K_T x)/(1 + K_R x)$.

Since Y and $1 - Y$ are proportional to oxygenated and deoxygenated forms, respectively, Q is an apparent equilibrium constant which expresses the oxygen affinity of oxygen carrier at a given x (or Y). $K_R - Q$ is the difference in affinity between the R state and the oxygen carrier at a given x; $Q - K_T$ is that between the oxygen carrier at the given x and the T state. Z is a ratio of these two quantities. Since $1 + K_T x$ and $1 + K_R x$ are the binding polynomials for the T and R states, respectively, $\log X$ is proportional to difference between the free energies of oxygenation for the two states.

Fig. 1 shows a model plot according to Eq.(5). Analysis according to this linear plot using representative oxygenation data for human hemoglobin (7) indicated that $N = 4$, just as expected.

Numerical method

Slope of the Hill plot at a given x is expressed as:

$$n = \frac{d\ln H}{d\ln x} = \left(\frac{P''}{P'} - NY\right)\Big/(1 - Y) \tag{6}$$

where $P'' = dP'/d\ln x$. When the oxygenation curve is symmetrical with respect to the half-saturation point, it follows that

$$L_0 = c^{-N/2} \text{ and } \alpha = c^{-1/2}. \tag{7}$$

Then, n at the half-saturation point is given by

$$n_{50} = 2[1 + (N - 1)(1 + c)/(1 + c^{1/2})^2] - N, \tag{8}$$

and, therefore,

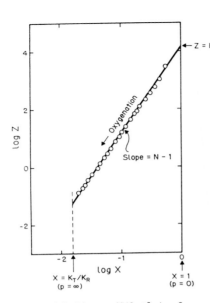

Fig. 1. A model linear MWC plot of oxygenation according to Eq.(5). p is partial pressure of oxygen.

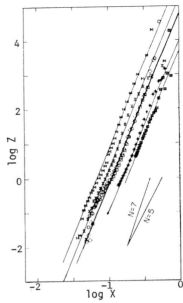

Fig. 2. Linear MWC plots for *Potamilla* Chl. See Table 1 for experimental conditions and meaning of symbols. The straight lines were calculated from Eq. (5) using N = 6 and the MWC parameters listed in Table 1. Duplicated from ref. 3.

Table 1. Oxygenation parameters for *Potamilla* Chl[a]

Plot symbol[b]	pH[c]	P_{50}	n_{50}	K_T	K_R	L_0[d]	c	N'
□	8.93	9.6	3.4	0.027	0.65	6.0×10^4	0.042	6.5
◇	8.42	11.5	3.9	0.018	0.55	6.5×10^4	0.033	7.0
⊠	7.89	20.6	4.9	0.0070	0.36	1.7×10^5	0.019	7.9
	7.39	49	5.2	0.0034	0.35	2.5×10^7	0.0097	7.2
⋈	7.32	62	4.7	0.0033	0.15	6.5×10^5	0.022	7.7
✳	6.76	186	2.5	0.0026	0.028	1.0×10^4	0.093	6.3
⊠	6.15	308	1.4	0.0025	0.015	5.0×10^3	0.17	3.3

a, Data except for N' values are from ref.(3). *b*, Refers to the symbols used in Fig 2. *c*, Other conditions: Chl concentration, 60 μM on a heme basis; 25°; in 0.05 M Tris buffer or bis-Tris buffer containing 0.1 M Cl⁻ and 10 mM MgCl₂. *d*, Evaluated for the case, N = 6 (Eq.(5)).

$$N' = (n_{50} + 2\lambda - 2)/(2\lambda - 1) \qquad (9)$$

where $\lambda = (1 + c)/(1 + c^{1/2})^2$ and N' substitutes for N so that it is distinguished from N obtained by the linear plot.

Eq.(9) that was derived by Colosimo et al. (12) indicates that the number of oxygen binding sites contained in allosteric unit can be inferred from the values of n_{50} and c. The n_{50} and c values of human hemoglobin (7) give N' values that are very close to 4 except under special conditions where the oxygenation curves are significantly asymmetric, e.g. in the presence of inositol hexaphosphate.

K_T, K_R, and P_{50} are given in $mmHg^{-1}$, $mmHg^{-1}$, and $mmHg$ throughout this paper.

RESULTS

Potamilla Chl

Oxygenation data for *Potamilla* Chl determined under a variety of solution conditions were analyzed by means of the linear MWC plot and N was found to be 6 (3). Linear plots at different pH values in the presence of 0.1 M Cl^- and 10 mM $MgCl_2$ are presented in Fig. 2, showing that the slope is definitely 5 (N = 6). The straight lines were calculated from Eq.(5) with N = 6. Values of the oxygenation parameters are listed in Table 1. Oxygenation data in the absence of $MgCl_2$ were also consistent with N = 6 (3). The analysis based on the linear MWC plot indicates that the allosteric unit for *Potamilla* Chl is the one containing six heme groups.

On the other hand, the N' values obtained from n_{50} and c (Eq.(9))

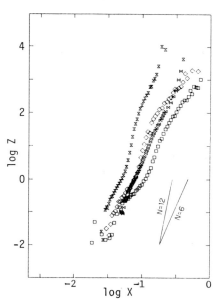

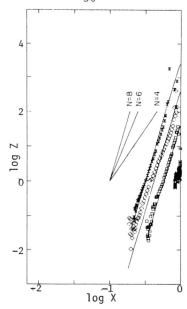

Fig. 3. MWC plots according to Eq.(5) for *Lumbricus* Ec. See Table 2 for experimental conditions and meaning of symbols.

Fig. 4. Linear MWC plots for *Lingula* Hr. See Table 3 for experimental conditions and meaning of symbols. The straight lines were calculated from Eq.(5) using N = 8 and the MWC parameters listed in Table 3.

Table 2. Oxygenation parameters for *Lumbricus* Ec[a]

Plot symbol[b]	pH[c]	P_{50}	n_{50}	K_T	K_R	c	N[d]	N'
□	8.88	1.6	4.9	0.052	3.7	0.014	8.0	7.3
	8.48	1.9	5.5	0.047	3.9	0.012	9.4	8.0
◇	8.10	2.6	6.4	0.031	3.7	0.0084	10.5	8.8
	7.73	3.9	6.9	0.032	3.5	0.0091	12.4	9.7
⊠	7.36	6.9	7.1	0.028	2.3	0.012	11.9	10.5
	6.98	11.6	4.2	0.027	1.0	0.027	9.7	7.2
⋈	6.58	15.5	2.7	0.025	0.62	0.040	6.2	4.8
	6.20	17.2	2.3	0.025	0.44	0.057	5.7	4.4

a, Data from Fushitani, K., Imai, K., and Riggs, A.F. (manuscript, in preparation). Also see the paper of them in this book. b, Refers to the symbols used in Fig. 3. c, Other conditions: Ec concentration, 60 μM on a heme basis; 25°C; in 0.05 M bis-Tris-propane buffer containing 0.1 M NaCl and 25 mM $CaCl_2$. d, Number of interacting oxygen binding sites as obtained from the maximum slope of the MWC plots in Fig.3.

are diverse and give no conclusive number of heme groups involved in heme-heme interactions (Table 1).

Lumbricus Ec

The experimental plots according to Eq.(5) are not simple straight lines; in most cases the slope tends to become smaller at both the ends (Fig. 3). Values of N obtained from the maximum slope at the middle portion are listed in Table 2. The N value depends on solution conditions and varies with pH in parallel with n_{50} within a range between N = 6 and N = 12.

The N' value obtained from n_{50} and c also shows behavior similar to that of N, but the former is somewhat and consistently smaller than the latter (Table 2).

Lingula Hr

The experimental plots at seven different pH values in 0.03 M bis-Tris-propane buffer containing 0.064 M Cl⁻ are essentially linear, giving an N value of 8 (Fig. 4). Values of L_0 were obtained by fitting the straight lines to the plots in Fig. 4 and are listed in Table 3 together with the values of K_T and K_R. Oxygenation data obtained using 0.07 M phosphate buffer of different pH values also gave the same result, although not presented here. Thus, the analysis based on the linear MWC plot indicates that the allosteric unit for *Lingula* Hr is the whole octameric molecule.

The N' values for this Hr range between 5 and 7 except for that at pH 6.53 (Table 3). The oxygenation data obtained in phosphate buffer showed more diverse values of N' (5.5 to 10). No consistent

Table 3. Oxygenation parameters for *Lingula* Hr[a]

Plot symbol[b]	pH[c]	P_{50}	n_{50}	K_T	K_R	L_0[d]	c	N'
□	8.18	3.5	1.43	0.15	0.46	4.4×10	0.33	6.7
	7.87	4.5	1.65	0.089	0.46	3.6×10^2	0.19	5.3
◇	7.64	5.9	1.77	0.066	0.37	4.3×10^2	0.18	5.7
	7.42	8.2	1.77	0.070	0.35	3.2×10^3	0.20	6.4
✕	7.32	10.2	1.71	0.057	0.28	3.1×10^3	0.20	6.1
	6.93	17.1	1.18	0.053	0.15	1.4×10^3	0.35	5.1
⋈	6.53	18.5	1.00	0.050	0.060	2.3	0.83	10.6

a, Data from ref. 4. b, Refers to the symbols used in Fig. 4. c, Other conditions: Hr concentration, 50 µM O_2 binding site; 25°C; in 0.03 M bis-Tris-propane buffer containing 0.064 M Cl^-. d, Evaluated for the case, N = 8 (Eq.(5)).

value of N' can be obtained from the numeric method.

DISCUSSION

Since the structure of *Potamilla* Chl is not known well, it is difficult to find exact correspondence between the functional and structural units. If the molecular structure of *Potamilla* Chl is the same as that of *Spirographis* Chl which has 72 heme groups per whole molecule (13), the submultiple, i.e. the structural unit having 1/12 molecular mass, would correspond to the allosteric unit that was inferred in the present study. Then, the heme-heme interactions would reside in each submultiple of *Potamilla* Chl; the number, six, of the interacting heme groups is sufficient to account for the observed cooperativity (maximum n of 5.82 (3)) of this Chl.

The allosteric unit for *Lumbricus* Ec inferred in the present study contains 12 heme groups under moderate pH conditions. According to previous studies by Rossi-Fanelli et al. (14) and Vinogradov et al. (15), Ec contains 144 to 146 heme groups per molecule. This would immediately imply that the submultiple corresponds to the allosteric unit. However, recent experiments give higher molecular mass for Ec ranging 3.8 to 4.0 MDa than that observed previously for annelid Ec's (16), suggesting that the number of heme groups can be as large as 192 per molecule (or 16 per submultiple). Then, it appears to be difficult to define a structural unit that corresponds to the allosteric unit. Further, the deviation of experimental plots from linearity (Fig. 3) implies that the allosteric unit for *Lumbricus* Ec is not constant in size; it tends to become smaller at extremes of oxygen saturation. The size of the unit also depends on solution conditions, becoming smaller as pH becomes either lower or higher than pH 8

374

(Table 2). The apparent variability of allosteric unit in *Lumbricus* Ec makes it difficult to correlate it with protein structure. It is noteworthy that cooperativity is substantially preserved in free subunits isolated from *Lumbricus* Ec that dissociated to various degrees (17; also see the paper by Fushitani, K., Imai, K., and Riggs, A.F. in this book).

The linear MWC plot provides a convenient means to infer the allosteric unit when K_T and K_R are already determined. When the size of the unit apparently depends on oxygen saturation, we may choose two alternative ways: one is to extend the original MWC model and the other is to reject that model.

The numerical method by Colosimo et al. (12) is simple and easy, but needs precautions. Since N' is determined from local properties of oxygenation curve, its value obtained from a single curve is not necessarily definitive. The N' value loses its meaning as the curve becomes highly asymmetric.

REFERENCES

1. Antonini, E., and Chiancone, E., Annu. Rev. Biophys. Bioeng. 6, 239-271 (1977).
2. Chung, M.C.M., and Ellerton, H.D., Prog. Biophys. Mol. Biol. 35 53-102 (1979).
3. Imai, K., and Yoshikawa, S., Eur. J. Biochem. 147, 453-463 (1985).
4. Imai, K., Takizawa, H., Tachiiri, Y., Handa, T., Yamamura, T., Satake, K., and Kihara, H., the 40th Annual Meeting of the Japanese Society of Physics, Kyoto (1985).
5. Imai, K., Morimoto, H., Kotani, M., Watari, H., Hirata, W., and Kuroda, M., Biochim. Biophys. Acta 200, 189-196 (1970).
6. Imai, K., Methods Enzymol. 76, 438-449 (1981).
7. Imai, K., Allosteric Effects in Haemoglobin, Cambridge University Press, Cambridge (1982).
8. Monod, J., Wyman, J., and Changeux, J.-P., J. Mol. Biol. 12, 88-118 (1965).
9. Wyman, J., J. Mol. Biol. 11, 631-644 (1965).
10. Wyman, J., J. Amer. Chem. Soc. 89, 2202-2218 (1967).
11. Decker, H., Savel, A., Linzen, B., and Van Holde, K.E., in Structure and Function of Invertebrate Respiratory Proteins (Wood, E.J., ed.), pp. 251-256, Harwood Acad. Pub., Chür/London/New York (1983).
12. Colosimo, A., Brunori, M., and Wyman, J., Biophys. Chem. 2, 338-344 (1974).
13. Antonini, E., Rossi-Fanelli, A., and Caputo, A., Arch. Biochem. Biophys. 97, 343-350 (1962).
14. Rossi-Fanelli, M.R., Chiancone, E., Vecchini, P., and Antonini, E., Arch. Biochem. Biophys. 141, 278-283 (1970).
15. Vinogradov, S.N., Schlom, J.M., Hall, B.C., Kapp, O.H., and Mizukami, H., Biochim. Biophys. Acta 492, 136-155 (1977).
16. Vinogradov, S.N., Schlom, J.M., Kapp, O.H., and Frossard, P., Comp. Biochem. Physiol. 67B, 1-16 (1980).
17. Kapp, O.H., Zetye, L.A., Henry, R.L., and Vinogradov, S.N., Biochem. Biophys. Res. Commun. 101, 509-516 (1981).

A COOPERATIVE MODEL FOR LIGAND BINDING AS APPLIED TO OXYGEN CARRIERS

Massimo Coletta, Enrico Di Cera, and Maurizio Brunori
CNR Center for Molecular Biology and Institute of
Chemistry, University of Rome "La Sapienza"
00185 Rome, ITALY

SUMMARY

This paper presents a thermodynamic model describing the cooperativity of ligand binding by a multimeric macromolecule. A multisubunit protein can be partitioned into a number of non interacting functional constellations, each one existing in two possible quaternary conformations. Each functional constellation is itself partitioned into a number of subsets, called "cooperons", where interactions occur according to an induced-fit mechanism.

This model is applied to the description of the functional properties of ligand binding to large oxygen-carrying proteins, such as hemocyanins and erythrocruorins.

INTRODUCTION

The thermodynamic and kinetic analysis of ligand binding to a multisite macromolecule often requires the use of a model which describes satisfactorily the physico-chemical behaviour of the molecule making use of a limited number of physically meaningful parameters. This need is mostly in order for those proteins which display a large number of homotropic sites (usually exceeding ∼4) acting in a cooperative fashion. In the last 20 years two theoretical formulations have been mo-

Invertebrate Oxygen Carriers
Ed. by Bernt Linzen
© Springer–Verlag Berlin Heidelberg 1986

re extensively employed and discussed, namely the allosteric two-state model (MWC) [1] and the sequential induced-fit model (KNF) [2].

Both models can account, to a first approximation, for the thermodynamic and kinetic properties of ligand binding to several oligomeric proteins, but for hemoproteins and hemocyanins the MWC model has been more widely used, although a rigorous and quantitative description of the data has required modifications of the original formulation [3,4, 5,6,7,8].

In this paper we report on a model, characterized by some of the features of both the MWC and the KNF models, which therefore represents an original integration of the two models.

THE MODEL

The model postulates that, in the absence of a ligand, a functional constellation can exist in only two quaternary states, T_0 and R_0, in equilibrium $L_0 = [T_0]/[R_0]$, as in the MWC model [1]. Within a functional constellation the subunits are segregated into smaller functional units, called "cooperons"; the number of subunits in a cooperon is called "valency" of the cooperon. Ligand binding to one subunit may affect the pairwise interactions only with the other subunits in the cooperon, as in the induced-fit model [2]. Furthermore, interaction between cooperons follows a concerted mechanism with reference to the quaternary structure of the whole functional constellation.

In the present version the following simplifications have been applied:
1) the macromolecule is in a single state of aggregation and is formed of m identical and non interacting functional constellations [6].
2) all the cooperons are dimeric (valency = 2).
3) functional equivalence of the subunits in a cooperon is assumed.
4) two possible tertiary structures are postulated for the T state (t_0 and t_1) as well as for the R state (r_0 and r_1).

5) both the $t_0 \rightarrow t_1$ and $r_0 \rightarrow r_1$ transitions are ligand-linked, following the induced-fit mechanism [2].

From these assumptions the binding polynomial of the functional
constellation in the T state is

$$P_T (x) = (\tau_{00} + 2\tau_{01} K_t x + \tau_{11} K_t^2 x^2)^z \qquad \text{(Equation I)}$$

and for the R state is

$$P_R (x) = (\varrho_{00} + 2\varrho_{01} K_r x + \varrho_{11} K_r^2 x^2)^z \qquad \text{(Equation II)}$$

where: x is the ligand activity, z is the number of cooperons which
form the functional constellation, K_t and K_r are the intrinsic affini-
ty constant of the T and the R states respectively. $\tau_{00}, \tau_{01}, \tau_{11}, \varrho_{00}$
ϱ_{01} and ϱ_{11} are the stabilization factors [2] of the dimer configura-
tions (0= empty site; 1= liganded site) within both quaternary states.

Therefore, the binding polynomial for the whole macromolecule is

$$P_M (x) = \left[(L_0(1+2K_T x+\delta_T K_T^2 x^2)^z + (1+2K_R x+\delta_R K_R^2 x^2)^z)/(1+L_0) \right]^{\bar{m}} \qquad \text{(Eq. III)}$$

and the saturation function $\bar{Y}$ becomes (Eq. IV)

$$\bar{Y} = \left[L_0 M_T (1+2K_T x+\delta_T K_T^2 x^2)^{z-1} + M_R (1+2K_R x+\delta_R K_R^2 x^2)^{z-1} \right]/(P_M(x))^{1/m}$$

where $M_i = K_i x (1+K_i \delta_i x)$ (i= T or R) and the five parameters of Eq. III
and IV are related to the stabilization factors in Eq. I and II according to the following relationships:

$$L_0 = (\tau_{00}/\varrho_{00})^z \quad ; \quad K_T = (\tau_{01}/\tau_{00}) K_t \quad ; \quad K_R = (\varrho_{01}/\varrho_{00}) K_r \quad ;$$
$$\delta_T = \tau_{00} \tau_{11}/(\tau_{01})^2 \quad ; \quad \delta_R = \varrho_{00} \varrho_{11}/(\varrho_{01})^2$$

Thus, each state is characterized by an observed equilibrium constant for the binding of the ligand (K_T or K_R) and an interaction constant (δ_T or δ_R) such that cooperativity may occur also without a quaternary transition, localized at the interface between the two subunits forming the cooperon.

Within this scheme, heterotropic ligands can display their effects on the functional properties of a macromolecule by altering the stabilization factors of the conformational arrangements in the cooperon. Moreover, it is worthwhile to point out that from the relationships given above it appears that the stabilization factors provide a linkage between tertiary and quaternary equilibrium constants, which is a special feature of this model.

APPLICATIONS OF THE MODEL

It is important to remark that for the cases considered in this paper, no need for $\delta_R \neq 1$ was required to fit the data, thus reducing the number of parameters which describe the system to 4 and implying that only within the T state cooperons display interaction phenomena.

Although the model can be applied also to oligomeric oxygen carriers, such as hemoglobin, it appears especially suited to describe the functional properties of very large multimeric macromolecules, like hemocyanins and erythrocruorins.

With the present model it has been possible to fit the whole set of oxygen binding isotherms of Helix pomatia β-Hcy at several pH values using a constant number of sites for a constellation (=16; z=8), while previous work [9] demanded a pH dependence of the number of sites. Although the pH range is insufficient to determine a pK_a value for the proton-induced variation of each parameter, a monotonic pH dependence of K_T, δ_T and L_0 is clear (see Table I). The heterotropic effect in the T state arises mostly from a pH effect on τ_{00}, thus from a mixed tertiary and quaternary effect (see above).

A second interesting example is represented by the hexameric hemocyanin from the crab Scylla serrata, which displays a small but definite cooperativity (n = 1.2) in binding carbon monoxide [10], at variance with many other hemocyanins investigated, which display a non cooperative CO binding behaviour [11,12]. The data on Scylla serrata Hcy cannot be described with the simple MWC model, since the value of

TABLE I

Thermodynamic parameters of ligand binding to <u>Helix pomatia</u> β-Hcy (O_2) and <u>Scylla serrata</u> Hcy (CO)

<u>Helix pomatia</u>

pH	$L_0(=[T_0]/[R_0])$	$K_T(mmHg^{-1})$	δ_T	$K_R(mmHg^{-1})$	δ_R
7.00	62.8 ± 12.8	0.24 ± 0.02	0.011 ± 0.01	0.92 ± 0.03	1.0
7.02	$1.77(\pm0.7)\times10^4$	0.13 ± 0.01	0.030 ± 0.02	0.89 ± 0.02	1.0
7.10	$2.28(\pm1.1)\times10^5$	0.10 ± 0.007	0.040 ± 0.03	0.79 ± 0.01	1.0
7.18	$2.31(\pm0.9)\times10^9$	0.08 ± 0.002	0.050 ± 0.006	0.81 ± 0.004	1.0
7.40	$2.71(\pm2.1)\times10^{12}$	0.05 ± 0.002	0.110 ± 0.02	0.73 ± 0.005	1.0
8.00	$3.75(\pm3.7)\times10^{16}$	0.03 ± 0.001	2.640 ± 0.26	0.69 ± 0.005	1.0
9.00	$1.56(\pm1.5)\times10^{18}$	0.03 ± 0.001	2.290 ± 0.2	0.68 ± 0.005	1.0

<u>Scylla serrata</u>

pH	L_0	K_T	δ_T	K_R	δ_R
8.00	5000	0.01	2.3	0.01	1.0

L_0 obtained from CO binding is very different from that resulting from oxygen binding [10]; thus a three state model was proposed [8]. Analysis of the CO binding data according to the cooperon model, assuming a functional constellation of 6 sites (z=3), is satisfactory. Furthermore, by comparison with the analysis carried out following the three state model [8], the cooperon model is able to provide a description of the data, making use of a single value of L_0, consistent with that obtained from the oxygen binding isotherm, simply implying a $\delta_T>1$ (see Fig. 1).

From this analysis, limited to hemocyanins, it is evident that this model can be considered a useful and logical extension of the MWC and KNF models, having the potentialities to describe more satisfactorily larger sets of data under several experimental conditions and with a

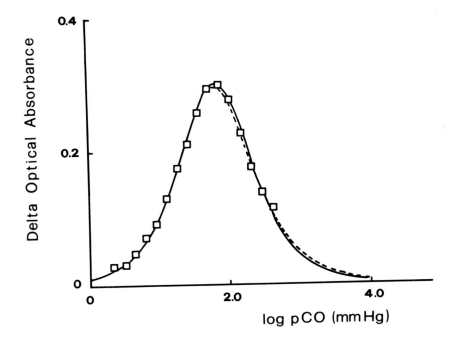

Fig. 1.
CO binding isotherm to Scylla serrata Hcy.
(Dashed line) Fitting according to the three-state
model [8].(Continuous line) Simulation according to
the model described in this paper, using the parameters reported in Table I.

limited number of parameters.

REFERENCES

1. Monod, J., Wyman, J., and Changeux, J.-P., J. Mol. Biol. 12, 88-
   118 (1965).
2. Koshland, D.E.jr., Nemethy, G., and Filmer, D., Biochemistry 5,
   365-385 (1966).
3. Szabo, A., and Karplus, M., J. Mol. Biol. 72, 163-197 (1972).

4. Ogata, R.T., and McConnell, H.M., Proc. Natl. Acad. Sci. USA 69, 335-339 (1972).
5. Minton, A.P., and Imai, K., Proc. Natl. Acad. Sci. USA 71, 1418-1421 (1974).
6. Colosimo, A., Brunori, M., and Wyman, J., Biophys. Chem. 2, 338-344 (1974).
7. Arisaka, F., and Van Holde, K.E., J. Mol. Biol. 134, 41-73 (1979).
8. Richey, B., Decker, H., and Gill, S.J., Biochemistry 24, 109-117 (1985).
9. Zolla, L., Kuiper, H.A., Vecchini, P., Antonini, E., and Brunori, M., Eur. J. Biochem. 87, 467-473 (1978).
10. Decker, H., Richey, B., and Gill, S.J., Biochem. Biophys. Res. Comm. 116, 291-296 (1983).
11. Brunori, M., Zolla, L., Kuiper, H.A., and Finazzi Agrò, A., J. Mol. Biol. 153, 1111-1123 (1981).
12. Bonaventura, C., Sullivan, B., Bonaventura, J., and Bourne, S., Biochemistry 13, 4784-4789 (1974).

Nesting – An Extension of the Allosteric Model and its Application to Tarantula Hemocyanin.

H. DECKER

Max Planck Institute for Medical Research

Department of Biophysics, Jahnstrasse 29, D-6900 Heidelberg, F.R.G.

C. H. ROBERT and S. J. GILL

Department of Chemistry and Biochemistry

University of Colorado, Boulder, Colorado 80309, U.S.A.

Summary

The coöperative oxygen-binding behavior of the 24-subunit tarantula hemocyanin (Hc) has been difficult to quantify in the context of a simple allosteric model [1]. For this reason and because of the obvious structural hierarchies present in tarantula and other Hc's, we have analyzed O_2 binding data using hierarchies of allosteric equilibrium-- a "nesting" model, originally conceived by Wyman [2]. The model can be easily generalized to account for a wide variety of structural observations as well as coöperative-binding data of Hc's or other large macromolecular assemblies.

Tarantula (Eurypelma californicum) oxygen—binding curves show strong asymmetry and high coöperativity. Attempts at fitting the binding data using a two-state MWC model with an allosteric unit of size 6, 12, or 24 sites have failed, and inclusion of another state is difficult to justify on structural grounds. When the size of the allosteric unit is allowed to vary, the curves can be fit by a two-state model, but the best-fit allosteric-unit size (~8) is not reconcilable with any known structural features [1].

When analyzed in the context of a nesting model, the macromolecule appears to be built from two dodecameric allosteric units. Each identical unit exhibits altered allosteric properties depending on the conformational state of the overall 24-mer.

Theory

In this section we will outline the idea of nesting as it applies to tarantula hemocyanin. We define nesting as the influence of macromolecular conformation on the ligand—binding properties of the composite subunits. Consider the first level of nesting, which is the influence of a macromolecular conformation on the ligand affinity of noncoöperative subunits. A good example of this concept is the allosteric model of Monod, Wyman, and Changeux [3], which postulates an allosteric molecule of n_0 binding sites which can exist in any of a number of conformational states. The macromolecule in each of these states binds ligand noncoöperatively but the subunits vary in affinity depending on the state,

Invertebrate Oxygen Carriers
Ed. by Bernt Linzen
© Springer–Verlag Berlin Heidelberg 1986

so that when conformational equilibria exist the ligand binding can be coöperative. The binding polynomial P_1 for the case of s major states is

$$P_1 = \sum_{i=1}^{s} v_i^o P_{0i}^{n_0} = \sum_{i=1}^{s} v_i^o (1 + \kappa_i x)^{n_0} \qquad (1)$$

where P_{0i} is the noncoöperative binding polynomial for a monomer in the ith allosteric state of the macromolecule, with affinity κ_i, x is the activity of ligand X, and v_i^o is the mole fraction of macromolecules present in state i in the absence of ligand. As viewed in the sense of nesting, the conformational environment controls the affinities of the "nested" subunits. In a more general approach the polynomial for the nested subunit (P_{0j}) could be formulated to provide for any specific situation by an Adair, KNF, or nonequivalent-site model. In this paper we will proceed following MWC principles.

Next we consider a macromolecule consisting of a number n_1 of these n_0-site allosteric units. The units are nested within the larger structure, which we assume can exist in different conformational states. The constants governing ligand—binding to each allosteric unit are subject to the particular conformational environment of the macromolecule as a whole. If there are t states available to the whole structure, then the resulting second—level binding polynomial P_2 is

$$P_2 = \sum_{j=1}^{t} v_j^o P_{1j}^{n_1} = \sum_{j=1}^{t} v_j^o \left(\sum_{i=1}^{s} v_{ij}^o (1 + \kappa_{ij} x)^{n_0} \right)^{n_1} \qquad (2)$$

where P_{1j} is the first—level (MWC) binding polynomial subject to the jth environment of the whole macromolecule. v_j^o has the analogous meaning to v_i^o in equation (1): here it refers to the mole fraction of macromolecules in the overall conformational state j. The j subscripts on v_i^o and κ_i indicate that the constants of the binding polynomial of each nested allosteric unit are subject to the overall conformational state. Here, also, one notes that in any given conformational environment the allosteric (coöperative) units enter independently of each other. As in the MWC model, the macromolecule can undergo a transition from a low-affinity to a high-affinity form as ligation proceeds, but in doing so it switches from one coöperative binding curve to another, thus extending the magnitude or functional range of the overall coöperative effect.

A number of high-level units like that described by equation (2) can themselves be nested in a higher-level structure, to produce a 3rd level of

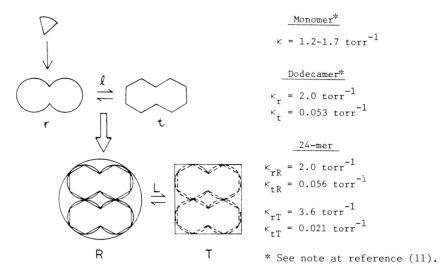

Monomer*

$\kappa = 1.2\text{-}1.7 \text{ torr}^{-1}$

Dodecamer*

$\kappa_r = 2.0 \text{ torr}^{-1}$
$\kappa_t = 0.053 \text{ torr}^{-1}$

24-mer

$\kappa_{rR} = 2.0 \text{ torr}^{-1}$
$\kappa_{tR} = 0.056 \text{ torr}^{-1}$

$\kappa_{rT} = 3.6 \text{ torr}^{-1}$
$\kappa_{tT} = 0.021 \text{ torr}^{-1}$

* See note at reference (11).

Figure 1. Binding constants for the monomer and dodecamer particle sizes are taken from references [9] and [10], respectively. Overlapping shapes represent the allosteric equilibrium of each dodecamer in the 24-site structure. Broken lines indicate the influence of conformational state (R or T) of the overall 24-site structure on the affinity and cooperativity of the nested dodecamers.

nesting. These too can be nested, and so on. In our present analysis, however, we will concern ourselves primarily with second-level nesting phenomena. Two dodecamer units (n_0=12) with two states (s=2), r and t, are nested to form a macromolecule with two states (t=2), R and T, thus giving rise to four possible oxygen—affinities κ_{rR}, κ_{tR}, κ_{rT}, and κ_{tT}. This model is shown schematically in Figure 1.

Results and Discussion

Oxygen-binding curves were conducted on tarantula Hc, pH 8.05, 0.1-M Tris buffer, 1mM Mg^{++}, 4mM Ca^{++}, 25°C. Data is shown as squares on a θ vs. $\ln(pO_2)$ plot in Figure 2 and in a Hill representation in Figure 3. The solid line shown in these plots is drawn from best-fit constants of the model just described. Species fractions for the various underlying conformations are shown in Figure 2 as well. Best-fit values of equilibrium constants are shown at the upper left in Figure 3.

In arriving at such a formulation we were guided by goodness of fit to the data of the various models we tested as well as by structural features of the Hc and reasonableness for the O_2 binding constants.

1)Goodness of fit. All data fitting was performed using the Marquardt algorithm for nonlinear least-squares optimization, modified to improve its convergence properties [4,5]. We conducted extensive simulations for each model tested in

order to arrive at good starting values for the parameters. We disregarded any
model that was unable to provide a statistically-good standard error of a point as
determined from a large number of starting guesses for the parameters.

2) Agreement with known structural features of Hc. Our model uses a dodecamer as
an allosteric unit, two of which are nested to form a 24—site structure. Tarantula
Hc is composed of two identical dodecamers, which are **not** made of identical
hexamers, owing to the seven different subunit types represented in the
physiological macromolecules [6]. The identification of the hexamers as
nonindependent entities in the dodecamer is supported by the data of Markl et al
[7], which shows both hexamers of the dodecamer sharing a dimeric ´linker´
subunit. It should be pointed out that we   first attempted a nesting scheme with
hexameric allosteric units. The model provided no satisfactory fit. We also tried
a high-level hybrid model, which does not invoke nesting, using two-state (R, T)
hexamers in the combinations RR, RT, and TT to form dodecamers. This model was
unsatisfactory as well.

3)Reasonableness of oxygen-binding constants. With a simple MWC two-state
analysis, it is often possible to associate either the R- or the T—state binding
constant of a macromolecule with that of a free subunit. For instance human HbA^o
in the oxygenated (R) form has about the same oxygen affinity as isolated α or β
subunits [8]. Tarantula Hc lends itself to comparisons of affinities at the
dodecamer level as well as at the monomer level, and binding curves on each of
these aggregation states have been performed [9,10]. The nesting model readily
permits comparison of whole-molecule constants with those applying to the
intermediate size particles. The values in Figure 1 show the affinities obtained
for the nesting model along with the affinities of the free monomers and R— and
T—state free dodecamers. One is struck by the persistence of oxygen affinities of
the building blocks as the molecule is assembled: monomers form the R—state of the
dodecameric half—molecules with no major change in affinity, and the affinities of
R— and T—state free dodecamers are preserved in the R—state of the 24-mer. Such
continuity gives credibility to the nesting model.

The Hill plot shown in Figure 3 shows how nesting permits extension of the
coöperative range of the nested subunits. The dodecamers, as MWC allosteric units,
produce coöperativity from the nested, noncoöperative single—site subunits by
allowing low—affinity subunits to become high-affinity ones through dodecameric
conformational changes. In the same way conformational changes in the 24-mer
permit conversion of nested, primarily low—affinity, coöperative dodecamers
(dotted line labelled tT at bottom) into into a mixture of high affinity ones
(upper dotted lines) as oxygen activity is increased. The result in this case is
an increase in the coöperative range accessible by the 24—mer (solid line and data
points) as compared to that of either free dodecameric form.

We have shown how a simple allosteric nesting model can be used to

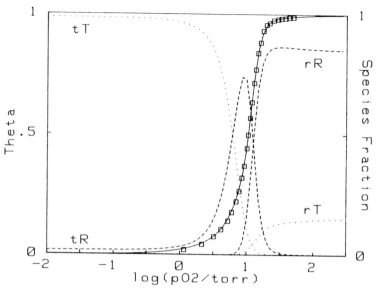

Figure 2. Fractional saturation of tarantula Hc with oxygen (squares and solid line). Other lines show species fractions: (----) Dodecamers in the R conformation of the 24-mer, with affinities rR and tR. (····) Dodecamers in the T conformation of the 24-mer, with affinities rT and tT.

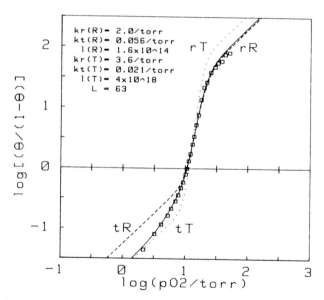

Figure 3. Hill plot of the data shown in Figure 2. As in Figure 2, the line representing theory is drawn from best-fit values of the parameters shown at upper left. (----) Hypothetical binding to pure R-state 24mer. (····) Hypothetical binding to pure T-state 24mer.

quantitatively explain the complex oxygen—binding curves of tarantula hemocyanin. In addition the model accounts for structural aspects of the macromolecule in a straightforward way. The model has potential applicability to the larger hemocyanins of arthropods and mollusks and large macromolecules in general.

Acknowledgement: We would like to express our gratitude to Jeffries Wyman for discussions and encouragement. This work was supported by National Institutes of Health grant HL22325 with travel to this meeting supported in part by NATO grant 132.82.

References

1. Decker, H., Savel. A., Linzen, B. and Van Holde, K. E., in Life Chemistry Reports Suppl. 1: Structure and Function in Invertebrate Respiratory Proteins, E. J. Wood, ed. (Harwood Academic Publishers, Chur, 1983), pp. 251-256.

2. Wyman, J. (1972) Current Topics in Cell. Reg. $\underline{6}$, 207-223.

3. Monod, J., Wyman, J., and Changeux, J. P., J. Mol. Biol. $\underline{12}$, 88-112 (1965).

4. Bevington, P. R., Data Analysis and Data Reduction in the Physical Sciences (McGraw Hill Book Co., New York, 1969), p 236.

5. Fraser, R. D. B., and Suzuki, E., in Physical Principles and Techniques of Protein Chemistry, S. J. Leach, ed. (Academic Press, New York, 1973), pp. 301-353.

6. Markl, J., Savel, A., Decker, H., and Linzen, B., Hoppe—Seyler's Z. Physiol. Chem. $\underline{361}$, 649-660 (1980).

7. Markl, J., Decker, H., Linzen, B., Schutter, W. G., van Bruggen, E. F. J., Hoppe—Seyler's Z. Physiol. Chem. $\underline{363}$, 73-87 (1982).

8. Imai, K., Allosteric Effects in Haemoglobin (Cambridge University Press, London, 1982), p. 174.

9. Decker, H., Markl, J., Loewe, R., and Linzen, B., Hoppe-Seylor's Z. Physiol. Chem. $\underline{360}$, 1505-1507 (1979).

10. Savel, A., Markl, J., and Linzen, B., this volume, pp. 399-402 (1986).

11. Equilibrium constants for monomers and dodecamers were determined at $20^{\circ}C$. Adjustments to $25^{\circ}C$ should be less than 20%.

Calorimetric Analysis of Oxygen Binding to Lobster Hemocyanin.

A. PARODY—MORREALE, C. H. ROBERT, G. A. BISHOP and S. J. GILL
Department of Chemistry and Biochemistry
University of Colorado, Boulder, Colorado 80309, U.S.A.

Background

Complete thermodynamic characterization and comparison of ligand—binding reactions of hemocyanins is motivated by the obvious structural similarities and hierarchies in these molecules. We present here a study on the thermodynamics of oxygen binding to lobster (<u>Homarus americanus</u>) hemocyanin, a dodecameric protein made up of two identical hexamers, and we compare our results to those of the well—characterized free—hexameric hemocyanin of the spiny lobster [1]. Our methods are based on the unifying principles of generalized binding phenomena, which have been introduced in detail elsewhere [2]. For this study we conducted two types of experiments. The first type was a simple binding curve, in which we recorded changes in the average saturation of the macromolecule as a function of oxygen activity. From this data we obtained values of the equilibrium constants governing the binding of oxygen. The second type of experiment was a calorimetric measurement of the average—enthalpy change of a hemocyanin solution as a function of oxygen activity. Analysis of this data gave ΔH's corresponding to the reactions involved in oxygen—binding.

Our analysis is based on the macromolecular binding—partition—function, or binding polynomial, which represents a specific equilibrium model for the system. In this paper we have used the two—state MWC model [3], which for an n—site macromolecule has the binding polynomial

$$P = \frac{1}{1+L_0} \left[(1 + \kappa_R x)^n + L_0 (1 + \kappa_T x)^n \right] \qquad (1)$$

where x is the activity of the ligand X; κ_R and κ_T are the intrinsic binding constants for ligand X to the low (R) and high (T) affinity forms of the macromolecule, and L_0 is the equilibrium constant between the R and T forms when no ligand is present. The temperature dependence of each equilibrium constant can be expressed in integrated van't Hoff form as $K = K^0 \exp[-\Delta H/R(1/T - 1/T^0)]$. The enthalpy changes for the respective reactions are ΔH_R, ΔH_T, and ΔH_{L_0}. In our analysis, the dodecamer is treated as two independent hexamers, so that the allosteric unit size n=6.

The average saturation of a macromolecule with ligand X is then given by

$$\overline{X} = \frac{\partial \ln P}{\partial \ln x} \qquad (2)$$

which is the expression for the familiar binding curve. In our experiments we measure change of optical signal (proportional to the average saturation) as we change X activity from x_{i-1} to x_i, and so represent our data by the following equation

$$\Delta OD_i / \Delta OD_{TOT} = \overline{X}_i - \overline{X}_{i-1} \qquad (3)$$

in which ΔOD_{TOT} is the optical—density change for the theoretical process of going from zero to infinite ligand activity, and must therefore be a parameter as well. Least—squares fitting to the data provides estimates of all of the parameters in equation (3).

In a way analogous to the average saturation, the average enthalpy of a macromolecule at equilibrium with ligand X can be expressed simply:

$$\overline{H} - \overline{H}_o = -R \, \frac{\partial \ln P}{\partial(1/T)} \qquad (4)$$

$\overline{H}_o$ is the reference—state enthalpy of the macromolecule, chosen here to be that of the unligated macromolecule. In this work we have used equation (4) to analyze calorimetric data obtained by titrating a solution of Hc with oxygen—saturated buffer. The heat evolved during a titration step as we move from ligand activity x_{i-1} to x_i at Hc concentration c is thus

$$q_i = c \left[(\overline{H} - \overline{H}_o)_i - (\overline{H} - \overline{H}_o)_{i-1} \right] \qquad (5)$$

We see that just as equation (2) enabled us to determine equilibrium constants for a specific model, so does equation (4) allow us to find the enthalpy changes for those equilibria.

Results and Discussion

A typical binding curve is shown at top in Figure 1. The actual data is pictured on the left—hand side: optical—density change as the pure oxygen atmosphere is diluted in logarithmic steps. Top right shows the data presented as a binding curve, $\overline{X}$ vs. $\ln(pO_2)$. The parallelism between average saturation and average enthalpy is evident from the typical calorimetric run shown in Figure 1, bottom. The data, heat absorbed as ligand titrates the macromolecule, is shown on the left and is plotted in the form of a "heat binding curve" on the right.

Using equilibrium constants obtained by fitting the delta—OD data, we analyzed the calorimetric data to determine the enthalpy ΔH_T for the MWC reaction $(Hc_T(O_2)_{j-1} + O_2 \rightleftarrows Hc_T(O_2)_j)$. However, for reasons discussed below, ΔH_R and ΔH_{L_o} were not determinable. The data permitted evaluation of ΔH_{TOT} for the total ligation process ($Hc + 6O_2 \rightleftarrows Hc(O_2)_6$) by summation of heats from each titration step, extrapolated to infinite ligand activity. It should be stressed that this value is model—independent; that is, its determination requires only a function capable of properly describing the area under a curve like that one shown in the bottom of Figure 1, left. We then obtained ΔH_{L_6}, for $R \rightleftarrows T$ of the fully ligated macromolecule, by the approximation $\Delta H_{TOT} = 6 \cdot \Delta H_T - \Delta H_{L_6}$.

The extent of the κ_R and L_o reactions proved to be highly correlated at all X activities and therefore their enthalpies could not be determined by the

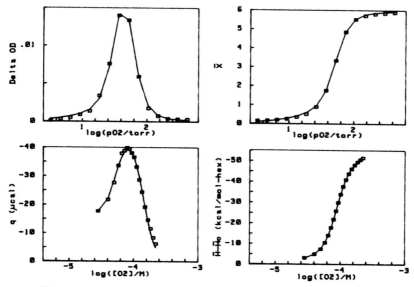

Figure 1.-

(Top) Oxygen equilibria experiment at 25°C of lobster hemocyanin in 20 mM CaCl$_2$, 0.1 M Tris buffer at pH 7.4, performed according the method of Dolman and Gill [7].

(Bottom) Calorimetric titration of a hemocyanin solution (0.31 mM in heme) in 20 mM CaCl$_2$, 0.1 M Tris, pH 7.4 with 1.05 mM O$_2$, 2.1% BSA in the same buffer. The experiment was done at 25°C using a microcalorimeter described elsewhere [8]. The experimental points show the heat effects of each injection (7 μl).

Solid lines show the best fits of the data.

calorimetric data. This difficulty appears to arise from characteristics of the MWC model itself. Equation (4) for the MWC binding polynomial can be rewritten simply: $H - H_0 = r(x) \cdot \Delta H_R + t(x) \cdot \Delta H_T + \ell(x) \cdot \Delta H_{L_0}$, where the coefficients $r(x)$, $t(x)$ and $\ell(x)$ depend on the binding constants and free—ligand activity alone and can be thought of as weighting factors for the associated reaction enthalpies. For

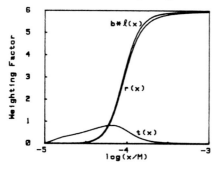

Figure 2.-
Scaling value b is -5.95.

the values of the binding constants used in Figure 1, Figure 2 reveals how highly correlated r and ℓ are, essentially differing only by a (negative) scaling factor. r and ℓ are thus related by a constant, say b, and equation (4) reduces to a form in which only the sum $(b\Delta H_R + \Delta H_{L_0})$ is resolvable.

By conducting calorimetric experiments in two different buffers, Tris and Bis—tris, we removed specific buffer—heat effects and calculated the number of protons absorbed, Δn_{H+}, for each reaction. As a rough check on one of these numbers, an additional binding experiment at pH 8.0 together with the pH 7.4 equilibrium constants permitted evaluation of Δn_{H+} for the overall reaction from $6 \cdot \partial \log p_m / \partial pH$. The value, -4.0, fortuitously agrees exactly with the result determined calorimetrically.

The standard—state thermodynamic parameters for the various reaction processes are summarized in Table 1. For the purposes of comparison, thermodynamic data for arthropod Hc's are scarce, with the exception of that for the well—characterized Panulirus interruptus (spiny lobster) system [1,4]. This Hc is present primarily in the 16S hexameric form at physiological conditions [5]. Van't Hoff studies at high pH indicate a value of -11 kcal/mole—O_2 for ΔH_T. This value is comparable to the H. americanus value (ΔH_T=-8kcal/mole—O_2). The P. interruptus results enable calculation of ΔH_{L_6}, -12 kcal/mole—hexamer, compared to the H. americanus value ΔH_{L_6}=-39 kcal/mole—hexamer found in the present study. Since when fully ligated the R form is favored in each of these systems the exothermic values for the transition as written (R⇌T) show that the reverse transition of the fully ligated macromolecule is strongly entropically driven.

The number of Bohr protons for H. americanus is similar to that for other hexameric systems [1,6]. The nearly zero value of Δn_{H+} for the T—state oxygen—binding also agrees with the relative invariance of κ_T with pH noted for P. interruptus [4].

The present study has demonstrated how titration calorimetric data can be used to determine 1) the overall heat of ligation, and 2) the underlying reaction heats within the framework of an appropriate molecular model.

Table 1.- Standard thermodynamic parameters for lobster hemocyanin oxygen-binding and related equilibria.

reaction	ΔG^o	ΔH^o	$-T\Delta S^o$	Δn_H+
$Hc_R(O_2)_{j-1} + O_2 \overset{K_R}{\rightleftharpoons} Hc_R(O_2)_j$	-7.0	-	-	-
$Hc_T(O_2)_{j-1} + O_2 \overset{K_T}{\rightleftharpoons} Hc_T(O_2)_j$	-4.7	-8	3	0.2
$Hc_R \overset{L_0}{\rightleftharpoons} Hc_T$	-8.4	-	-	-
$Hc + 6O_2 \overset{\beta_6}{\rightleftharpoons} Hc(O_2)_6$	-33	-9	-24	-4.0
$Hc_R(O_2)_6 \overset{L_6}{\rightleftharpoons} Hc_T(O_2)_6$	5.1	-39	44	5

Standard state is 1 M.
Units for ΔG^o, ΔH^o and $T\Delta S^o$ are kcal mole^{-1}.
Units for Δn_H+ are mole of proton per unit reaction.

We would like to acknowledge the counsel of Jeffries Wyman and assistance from National Institutes of Health grant HL22325 and NATO grant 132.82.

References

1. Antonini, E., Brunori, M., Colosimo, A., Kuiper, H. A., Zolla, L., Biophys. Chem. 18, 117-124 (1983).

2. Gill, S. J., Richey, B., Bishop, G., and Wyman, J., Biophys. Chem. 21, 1-14 (1985).

3. Monod, J., Wyman, J., and Changeux, J. P., J. Mol. Biol. 12, 88-112 (1965).

4. Kuiper, H. A., Coletta, M., Zolla, L., Chiancone, E., and Brunori, M., Biochim. Biophys. Acta 626, 412-416 (1980).

5. Markl, J., Hofer, A., Bauer, G., Markl, A., Kempter, B., Brenziger, M., and Linzen, B., J. Comp. Physiol. 133, 167-175 (1979).

6. Arisaka, F., and Van Holde, K. E., J. Mol. Biol. 134, 41-73 (1979).

7. Dolman, D., and Gill, S. J., Anal. Biochem. 87, 127-134 (1978).

8. McKinnon, I. R., Fall, L., Parody-Morreale, A., and Gill, S. J., Anal. Biochem. 139, 134-139 (1984).

STRUCTURAL AND FUNCTIONAL STUDIES ON THE HEMOCYANIN OF THE MANGROVE CRAB (SCYLLA SERRATA)

H. Decker[§], B. Richey[*] and S.J. Gill[*]

[§]Max Planck Institute for Medical Research, Jahnstr. 29, D-69 Heidelberg,F.R.G.and [*]Dept. of Chemistry, University of Colorado, Boulder,Co. 80309, U.S.A.

Arthropod hemocyanins are macromolecules composed of 6,12,24 or 48 subunits (1). Each subunit reversibly binds one oxygen molecule bridged between two copper atoms. The oligomers exhibit strong homotropic and heterotropic interaction. Another ligand,carbon monoxide can also bind to the same site, but in a different way, since it does not bridge the two coppers (2,3). Application of both ligands together in competition experiments on the Hc of Scylla serrata was useful in analysing the allosteric mechanism (4). This paper presents more basic information about the structural and functional properties of the hemocyanin from Scylla serrata.

Material and methods

S. serrata Hc was prepared as described previously (5). Binding studies and computer fitting were performed as described by Richey et al.(1985). SDS PAGE (polyacrylamide gel electrophoresis) and crossed immuno electrophoresis were performed as described elsewhere (6).

Results

The hemolymph proteins, salted out in ammonium sulfate, were dialysed against 0.1 M TRIS/HCl buffer, pH 8.0 in the presence of 20 mM $CaCl_2$ for 12 hrs. changing the buffer three times. The solution contained a single component (95%), sedimenting with 21 S (apparent). Gel column chromatography also revealed one peak when recorded at two different wavelengths (280 nm and 340 nm). The ratio of the absorbance at these wavelengths was 4.2 . Samples dialysed overnight against the same buffer but with 10 mM EDTA sedimented as one component with 11 S (apparent). The Hc completely dissociates into the subunits after dialysis against 0.05 M glycine/NaOH buffer, pH 9.6 in the presence of 10 mM EDTA for two days. The mixture sedimented with 4.7 S (apparent). The analysis of this mixture by SDS PAGE revealed six different bands. In a PAGE sytem without SDS hexamers, no dodecamers could be seen. This may have been due to the lack of Ca^{2+} ions in the gel system, as the ultracentrifugation studies showed that Ca^{2+} stabilises the dodecamer. The dissociation/association behaviour can be summarized schematically:

$$\text{dodecamers} \underset{\text{+EDTA}}{\overset{+Ca^{2+}}{\rightleftharpoons}} \text{hexamers} \underset{\text{+EDTA, pH} > 9}{\overset{\text{pH} < 9}{\rightleftharpoons}} \text{monomers}$$

Invertebrate Oxygen Carriers
Ed. by Bernt Linzen
© Springer–Verlag Berlin Heidelberg 1986

<u>Immunological Studies</u>: It is thought that crustacean Hc's are composed of two classes of subunits, a conservative one, α , being responsible for the interhexamer contacts, and a variable one, ß , located at the periphery (7,8,11). The Hc of <u>S</u>. <u>serrata</u> also contains these two subunit types since they were recognized by antibodies against the subunit mixture of <u>Callinectes</u> <u>sapidus</u> (Fig.1). An additional subunit, γ , is also recognized, which shares some antibody determinants with α and thus γ is thought to be antigenically deficient (E. Precht and J. Markl, pers. commun.)

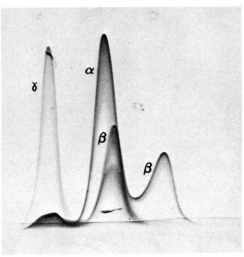

Fig. 1: Crossed immunogelelectrophoretic pattern of the dissociated Hc from <u>Scylla</u> <u>serrata</u> against antibodies made from <u>Callinectes</u> <u>sapidus</u>. The anode of the first dimenson was on the left. The immunologically destinct subunits are designed as α , ß ,and γ .

<u>Binding Studies</u>: Binding of oxygen and carbon monoxide to the dodecamers was measured at different pH (Fig.2, Tab.1). The Bohr effect is normal for both ligands (Δ log $p_{50}(O_2)$/ ΔpH=-0.89; Δ log $p_{50}(CO)$/ ΔpH=-0.35) and for both the binding is cooperative over the pH range measured. The binding behaviours of dodecamers, hexamers and monomers were compared (Tab.2), although the solution conditions were not identical. While dodecamers bind both ligands cooperatively, hexamers show cooperative binding only for oxygen: The CO binding might be cooperative but the CO affinity too low to ligate the hexamer completely. The oxygen and CO binding of the monomer mixtures are characterized by a Hill coefficient of 1.0 . Thus the subunit mixture does not show any heterogeneity in binding. In all cases the affinity for oxygen was much higher than for CO.

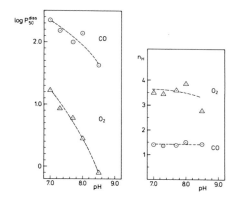

Fig 2. pH dependence of the binding affinity (a) and cooperativity (b) of the dodecameric Hc to oxygen and carbon monoxide. The conditions were the same as in table 1.

Discussion

The close relationship of the two crabs Scylla serrata and Callinectes sapidus, both members of the family Portunidae, is demonstrated by crossed immunoelectrophoresis (Fig.1). As many crustaceans the native form of the Hc is the dodecamer. The dimerisation of the two hexamers is stabilized by Ca^{2+} as reported extensively for the Hc of Homarus americanus (9). The reason for the structural heterogeneity of the monomers may be seen in their arrangement within the oligomers as reported for other Hc´s (10). The α subunit might be involved in the interhexameric contact as reported for other crustacean hemocyanins (11). The oxygen and CO binding behaviour of the heterogeneous monomer mixture is homogeneous and not cooperative. The association of the subunits to dodecamers and hexamers establishes homotropic and heterotopic properties. The high CO affinity of the Hc of S. serrata as compared with other Hcs allows one to obtain complete binding curves, and a weak cooperativity for CO binding has been shown definitely (4,5). According to Richey et al.(1985) the comparison of the ratio of the binding affinities gives information about the local structure around the binding sites of the subunits. The pH dependence of this ratio (Tab.1) indicates a ligand-dependent interaction of protons at the active site during binding. Competition experiments with both ligands may help to reveal the allosteric mechanism of these highly cooperative macromolecules. For the Hc of S. serrata it could be demonstrated that the two state MWC model (12) cannot be applied (5). Therefore a more appropiate three state model has been developed (4).

Acknowledgement: We thank Drs. J. Markl, W. Stöcker and E. Precht for the immunological experiments and helpful discussions. This work was supported by an NIH grant HL22325.

References:
1.Miller K. and van Holde K.E. (1982), Quart. Rev. of Biophys.15:1-129.
2.Bonaventura C., Sullivan B., Bonaventura J. and Bourne S. (1974), Biochemistry 13:4784-4789.
3.Brunori M., Zolla L., Kuiper H.A. and Finazzi Agro A.(1981), J. Mol. Biol. 153:1111-1123.
4.Richey B., Decker H. and Gill S.J. (1985), Biochemistry 24:109-117.
5.Decker H., Richey B. and Gill S.J. (1983), Biochem. Biophys. Res.

Commun.116:291-296.
6 Markl J., Savel A., Decker H. and Linzen B. (1980), Hoppe-Seyler's
    Z. Physiol. Chem. 361:649-660.
7.Markl J. and Kempter B. (1981),in Invertebrate Oxygen Binding
    Properties (Lamy J. and Lamy J. eds., Marcel Dekker, N.Y.).
8.Markl J., Stöcker W., Runzler R. and Precht E., this volume, pp. 281-292.
9.Morimoto K. and Kegeles G. (1971), Arch. Biochem. Biophys.
    142:247-257.
10.Markl J., Decker H., Linzen B., Schutter W.G. and van Bruggen E.F.
    J. (1982), Hoppe-Seyler's Z. Physiol. Chem. 363:73-87.
11.Stöcker W., Raeder U., Bijlholt M.M.C., Schutter W.G., Wichertjes T. and
Markl J.,this volume, pp. 213-216.
12.Monod J., Wyman J. and Changeux J.-P. (1965),J. Mol. Biol. 12:88-
    118.

TABLE 1: pH dependence of the oxygen and carbon monoxide binding of
          dodecamers. 0.1 M TRIS/HCl buffer, pH 8.0, 20 mM Ca^{2+}

pH	7.0	7.3	7.7	8.0	8.5
oxygen:					
P_{50}(Torr)	17.0	8.5	6.2	2.8	0.8
n_H	3.50	3.45	3.60	3.85	2.80
carbon monoxide:					
P_{50}(Torr)	227.3	151.5	125.0	138.9	42.6
n_H	1.4	1.35	1.38	1.50	1.42
k_1	0.004	0.006	0.006	0.006	0.019
k_2	0.006	0.008	0.013	0.009	0.030
P_{CO}/P_{O_2}	13.4	17.8	20.2	49.6	53.3

k_2, k_2 were obtained by fitting the Adair equation for two binding sites

TABLE 2: Comparison of the oxygen and CO binding behaviour of the
          dodecamers, hexamers and monomers at T=25°C. Buffer condition for
dodecamers: 0.1 M TRIS/HCl, pH 8.0, 20 mM $CaCl_2$
hexamers: 0.1 M TRIS/HCl, pH 8.0, 10 mM EDTA
monomers: 0.05 M glycine/NaOH, pH 9.6, 10 mM EDTA

Aggregation level	oxygen		carbon monoxide		ratio
	P_{50}	n_H	P_{50}	n_H	P_{CO}/P_{O_2}
dodecamers	2.8	3.8	138.9	1.5	49.6
hexamers	16.0	3.1	209.9	1	13.1
monomers	11.1	1.0	47.6	1.0	4.3

THE SPATIAL RANGE OF ALLOSTERIC INTERACTION IN A 24-MERIC ARTHROPOD HEMOCYANIN

A. Savel, J. Markl and B. Linzen
Zoologisches Institut, Universität München, Luisenstr. 14, 8000 München 2, F.R.G.

In the 37 S hemocyanin (Hc) of the tarantula, _Eurypelma californicum_, all the specific oxygen binding features arise by subunit interaction. Native tarantula Hc shows a relatively low oxygen affinity (P_{50} = 25 Torr at pH 7.5 and 25 ^{O}C), a high Bohr effect and extremely high cooperativity (n_{50} running up to > 9) which is also pH dependent (1). Isolated subunits have high oxygen affinity and no Bohr effect (2). The quaternary structure of this Hc is well known (3). It can be described as an assembly of two pairs of unequal hexamers. Each hexamer comprises five distinct monomers, a, d, e, f and g, and one half of a heterodimer, bc. So, while the whole molecule is symmetric, the half-molecule is not, strictly speaking. It may be regarded as symmetric only in so far as it is composed of two hexamers.

The present work addresses the functional interaction of the subunits, in particular the question whether the distinct morphological substructures of the native Hc - the hexamers and the two-hexamers (= half-molecules) - may be considered also as functional units which, if isolated, could by themselves bring about the observed allosteric effects. In other words: what is the range of allosteric interaction within the native hemocyanin molecule?

One approach to this problem was provided by the observation that tarantula Hc does not dissociate, under alkaline conditions, in a all-or-none fashion, but that it is possible to isolate dissociation intermediates which are stable in Tris buffers of pH 7-9 and therefore amenable to further study (4). We have previously shown that treatment of tarantula Hc with 4 M urea results in hexamers which have a very low cooperativity in O_2 binding but exhibit the full Bohr effect (5). These studies have been extended now to include the higher dissociation intermediates, viz. the heptamer, dodecamer and 19-mer. Aggregates with less than six subunits have not yet been isolated.

The methods employed to isolate and to handle the different aggregates have either been published (4) or will be described in detail elsewhere. Briefly, the 19-meric and heptameric aggregates were obtained by dissociating the Hc for short periods in dithiothreitol or under mildly alkaline conditions, and subjecting the mixture to slab gel electrophoresis and eluting the desired bands from the gel. The 19-mer was also prepared by incubating native Hc overnight in 20 % sucrose. The hexamers were prepared as previously (5) by treatment of the Hc with 4 M

urea and separating the undissociated 37 S from the 16 S material by gel filtration. The greatest difficulties were encountered with the half-molecules. These
were prepared by treating the Hc with a reducing agent (dithiothreitol, cysteine)
at neutral or acid (4.2) pH. Isolation was again by gel electrophoresis and
elution of the appropriate band. However, under the conditions of O_2 binding
measurement there was strong reassociation. This could only be prevented by
carboxymethylation of the protein. In order to examine to which extent O_2 binding
is affected by this procedure, native 37 S Hc was also carboxymethylated; P_{50} and
n_H were altered negligibly.

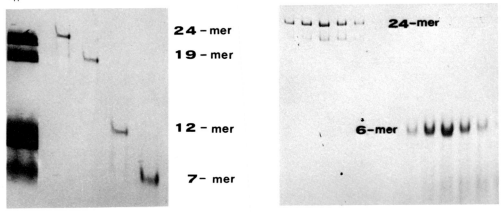

Fig. 1 (left). Polyacrylamide electrophoresis of <u>Eurypelma</u> hemocyanin after partial dissociation (0.1 M Tris/HCl, pH 7.0, containing 40 mM dithiothreitol, 30
min at $37°C$). The left lane is from a preparative gel. The four lanes to the
right are from an analytical gel showing the isolated fractions. Gel concentration
3.7 %, buffer system no. 6 in the list of Maurer (6).

Fig. 2 (right). Separation of remaining 37 S (= 24-meric) <u>Eurypelma</u> hemocyanin
and 16 S (hexameric) dissociation product after treatment with 4 M urea. The mixture was passed through a Biogel A 5m column (1.5 x 150 cm, elution rate 3.3
ml/h) and 2.5 ml-fractions taken. Aliquots were analyzed by slab gel polyacrylamide electrophoresis (conditions as in Fig. 1). The lanes correspond (from left
to right) to the fractions 7-35 (only each second fraction being analyzed).

Figs. 1 and 2 document the isolation of the products of partial dissociation. It
must be noted that the O_2 binding characteristics of Hc are affected by most operations to a variable extent, resulting in higher affinity and lower cooperativity.
Therefore it was imperative to set the conditions for dissociation in such a way
that part of the starting material remained intact, serving as an internal control
to which the data of the dissociation products were related. Cooperativity in
these control 24-mer preparations was invariably low, often falling below n_H = 4
at pH 8, but if expressed on a percentage basis ($n_{control}$ = 100 %), the n_H values
of the dissociation intermediates showed surprisingly little variation.

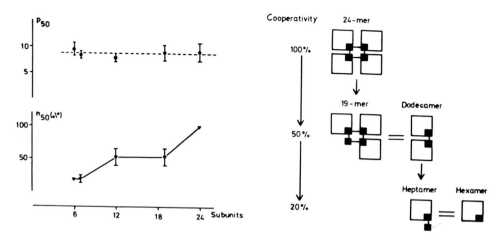

Fig. 3 (left). P_{50} and $n_{H,50\%}$ values of dissociation intermediates of Eurypelma californicum hemocyanin. The Hill coefficient is expressed as percent of the control (remaining undissociated 37 S = 24-meric) hemocyanin subjected to the same operations, e.g. electrophoresis, elution, standing overnight, as the corresponding intermediate. Vertical bars represent the S.E.M.; in case of the hexamer this was as small as the symbol of the figure.

Fig. 4 (right). Interpretation of the drop in cooperativity during dissociation of the four-hexamer Eurypelma hemocyanin. The drop occurs in two steps, each step being associated with the loss of symmetry. The squares represent the hexamers, the black corners represent the subunits b-c which form two very stable heterodimers in the centre of the native molecule.

Oxygen binding curves were measured by the fluorimetric-polarographic method of Loewe (1) with Hc concentration ranging between 5 and 150 µg/ml. Without this very sensitive method, the study would not have been feasible. The temperature was held at 20 °C, and the pH was 8. In some experiments, other pH values were chosen in addition.

The results are summarized in Fig. 3 and interpreted in Fig. 4. The oxygen affinity is the same in 37 S hemocyanin and in all smaller aggregates down to the hexamer. The Bohr effect of the hexamer has previously (5) been found to be identical with the Bohr effect of the control four-hexamer Hc. For the aggregates between hexamer and 24-mer it was therefore examined only by measurements at 3 pH values: 7.7, 8.0 and 8.4. Thereby the constancy of the Bohr effect was confirmed. It is concluded that for heterotropic modulation by protons, the hexamer represents the allosteric unit.

Cooperativity, in contrast, increases in two steps, going from the hexamer through the heptamer, dodecamer, 19-mer to the whole molecule. The highest values of n_H, significantly above those of the 19-mer, are obtained only in the four-hexamer. This shows clearly that for homotropic interaction, the whole, four-hexamer Hc molecule is the allosteric unit. The Hill coefficient is virtually independent of pH in the hexamer and heptamer, but its modulation by changing pH becomes stronger with progressive aggregation, being maximum only in the whole molecules (data not shown). It is interesting to note that the hexamer and heptamer cannot be distinguished by their mode of oxygen binding, nor can the dodecamer and 19-mer. It appears thus that in order to increase cooperativity beyond the value obtained for a defined aggregate structure, an equivalent aggregate must be added to it. In other words, only when a new symmetry is established, a new (higher) degree of cooperativity will be obtained. This seems to substantiate ideas of "subsets of allosteric interaction", as proposed by Colosimo et al. (7) and developed by Decker et al. (8) and Coletta et al. (9).

In conclusion, the present work has shown that (1) oxygen affinity and Bohr effect are not altered by increasing the state of aggregation beyond the hexamer; (2) cooperativity increases from the hexameric to the four-hexameric state; (3) only the four-hexameric, 'whole' molecule can be regarded as the allosteric unit; (4) within this unit "subsets" can be discerned which show a lower degree of cooperativity; (5) these subsets are "regular" structures, i.e. symmetrical, at the level of quaternary structure.

1. Loewe, R., J. Comp. Physiol. 128, 161-168 (1978).
2. Decker, H., Markl, J., Loewe, R., and Linzen, B., Hoppe-Seyler's Z. Physiol. Chem. 360, 1505-1507 (1979).
3. Markl, J., Kempter, B., Linzen, B., Bijlholt, M.M.C., and van Bruggen, E.F.J., Hoppe-Seyler's Z. Physiol. Chem. 362, 1631-1641 (1981).
4. Markl, J., Savel, A., and Linzen, B., Hoppe-Seyler's Z. Physiol. Chem. 362, 1255-1262 (1981).
5. Savel, A., Markl, J., and Linzen, B., in "Structure and function of invertebrate respiratory proteins" (Wood, E.J., ed.). Life Chemistry Reports, Suppl. 1. Harwood Acad. Publishers, Chur-London-New York, pp. 265-266 (1983).
6. Maurer, W., Disc-Elektrophorese. De Gruyter, Berlin (1968).
7. Colosimo, A., Brunori, M., and Wyman, J., Biophys. Chem. 2, 338-344 (1974).
8. Decker, H., Robert, C.H., and Gill, S.J., this volume, pp. 383-388 (1986).
9. Coletta, M., di Cera, E., and Brunori, M., this volume, pp. 375-381 (1986).

Acknowledgement. The financial support given by the Deutsche Forschungsgemeinschaft (Li 107/24-5+6) is gratefully acknowledged.

MERCURY IONS – A TOOL TO STUDY THE SPECIFIC ROLE OF INDIVIDUAL SUBUNITS
IN THE ALLOSTERIC INTERACTION OF ARTHROPOD HEMOCYANINS

Jürgen Markl, Anette Savel, Birgit Knabe, Heide Storz, Thomas Krabbe,
Stephan Abel and Barbara Markl
Zoologisches Institut der Universität München,
Luisenstrasse 14, 8000 München 2, F.R.G.

INTRODUCTION
Native 37S hemocyanin of the tarantula Eurypelma californicum shows a complex
oxygen binding behavior with low oxygen affinity, strong normal Bohr effect,
and high cooperativity (Hill coefficient beyond 7). All three effects are
created by an interaction of the 24 constituent subunits, because isolated
subunits have neither Bohr effect nor cooperativity, and their oxygen affinity
is high (1). The oxygen binding behavior of the native 24-mer establishes
itself by an interaction of two dodecamers which, on their part, consist of
two interacting hexamers (2).
A challenging question is, how each of the six subunit types (the monomers a,
d, e, f, g, and the heterodimer bc) does partake in the manifestation of the
native function? On the basis of a known oligomeric subunit topology (3) we
try an approach by analysing the oxygen binding of reassembled 37S hemocyanin
after removal, modification, and reincorporation of one individual subunit
type. Eurypelma hemocyanin is particularly suitable for this, because it can
be easily dissociated into subunits, reassembles to 37S particles in high
yield, and the reassembly process is very specific: each subunit type is
required to form the 37S structure, and the subunit composition of reassembled
37S equals that of native 37S (4,5).
Brouwer and coworkers have reported that mercury ions, undialysably bound to
the HS-residue of cysteine, forestall the interaction of the subunits in the
native 60S hemocyanin particle of Limulus polyphemus, without however
influencing the inherent ability of the subunits to bind oxygen (6). Our idea
was to use mercury ions to specifically inhibit the allosteric interaction of
a particular subunit within reassembled 24-mers. This is a preliminary report.

RESULTS AND DISCUSSION

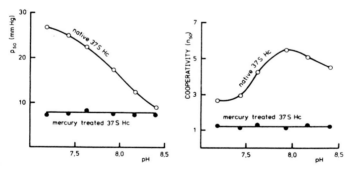

FIGURE 1: Oxygen binding of mercury-labelled 37S Eurypelma hemocyanin (Hc).
37S hemocyanin (1 mg/ml) was treated 1 hour with divalent mercury ions (0.1M
Tris/HCl buffer, pH 7.5, 1 mM HgCl$_2$), followed by extensive dialysis against
mercury-free buffer. The undialysably, most probably to cysteine, bound
mercury amounted to 1.4 atoms per subunit as measured by atomic absorption
spectrophotometry. Sedimentation analysis and electron microscopy showed that
still 37S units were present. Oxygen binding curves were recorded by
fluorometry (7). The labelling resulted in an almost total loss of the subunit
interaction phenomena: the 37S unit behavesvery much like isolated subunits.
Similar results were obtained by Brouwer et al. (6) with Limulus hemocyanin.

Invertebrate Oxygen Carriers
Ed. by Bernt Linzen
© Springer–Verlag Berlin Heidelberg 1986

404

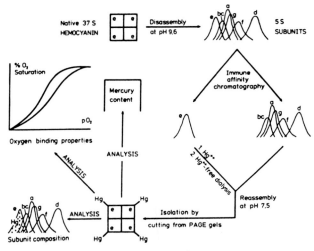

FIGURE 2: Experimental procedures and controls.
According to the results with whole mercury-labelled 37S Eurypelma hemocyanin,
it can be expected that mercury-labelled subunits still bind oxygen, but are
unable to partake in allosteric interaction processes. For the following
experiments, an important requirement was that the 37S starting material was
functionally in perfect condition. Isolation of native 37S hemocyanin from
hemolymph samples by gel chromatography (9) caused a considerable drop of
cooperativity (up to 50% of the original cooperativity of n50 = 7 was lost).
In contrast, isolation by ultracentrifugation ("pelleting") retained the
original cooperativity. A gentle procedure was to remove the Non-Respiratory
Protein (9) by immune affinity chromatography using a rabbit anti-NRP
antiserum (according to (10), IgG was bound to CNBr-activated Sepharose 4B);
this isolation procedure was frequently applied here. Also to free the mixture
from one subunit type as gently as possible, immune affinity chromatography
was carried out as described above, but with subunit-specific rabbit IgG.
Since the removed subunit could not be recovered from the matrix undenatured,
single subunits were isolated by immune affinity chromatography with 5 of the
6 IgG specifities bound to the matrix. To label a subunit with mercury, the
procedure described in Figure 1 was applied, but with glycine/NaOH buffer, pH
9.6, I = 0.05. Each subunit type undialysably bound 1-2 mercury atoms.
The procedure of removal, modification, and reincorporation is illustrated for
subunit e as an example, and was carried out in the same way also for the
other subunits. Until now we have modified and reincorporated the subunits a,
bc, d, e, and f; experiments with subunit g are planned. In case of a, d, e,
and f, considerable amounts of reassembled 37S particles could be isolated.
When bc was labelled with mercury, however, only traces of 37S were obtained.
Most of the material was present as 24S half-molecule. This dodecamer was also
isolated by gel cutting (8) and studied. Obviously, the binding of mercury
prevents the formation of the tetrameric bc-bc core, which is responsible for
the connection between dodecamers (see model, Figure 3). It was demonstrated
that the isolated 37S reassembly products, as well as the 24S particle,
contained all subunit types. The molar proportions were similar as in native
hemocyanin. Oxygen binding curves were recorded continuously by fluorometry
(7). Each experiment was controlled by a reassembled 37S reference, which had
undergone the same procedure except of the treatment with mercury.
Since one experiment lasts about 2 weeks, it was verified that isolated 37S
hemocyanin of Eurypelma (1mg/ml in 0.1M Tris/HCl buffer, pH 8.0) can be stored
at 20 degree Celsius for 8 days, or at 4 degree for at least 30 days: within
these periods, neither oxygen affinity nor cooperativity was affected.
Subunits (1mg/ml, glycine/NaOH, pH 9.6, I=0.05) can be stored correspondingly.

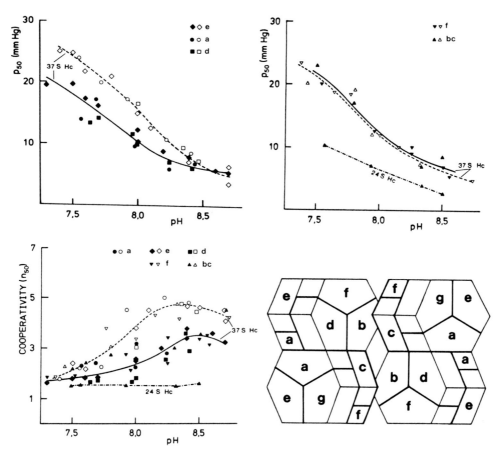

Figure 3: Oxygen binding of hemocyanin with one subunit type mercury-labelled. In the diagramms, CLOSED SYMBOLS mark 37S molecules with mercury-labelled subunits incorporated, whereas OPEN SYMBOLS mark reference 37S molecules. The topologic model of quaternary structure (3) is shown for comparison. There was no significant difference detected in the results from subunits a, d, and e. To a certain degree, all three of them help to establish the low oxygen affinity and the high cooperativity of the 37S structure, and slightly influence the Bohr effect. Reassembled 37S with the subunits f or bc labelled showed a similarly reduced cooperativity but, in contrast to the behavior of a, d, and e, in these cases oxygen affinity and Bohr effect equaled that of the 37S references. This might indicate a minor influence of these particular subunits on the overall function; however, at least for bc this is certainly not true:

Most of the mercury-labelled bc molecules were uncapable to organize the four-hexamer (37S); the reassembly process stopped at the level of the two-hexamer (24S). We believe that "efficient" bc is incompletely labelled; probably only one of the presumably two mercury binding sites is occupied. Cooperativity of the 24S reassembly product with mercury-bc was unexpectedly low – in the range of hexameric 16S intermediates (2); obviously, the interaction between the two hexamers was blocked. Also Bohr effect and p50 were strongly reduced compared to 37S hemocyanin (in contrast, 16S and 24S intermediates equal native 37S in both respects (2)). These results indicate an important functional role of subunit bc in the allosteric interaction.

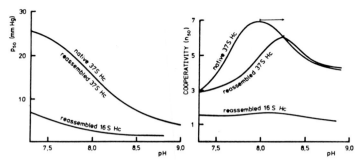

FIGURE 4: Oxygen binding of reassembled 37S and 16S Eurypelma hemocyanin (Hc). In the experiments described in Figure 3, the functional properties of reassembled 37S references were remarkably constant from preparation to preparation. It should be pointed out, however, that native 37S hemocyanin behaves different in some respects:

Dissociated hemocyanin (1 mg/ml) was reassociated to 37S by 24 hours dialysis against 0.1M Tris/HCl buffer of pH 7.5. The reassembled molecules were isolated from PAGE gels according to (8), and the oxygen binding properties analysed. Compared to native 37S hemocyanin, in reassembled 37S oxygen affinity and Bohr effect were completely restored. Cooperativity, however, was only fully recovered at the more alkaline pH values. This caused a clear shift of the curve maximum from pH 8.0 to pH 8.25.

To closer analyse the role of subunit bc, we studied a 16S fraction reassembled from a + d + e + f + g, omitting the heterodimer. Cooperativity of these molecules (n50 = 1.6) was like in 16S dissociation intermediates, which contain either b or c (2,8). However, oxygen affinity of reassembled 16S was much higher (p50 at pH 7.5 was 5 Torr compared to 25 Torr), and the Bohr effect was negligible . This again shows that not only inter- but also intra-hexamer allosteric interactions are clearly enhanced by b and c.

CONCLUSION

In native 37S hemocyanin of Eurypelma, to display the low oxygen affinity and the strong normal Bohr effect (which are completely present already in the hexameric substructure (2)), f plays a minor, a, d, and e a moderate, and bc a substantial role. To establish a low basic cooperativity within each hexamer, all subunits contribute equally; interactions between hexamers, which amplify cooperativity to the high final value, are essentially achieved by subunit bc.

Supported by the Deutsche Forschungsgemeinschaft (Ma 843/2-3)

LITERATURE
(1) Decker H., Markl J., Loewe R. & Linzen B. (1979): Hoppe-Seyler's
    Z. Physiol. Chem. 360, 1505 - 1507.
(2) Savel A., Markl J. & Linzen B. (1986): This volume. pp. 399-402.
(3) Markl J., Kempter B., Linzen B., Bijlholt M.M.C. & van Bruggen E.F.J.
    (1981): Hoppe-Seyler's Z. Physiol. Chem. 362, 1631 - 1641.
(4) Decker H., Schmid R., Markl J. & Linzen B. (1980): Hoppe-Seyler's
    Z. Physiol. Chem. 361, 1707 - 1717.
(5) Markl J., Decker H., Linzen B., Schutter W.G. & van Bruggen E.F.J (1982):
    Hoppe-Seyler's Z. Physiol. Chem. 363, 73 - 87.
(6) Brouwer M., Bonaventura C. & Bonaventura J. (1983): Biochemistry 22,
    4713 - 4723.
(7) Loewe R. (1978): J. Comp. Physiol. 128, 161 - 168.
(8) Markl J., Savel A. & Linzen B. (1981): Hoppe-Seyler's Z. Physiol. Chem.
    362, 1255 - 1262.
(9) Markl J., Schmid R., Czichos-Tiedt S. & Linzen B. (1976): Hoppe-Seyler's
    Z. Physiol. Chem. 357, 1713 - 1725.
(10) Johnson G. & Garvey J.S. (1977): J. Immunol. Meth. 15, 29 - 37.

ACTIVE-SITE HETEROGENEITY AS REVEALED BY PEROXIDE AND MERCURY: INTERACTIONS WITH PURIFIED SUBUNITS OF <u>LIMULUS</u> HEMOCYANIN.

R. W. Topham, S. Tesh, C. Bonaventura, and J. Bonaventura
Marine Biomedical and Biotechnology Center, Duke University Marine Laboratory
Beaufort, NC 28516 USA

ABSTRACT

Previous work has shown that the active sites of mollusc and arthropod hemocyanins may be differentiated by their interactions with hydrogen peroxide and with mercury. In contrast to mollusc hemocyanins, the arthropod hemocyanins are typically oxidized by hydrogen peroxide and are susceptible to loss of copper after mercury incubation. We have undertaken studies of peroxide and mercury interactions with native hemocyanins and their fractionated subunits to gain a better understanding of the extent and significance of active-site heterogeneity. We selected whole, stripped and fractionated forms of <u>Limulus</u> hemocyanin for detailed study, concentrating on Ip IIIA because of its unique characteristics. Ip IIIA, unlike the other subunits of <u>Limulus</u> hemocyanin and other arthropod hemocyanins in general, has the ability to catalyze the decomposition of hydrogen peroxide without loss of 340 nm absorbance. Also, Ip IIIA upon long-term storage slowly loses 340 nm absorbance, an aging process that can be reversed by hydrogen peroxide. It does not undergo extended self-aggregation in the presence of mercury as do the other subunits, and it loses more 340 nm absorbance than other subunits upon mercury treatment and subsequent EDTA dialysis. These exceptional characteristics make subunit IIIA of <u>Limulus</u> hemocyanin an outstanding candidate for detailed active-site study.

<u>Acknowledgments</u>

This work was supported in part by Grant ESO 1908 from the National Institutes of Health, Grant DMB 8309857 from the National Science Foundation, and Contract N00014-83-K-0016 from the United States Office of Naval Research.

INTRODUCTION

Hemocyanins, high-molecular-weight copper proteins, serve as oxygen carriers in many arthropods and molluscs. The physical and chemical properties of these proteins and the nature of the oxygen-binding site have been the subject of numerous reviews (1-6). A feature shared by all hemocyanins is a binuclear copper center where oxygen is reversibly bound. Spectroscopic studies have shown similarities and differences between the copper centers of arthropod and mollusc hemocyanins. A commonality is that in both types of molecules the copper center contains both an endogenous (protein) and an exogenous (dioxygen) ligand bridge (9). EXAFS (extended x-ray absorption fine structure) analysis (10,11) and resonance Raman data (12) strongly suggest the presence of imidazole nitrogens as the primary copper ligands. The high molecular weights of the hemocyanins and the existence of subunit diversity have hindered researchers in their attempts to specify more completely the manner in which the coppers are bound. Very recently, however, the intensive efforts of a number of research groups undertaking structural analysis of arthropod hemocyanins have yielded x-ray and amino-acid sequence information indicating that six histidine residues ligate the two oxygen-binding copper atoms (13).

Several authors have addressed the question of active-site heterogeneity among the hemocyanins (1, 5-9). Studies of Solomon and coworkers have provided spectroscopic evidence of active-site heterogeneity (7-9, 14), with one conclusion being that the active site in Limulus hemocyanin is not typical of the arthropods, being more like that of mollusc hemocyanins in several respects.

For purposes of further evaluating the extent and significance of active-site heterogeneity among hemocyanins, we have undertaken an analysis of the interactions of hydrogen peroxide and organic peroxides with whole and fractionated hemocyanins. Preliminary results of these studies have been presented elsewhere (15, 16). As a secondary probe, we have examined peroxide interactions after treatment with mercury. The rationale for these probes of active-site heterogeneity is that the hemocyanins of arthropods and molluscs have long been recognized as differing in their hydrogen peroxide interactions (17-21). Mercury as a secondary probe was suggested by the fact that mollusc and arthropod hemocyanins differ significantly in their response to mercury treatment (22).

The results presented in the following discussion focus on peroxide and mercury interactions with whole and fractionated Limulus hemocyanin and the exceptional properties of subunit Ip IIIA. The functional, immunological and electrophoretic heterogeneity of the subunits of Limulus hemocyanin has been the subject of detailed study (23-25). The active-site heterogeneity revealed in the studies reported here was not predicted by earlier data, but may be related to the fact that of the Limulus subunits, only Ip IIIA is subject to slow methemocyanin formation as a result of aging (26). This behavior is reminiscent of that observed for a number of mollusc hemocyanins that, like Ip IIIA, are able to catalyze the decomposition of hydrogen peroxide without loss of the 340 nm copper-oxygen absorbance band.

MATERIALS AND METHODS

Subunits of <u>L. polyphemus</u> hemocyanin, isolated as described by Brenowitz et al. (25-28), were used immediately after preparation or stored at 4°C. Oxygen-equilibrium experiments were performed with the tonometric method of Riggs and Wolbach (29). All experiments were conducted at 20°C unless otherwise indicated. Constant ionic-strength tris-HCl buffers were prepared as described by Bates (30). Subunits were investigated as monomers (favored by high pH and 10mM EDTA) or in aggregated states (favored by neutral pH and 10mM CaCl$_2$). The reaction of hydrogen peroxide with whole, stripped, and fractionated forms of <u>Limulus</u> hemocyanin was determined by continuously monitoring changes in the absorbance at 340 nm with a Hewlett-Packard Model 8451A Diode Array Spectrophotometer. Unless otherwise specified, the concentration of hemocyanin was adjusted with the appropriate tris buffer to an absorbance at 340 nm of the oxy form of 1.3 (4.9 mg protein/ml).

In the mercury studies, EDTA-free samples of the <u>Limulus</u> hemocyanin zones and subunits at pH 7.0 were treated with 1mM HgCl$_2$ and incubated 24 hr. The HgCl$_2$ treatment was stopped by the addition of 10mM EDTA. The samples were extensively dialyzed against tris buffer at pH 7.0, then pH 9.0, both containing 10mM EDTA.

RESULTS AND DISCUSSION

It has long been recognized that <u>Limulus</u> hemocyanin, like that of other arthropods, is susceptible to oxidation by hydrogen peroxide as evidenced by its loss of 340 nm absorbance (17). A reexamination of this phenomenon has led us to new insights regarding active-site heterogeneity among the <u>Limulus</u> subunits. Figure 1 shows the pH dependence of the interaction of hydrogen peroxide with unfractionated <u>Limulus</u> hemocyanin that has been stripped of divalent cations by extensive dialysis against 10 mM EDTA at pH 9. When treated with 100-fold excess of hydrogen peroxide, an air-equilibrated solution of stripped <u>Limulus</u> hemocyanin loses 340 nm absorbance more rapidly at pH 9 than at pH 7. As shown in previous studies (18), if the hydrogen peroxide treatment is carried out with a deoxygenated sample, the preparation is much more readily oxidized. These two effects may be attributed to the fact that the active site is resistant to peroxide-induced oxidation when it is oxygenated. Stripped solutions with lower oxygen affinity, such as the complex at pH 9, consistently show higher rates of peroxide-induced oxidation than the same materials under conditions where a higher oxygen affinity leads to a proportionately lower availability of deoxygenated copper sites. This conclusion is born out by analysis of the rate of the loss of 340 nm absorbance by peroxide-treated <u>Limulus</u> hemocyanin components that have varying oxygen affinities. Figure 2 shows that the initial rate of 340 nm absorbance decrease upon peroxide treatment is inversely related to the oxygen affinity of the preparation.

As shown in Figure 1 inserts, the absorbance characteristics of stripped hemocyanin after prolonged treatment with hydrogen peroxide are indicative of incomplete oxidation, with a more noticeable 340 nm absorbance maintained at pH 9

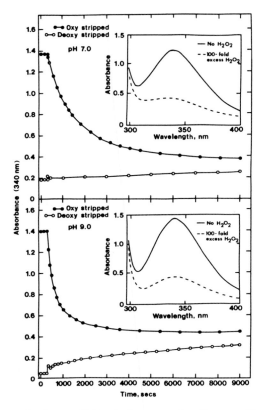

FIGURE 1. Reaction of air-equilibrated oxy and nitrogen-flushed deoxy stripped _Limulus_ hemocyanin, pH 7 (top) and pH 9 (bottom), with 100-fold excess hydrogen peroxide over copper. The insets show the spectra of the air-equilibrated samples before treatment and after 2.5 hr. Experiments were carried out at 20°C in I=0.1 tris, 10mM EDTA.

than at pH 7. It is also shown (Fig.1) that the low absorbance seen for a deoxy sample after treatment with hydrogen peroxide is subject to a slow increase that is more marked at pH 9 than at pH 7. This regeneration of 340 nm absorbance after peroxide-induced oxidation is reminiscent of the behavior of molluscan hemocyanins and a clear indication that, for this particular arthropod hemocyanin, there is some capability for oxidized active sites to become regenerated to the oxy state by hydrogen peroxide treatment.

Experiments like those shown in Figure 1 were repeated with the five major chromatographic fractions of _Limulus_ hemocyanin. As shown in Figure 3, Zone III at pH 9.0 can be more than 60% regenerated to the oxy state, while for the other zones under this condition regeneration was 20% or less. Figure 3 clearly shows that the ability of the oxidized hemocyanin fractions to be regenerated by prolonged hydrogen peroxide treatment is not correlated with their oxygen affinities. Figure 4 shows the absorbance characteristics of Zone III _Limulus_

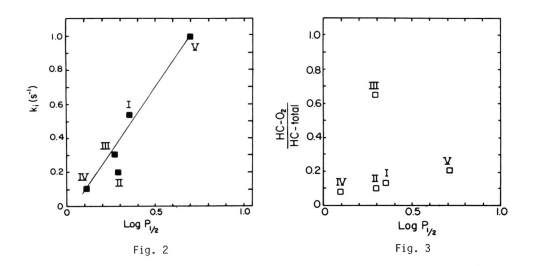

Fig. 2

Fig. 3

FIGURE 2. The initial rate of decrease of the absorbance at 340 nm upon peroxide addition is shown for the five major zones of _Limulus_ hemocyanin. Conditions as in Figure 1, pH 9.

FIGURE 3. Per cent regeneration is shown $\underline{vs}$ PO_2 at half saturation for the five major zones of _Limulus_ hemocyanin. Conditions as in Figure 1, pH 9.

hemocyanin after treatment with ethylhydroperoxide or with hydrogen peroxide. Treatment of the oxy or deoxy sample with the organic peroxide leads to a complete disappearance of the 340 nm absorbance band. The oxidation proceeds more rapidly if the hemocyanin is deoxygenated (several hours $\underline{vs}$ several days). If the sample is then treated with hydrogen peroxide, the reductive capabilities of hydrogen peroxide are manifest and the oxidized sites of Zone III are able to regain a large extent of the 340 nm absorbance. The 340 nm absorbance after this sequence is the same as that observed when oxygenated Zone III was subjected to treatment with hydrogen peroxide. These results fit the reaction scheme shown in Figure 5, a modified version of that proposed for the interaction of hydrogen peroxide with mollusc hemocyanin (17-21). Because hydrogen peroxide was not able to return oxidized Zone III to its fully oxygenated state, the next step was to examine the Zone III subunits, Ip IIIA, Ip IIIB, and the minor contributor to this fraction, Ip IIIB'. Figure 6 shows the result of experiments similar to those shown in Figure 1 for hydrogen peroxide treatment of oxy and deoxy samples of these three purified hemocyanin subunits. The failure of Zone III to regain completely its fully oxygenated state is clearly a result of subunit heterogeneity. Subunit Ip IIIA is singular among the _Limulus_ subunits in that it can catalyze the decomposition of hydrogen peroxide without loss of the copper-oxygen absorbance band. The regeneration of the oxy state from oxidized Ip IIIA occurs completely with addition of 100-fold excess of hydrogen peroxide at pH 9, but incompletely at pH 7, where a final 340 nm absorbance of 74% of the initial 100% oxy absorbance is achieved. An additional distinguishing feature of Ip IIIA is that, unlike other subunits, it tends to become oxidized upon aging, a process that can be reversed by treatment with hydrogen peroxide. Similar behavior has been observed for a number of mollusc hemocyanins.

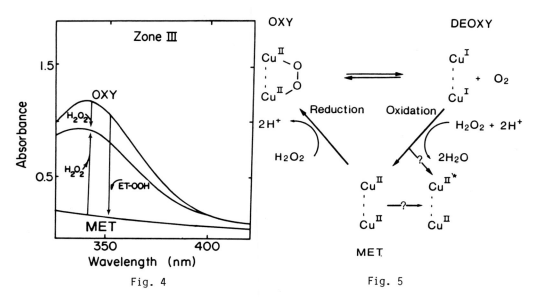

Fig. 4

Fig. 5

FIGURE 4.   Spectral changes of Zone III of *Limulus* hemocyanin brought about by addition of 100-fold excess of hydrogen peroxide. Conditions as in Figure 1, pH 9.

FIGURE 5.   A diagrammatic representation of the interaction of the binuclear copper center of hemocyanin with oxygen and hydrogen peroxide, showing the possibility of the formation of a type of oxidized hemocyanin that cannot readily be regenerated, along with the form, predominant in Ip IIIA, that can be readily regenerated to the oxy form by hydrogen peroxide addition.

Ip IIIA was then used to test further the scheme shown in Figure 5. Ethylhydroperoxide was used to oxidize Ip IIIA and then removed by extensive dialysis. The rate of regeneration of oxy hemocyanin was then followed by measuring the initial rate of absorbance change when an air-equilibrated solution was mixed with varying concentrations of hydrogen peroxide. According to the scheme of Figure 5, the regeneration of the 340 nm absorbance band should be a second order process, dependent upon the concentration of hydrogen peroxide and oxidized active sites. When experiments were carried out under pseudo-first order conditions (peroxide in at least 10-fold excess over hemocyanin active sites), the expected linear relationship with hydrogen peroxide concentration was observed (Figure 7).

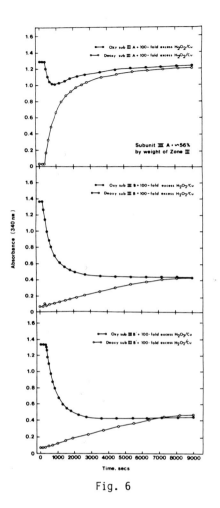

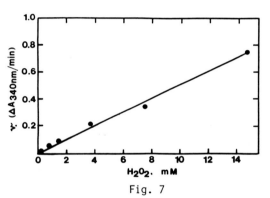

Fig. 6

Fig. 7

FIGURE 6. The reaction with hydrogen peroxide of oxy and deoxy forms of the Limulus subunits comprising chromatographic Zone III. Only Lp IIIA shows the capability of complete regeneration under these conditions (as in Fig. 1, pH 9).

FIGURE 7. The initial rate of 340 nm absorbance change (V_O) is shown for oxidized Lp IIIA after addition of varied concentrations of hydrogen peroxide. The purified subunit was oxidized by ethylhydroperoxide and then extensively dialyzed prior to the reaction. Experiments were conducted in I=0.1 tris buffer at pH 9, 20 °C.

As noted in the Introduction, mollusc and arthropod hemocyanins react differently to mercury. As an additional probe of active site heterogeneity among the Limulus hemocyanin subunits, we have initiated studies on the effects of mercury treatment. The results shown in Table 1 for the five major zones of Limulus hemocyanin and for subunits IIIA and IIIB isolated from Zone III provide a clear indication that the subunits are indeed heterogeneous with respect to mercury treatment. Visual observation of the five zones after addition of mercuric chloride provides a clear indication of subunit heterogeneity. All fractions except Zone III show extensive self-aggregation, resulting in highly turbid samples. Zone III alone remains clear. The treated samples can be clarified by the addition of EDTA, with little loss of 340 nm absorbance. Extensive dialysis against EDTA causes a change in absorbance that differs among the fractions. Zone I shows no loss of absorbance. Zone III and its constituent subunits show by far the highest degree of loss of absorbance. Subunit IIIA was selected for further study. A mercury-treated sample of Lp IIIA had, for this series of experiments, an absorbance at 340 nm of approximately 50% of that of the untreated sample. Treatment of this oxygenated material with 100-fold excess hydrogen peroxide did not regenerate the full 340 nm absorbance, but resulted in a further decrease in 340 nm absorbance. The absorbance at 340 nm was approximately 30% of the original value after 20 hours. From this experiment we must conclude that the loss of 340 nm absorbance brought about by mercury treatment was not equivalent to the oxidation brought about by peroxide and may result from loss of copper from the active site. It is not yet clear whether the treatment with mercury and subsequent EDTA dialysis, which results in loss of 50% of the absorbance band for this purified subunit, is a kinetic phenomenon or one related to active-site heterogeneity even within this purified subunit.

Table 1. Effect of 1mM $HgCl_2$ on Separated Limulus Hemocyanin Zones

Limulus sample	Oxygen-ation state	A_{340} nm before 1mM $HgCl_2$	A_{340} nm after 1mM $HgCl_2$	A_{340} nm after $HgCl_2$ and EDTA dial.	% loss of A_{340} nm
Zone I	oxy	1.15	1.11	1.15	0
Zone II	oxy	1.17	1.12	0.860	26.5
Zone III	oxy	1.14	1.11	0.404	64.5
Zone IV	oxy	1.19	1.11	0.882	25.9
Zone V	oxy	1.13	0.865	0.633	44.0
Sub IIIA	oxy	1.24	1.23	0.528	57.4
Sub IIIB	oxy	1.13	1.06	0.208	81.6

In summary, we find among the subunits of Limulus hemocyanin a remarkable diversity in their interactions with hydrogen peroxide and mercury. All of the Limulus subunits have some capability for catalase-like activity, being able to cycle repetitively between oxidized and reduced forms when exposed to hydrogen peroxide. However, the oxidized form of Lp IIIA is the only one that will undergo complete reductive regeneration in the presence of hydrogen peroxide. The regeneration of the oxy form is markedly pH dependent, occurring less readily at pH 7 than at pH 9. This pH sensitivity, not seen for molluscan hemocyanins, is being further analyzed since it may provide additional insight into factors influencing the geometry of the active site. It may be inferred from these results that different active-site geometries exist for the oxidized Limulus subunits. Further experimentation and structural analysis will be required to clarify the basis of active-site heterogeneity. It appears that comparisons between the various Limulus subunits and Lp IIIA in particular may prove to be particularly valuable in unraveling the structural basis for the disparity in catalase activity exhibited by hemocyanins of varied species.

REFERENCES

1. Van Holde, K.E., and Miller, K.I., Quart.Rev.Biophys. 15, 1-129 (1982).
2. Ellerton, H.D., Ellerton, N.F., and Robinson, H.A., Prog. Biophys. Molec. Biol. 41, 143-248 (1983).
3. Bonaventura, C., and Bonaventura, J. inThe Mollusca, (Wilbur, K., ed.), vol. 2, pp. 1-50, Academic Press, New York (1983).
4. Snyder, G.K., and Mangum, C.P. in Physiology and Biology of Horseshoe Crabs, (Bonaventura, J., Bonaventura, C., and Tesh, S., eds.), pp. 173-188, Alan R. Liss, New York (1982).
5. Lontie, R., and Witters, R., inMetal Ions in Biological Systems, (Sigel, H., ed), pp. 229-246, Marcel Dekker, New York (1981).
6. Lontie, R., Gielens, C., Groeseneken, D., Verplaetse, J., and Witters, R., inOxidases and Related Redox Systems, (King, T.E., Mason, H.S., and Morrison, M., eds.), pp.245-261, Pergamon Press, New York, 245-261 (1982).
7. Solomon, E.I., inCopper Proteins, (Spiro, T.G., ed.), pp. 41-108, John Wiley and Sons, New York (1981).
8. Eickman, N.C., Larrabee, J.A., Solomon, E.I., Lerch, C., and Spiro, T.G., J.Am.Chem.Soc. 100, 6529-6531 (1978).
9. Himmelwright, R.S., Eickman, N.C., Lublen, C.D., and Solomon, E.I., J.Am.Chem.Soc. 102, 5378-5382 (1980).
10. Brown, J.M., Powers, L., Kincaid, B., Larrabee, J.A., and Spiro, T.G., J.Am.Chem.Soc. 102, 4210-4216 (1980).
11. Co, M.S., Hodgson, K.O., Eccles, T.K., and Lontie, R., J.Am.Chem.Soc. 103, 984-986 (1981).
12. Larrabee, J. A., and Spiro, T. G., J.Am.Chem.Soc. 102, 4217-4223 (1980).
13. Linzen, B., Soeter, N. M., Riggs, A.F., Schneider, H.J., Schartau, W., Moore, M.D., Yokota, E., Behrens, P.Q., Nakashima, H., Takagi, T., Nemoto, T., N., Vereijken, J.M., Bak, H.J., Beintema, J.J., Volbeda, A., Gaykema, W.P.J., and Hol, W.G.J., Science, 229, 519-524 (1985).
14. Solomon, E.I., Eickman, N.C., Himmelwright, R.S., Hwang, Y.T., Plon, S.E., and Wilcox, D.E., inPhysiology and Biology of Horseshoe Crabs (Bonaventura, J., Bonaventura, C., and Tesh, S., eds.), pp.189-230, Alan R. Liss, New York (1982).
15. Bonaventura, C., and Bonaventura, J., Biophys.J.47, 375a (1985).
16. Bonaventura, J., and Bonaventura, C., in Respiratory pigments in animals (Lamy, J.N., Truchot, J.P., and Gilles, R., eds), pp. 21-34, Springer-Verlag Berlin (1985).
17. Felsenfeld, G., and Printz, M.P., J.Am.Chem.Soc. 81, 6259-6264 (1959).
18. Verplaetse, J., Declercq, P., Deleersnijder, W., Witters, R., and Lontie, R., inInvertebrate Oxygen-Binding Proteins (Lamy, J., and Lamy, J., eds), pp. 589-596, Marcel Dekker, New York (1981).
19. Lijnen, H.R., Witters, R., and Lontie, R., Comp.Biochem. Physiol., 63B 35 (1978).
20. Witters, R., Verplaetse, J., Lijnen, H. R., and Lontie, R.,

416

inInvertebrate Oxygen-Binding Proteins (Lamy, J., and Lamy, J., eds.), pp.597-602, Marcel Dekker, New York (1981).

21. Ghiretti, F., Arch.Biochem.Biophys. 63, 165 (1956).
22. Brouwer, M., Bonaventura, C., and Bonaventura, J., Biochemistry 22, 4713-4723 (1983).
23. Sullivan, B., Bonaventura, J., and Bonaventura, C., and Godette, G., J.Biol.Chem. 251, 7644-7648 (1976).
24. Bonaventura, J., Sullivan, B., Bonaventura, C., and Bourne, S., Biochemistry 13, 4784-4789, (1974).
25. Brenowitz, M., Bonaventura, C., and Bonaventura, J., Arch. Biochem. Biophys. 230, 238-249 (1984).
26. Brenowitz, M., Bonaventura, C., Bonaventura, J., and Gianazza, E., Arch.Biochem.Biophys. 210 748-761 (1981).
27. Brenowitz, M., and Moore, M., in Physiology and Biology of Horseshoe Crabs (Bonaventura, J., Bonaventura, C., and Tesh, S., eds.), pp. 257-267, Alan R. Liss, New York (1982).
28. Brenowitz, M., Bonaventura, C., and Bonaventura, J., Biochemistry 22, 4707-4713 (1983).
29. Riggs, A., and Wolbach, R.A., J.Gen.Physiol. 3, 585-605 (1956).
30. Bates, R.G., Determination of pH, Theory and Practice, John Wiley and Sons, New York (1973).

OXYGEN-LINKED DISSOCIATION AND OXYGEN BINDING BY SUBUNITS OF OCTOPUS DOFLEINI
HEMOCYANIN

Karen I. Miller and K.E. van Holde
Oregon State University, Corvallis, OR 97331, USA

The hemocyanin of Octopus dofleini consists of a single component with a
sedimentation coefficient of 51S. It can be reversibly dissociated in 10 mM
EDTA at pH 8.0 to yield 10 identical 11S subunits (Miller & van Holde, 1982; van
Holde & Miller, 1985). At pH near 7.0 these subunits form a ~20S putative
dimer. We have already examined the oxygen binding of the intact molecule in
considerable detail (Miller, 1985). We here report on oxygen linked dissociation in the presence of 10 mM EDTA and on oxygen binding by subunits under
these conditions.

Experimental Methods

Hemocyanin was purified from Octopus dofleini hemolymph by gel filtration
(van Holde & Miller, 1982). It was dialyzed against .1 I Tris with 10 mM EDTA
to produce subunits. Oxygen binding was performed tonometrically as described
previously (Miller, 1985). Sedimentation experiments were carried out in a
Beckman Model E ultracentrifuge equipped with scanner optics. Oxygenated
samples were loaded into the cells in air, deoxygenated samples were loaded
under N_2 in a glove bag. When partially deoxygenated samples were desired, they
were prepared in a tonometer by evacuation followed by addition of the desired
partial pressure of oxygen (as in O_2 binding). After the sample had
equilibrated several hours, the tonometer was held vertically in a N_2 filled
glove bag (to minimize loss of O_2) and a sample was withdrawn with a long needle
to fill the ultracentrifuge cell. Absorption spectra of the sample were taken
both before and after this process to be certain of the exact fraction oxygen
saturation of the loaded sample.

Results & Discussion

The transition between the 51S molecule and its subunits occurs rather
abruptly at around pH 7.0 (10 mm EDTA) in material equilibrated with air.
However, we have observed that when the subunits are deoxygenated,   transition
occurs gradually between pH 7.0 and 8.0 (Fig. 1). It should be noted that
Ricchelli et al. (1984) have detected two conformational changes in the 51S hemocyanin of O. vulgaris - one occurs just below pH 7.0, the second just above 8.0.
Thus, the pH range between 7.0 and 8.0 is characterized by a unique structure in
the decamer. The maximum difference between the association-dissociation

Invertebrate Oxygen Carriers
Ed. by Bernt Linzen
© Springer–Verlag Berlin Heidelberg 1986

equilibria for oxy and deoxy hemocyanin occurs at ~pH 7.2; at this pH in the oxygenated form we see a mixture of 11S and 20S. (The precise amounts of each depend on protein concentration, dilution favoring dissociation to monomers). The deoxy form is a mix of 51S and 20S. In order to define the nature of this O_2 dependent transition, we measured in the analytical ultracentrifuge the amounts of 51S decamer present at several partial pressures of oxygen inter- mediate between deoxygenated and fully oxygenated states. Fig. 2 shows the results of this experiment when scans were measured at 280-300 nm, the protein absorption band. In this wavelength region absorption by the copper-oxygen chromophore is negligible, so that the apparent relative amounts of decamer and subunits are not influenced by oxygenation. We find about 70% decamer at oxygen partial pressures ranging from 0 to about 35 mm Hg, corresponding to about 30% saturation of O_2 binding sites. From this point we see a steady decrease in decamer until by 200 mm Hg (above air saturation) the hemocyanin is 75-80% saturated and all decamer has dissociated. Thus, these experiments allow direct investigation of the linkage between oxygenation and dissociation under con- ditions where pH and ionic composition are fixed.

Fig. 1   Fig. 2

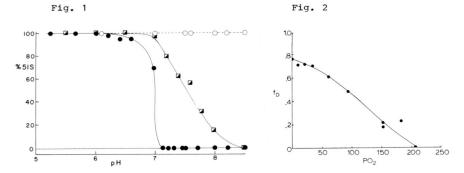

Figure 1. The effect of pH, presence of divalent cations and level of oxygena- tion on aggregation state of <u>Octopus dofleini</u> hemocyanin at 20°C. Below pH 7.0 the buffer was 0.1 M PIPES, above pH 7.0 we used 0.1 I Tris. O, 50 mM MgCl₂, oxygenated; ●, 10 mM EDTA, oxygenated; ◪, 10 mM EDTA, deoxygenated

Figure 2. The effect of PO₂ on association-dissociation equilibria of <u>Octopus</u> hemocyanin. All samples were in .1 I TRIS pH 7.2 with 10 mM EDTA, 20°C.

When we scan at 345 nm (the peak of the difference spectrum between oxy and deoxy hemocyanin) and compare the results with scans at 280 nm we observe dif- ferences in the relative saturation of the decameric hemocyanin and its subunits. We find that subunits bind oxygen with higher affinity than the intact hemocyanin under these conditions, as would be expected from the fact that oxygenation promotes dissociation. Calculations are in progress which should allow us to reconstruct from such data approximate binding curves for the two components under identical conditions.

The quality of the data in Figure 2 is such that it should be possible to calculate equilibrium constants for the reaction, and thus investigate the thermodynamics of the linkage. However, the existence of a monomer-dimer equilibrium, in addition to the equilibrium with decamer, complicates the reaction. Therefore, a detailed analysis must await investigation of the monomer-dimer reaction.

We have compared the oxygen binding of <u>Octopus</u> hemocyanin at pH 7.2 under two conditions: (a) "physiological saline," wherein the molecules remain in the 51S decameric form throughout the binding, and (b) 10 mM EDTA, in which dissociation occurs during binding as shown in Figure 2. In the latter case the experiments were conducted very slowly, with an equilibration period of at least 1 hour between O_2 additions, to allow for dissociation re-equilibration. The results are shown in Figure 3a. The curve in physiological saline is typical of those we observed in earlier studies of the 51S component; the molecule exhibits strong cooperativity. The results under dissociating conditions are complex. At low PO_2, where the hemocyanin exists as a mixture of 20S and 51S particles, the line has a slope <1, suggesting heterogeneity in binding. After dissociation begins, ($f \cong 0.30$), the curve abruptly becomes steeper, as would be expected if the dissociated material binds oxygen more strongly. The data also suggest that the 51S particles present in 10 mM EDTA at pH 7.2 do not bind O_2 with the same affinity as those present in physiological saline at the same pH.

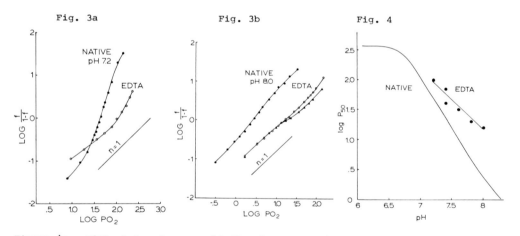

Fig. 3a   Fig. 3b   Fig. 4

Figure 3a. Hill plots of oxygen binding by <u>Octopus</u> hemocyanin at pH 7.2, 20°C.
     O, 0.1 I Tris, 10 mM EDTA, monomers + dimers at low PO_2, dimers + decamers at high PO_2; ●, 0.1 M HEPES, phys. saline, decamers
Figure 3b. Hill plots of oxygen binding by <u>Octopus</u> hemocyanin at pH 7.8 and 8.0, 20°C.
     O, 0.1 I Tris, 10 mM EDTA, monomers, pH 8.0; ▲, 0.1 I Tris, 10 mM EDTA, monomers, pH 7.8; ●, 0.1 M HEPES, phys. saline, decamers

Figure 4. Bohr effect of <u>Octopus</u> hemocyanin subunits in 10 mM EDTA, compared with native hemocyanin in buffered physiological saline.

Similar binding curves are also observed at pH 7.4 and 7.6 (not shown). At pH 7.8 and above, where the molecule is almost entirely in the 11S form (without any large amounts of 20S material) in both the oxy and deoxy states, the binding curves show slopes near 1.0, but exhibit regions with slope <1.0, indicating a heterogeneity of function among domains (Fig. 3b).

One of the most notable features of oxygen binding by native <u>Octopus</u> hemocyanin is the very pronounced Bohr effect. It is interesting therefore to observe that there is a substantial Bohr effect for subunits examined in 10 mM EDTA. The value of ($\Delta\log p50/\Delta pH$) is -1.0, compared to -1.7 for native hemocyanin (Fig. 4). Therefore, the Bohr effect does not require associated material. The allosteric behavior, as we might expect, is very different in the absence of divalent cations.

This research was supported by a grant (PCM82 12347) from the National Science Foundation.

References

Miller, K.I. (1985) Biochemistry (in press).
Miller, K.I. and K.E. van Holde (1982) Comp. Biochem. Physiol. 7313; 1013-1018.
Riccheli, F., Jori, G., Tallandini, L., Zatta, P., Beltramini, M., and Salvato, B. (1984) Arch. Biochem. Biophys. <u>235</u>, 461-469.
van Holde, K.E. and K.I. Miller (1985) Biochemistry (in press).

CONFORMATIONAL DIFFERENCES BETWEEN OXY- AND DEOXY-HEMOCYANIN

L. ZOLLA[*], P. THYBERG[o], R. RIGLER[o], M.BRUNORI[*]

* CNR Centre of Molecular Biology, Institute of Chemistry,
  Faculty of Medicine, University of Rome "La Sapienza"
  Rome Italy and
o Department of Medical Biophysics, Karolinska Institute,
  Stockholm, Sweden

INTRODUCTION

As in the case of hemoglobin (1) the classic two-state allosteric
model (2), based on the presence of two functionally relevant
quaternary states of the macromolecule in chemical equilibrium,
is often adequate to describe cooperative effects in the binding
of O_2 by hemocyanin (3,4). Nevertheless direct physical evidence
for the presence of different conformational states of
hemocyanin in equilibrium is (at best) very limited, with the
possible exception of some observations (5,6). Recent
spectroscopic studies have shown that upon binding of O_2 at the
active site of hemocyanin the Cu-Cu distance and the number of
ligands (7) bound to the metal atoms change, with a strong
involvement of the endogeneous protein bridge. This probably
triggers the quaternary conformational change(s) involved in the
cooperative binding of O_2. In order to aquire direct experimental
information on the ligand linked conformational transition of
hemocyanins we have applied autocorrelation spectroscopy to probe
the internal mobility and the hydrodynamic properties of oxy and
deoxy Helix p. ß -hemocyanin (M.W.9 millions).

MATERIALS AND METHODS

Helix p. ß -hemocyanin was isolated, stored and regenerated as
previously described (3); its state of aggregation was controlled
by analytical ultracentrifugation.The apo-form of ß -Helix p. was
prepared removing copper from the protein by treatment with KCN.
For the excitation of tryptophan fluorescence a mode-locked and
cavity dumped dye laser (Rhodamine 6G) was used which is
frequency doubled by a KDP crystal (λ exc =300 nm). The emitted
fluorescence is observed through cut-off filters and measured by
time correlated single photon counting. In the present instrument
an overall response time of 40 ps towards a laser pulse of 5 ps
was obtained (8). Fluorescence intensities polarized parallel to
the excitation polarization I (T) and at the magic angle I (T)
were measured together with the laser pulse in a double beam
spectrometer, and were collected in 2000 channels each.

Invertebrate Oxygen Carriers
Ed. by Bernt Linzen
© Springer–Verlag Berlin Heidelberg 1986

RESULTS

The translational diffusion coefficient of deoxy-and oxy-hemocyanin has been determined by measuring the autocorrelation function of the scattered light by dynamic laser light scattering. The oxy and deoxy derivatives display the same diffusion coefficient of 8.5×10^{-8} cm^2 s^{-1} when oxygen is removed by flushing with pure N$_2$. It may be mentioned that if oxygen was removed by addition of dithionite, a lower value was obtained probably due to aggregation of the deoxyprotein promoted by dithionite or its by products.

The fluorescence lifetime and the anisotropy decay of tryptophans in the presence and absence of oxygen differ significantly.

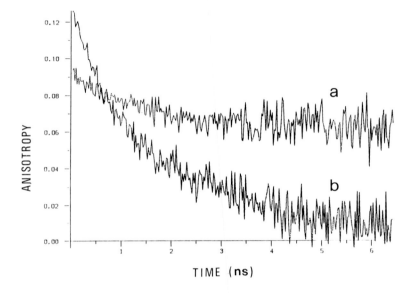

TIME (ns)

FIG. 1   Time course of anisotropy decay of <u>Helix p.</u>
ß-hemocyanin recorded in the presence and in
the absence of oxygen. Excitation at 295 nm;
emission above 370 nm, a. Deoxy Hcy, b. Oxy Hcy.

Evaluation of the time dependent fluorescence intensity (I) by non linear least squares parameter fitting and appropriate deconvolution routines yields a spectrum of decay times which is represented by at least 3 different decay processes. In the presence of oxygen, which is known to quench the stationary fluorescence of hemocyanin (9), all decay times are significantly shortened and the population strongly shifted to the shortest times.

Evaluation of rotational relaxation times (Fig. 1) from anisotropy decay curves reveals a large increase of the rotational motion of tryptophans emitting above 370nm, as seen by the complete depolarization of emission on a time scale where the large hemocyanin is practically immobile. Analysis of data indicates that τ_R (the anisotropy decay time) corresponds to 2.0+.16 ns for oxy-hemocyanin. It is interesting to underline that oxyhemocyanin displays a fast decay only at emission wavelenghts above 370 nm, while at lower wavelenghts the anisotropy decay of oxy- and deoxy are more similar. Static emission spectra of oxy-hemocyanin is broad, with a tail over 400 nm.

In order to collect more information, we have also carried out measurements of the anisotropy decay of apo-hemocyanin in the presence and in the absence of oxygen, under the same experimental conditions; in this case the decay process is oxygen independent.

DISCUSSION

The observations reported above show unequivocally that oxygen binding to Helix p. ß-hemocyanin is associated to a conformational change of the protein. In fact the anisotropy decay curve of the apo-protein is not affected by the presence of oxygen, and therefore the increase in rotational mobility of tryptophans emitting above 370nm and induced by O_2 cannot be attributed to a quenching effect of oxygen, but is a monitor of a ligand-linked conformational change of the protein. On the other hand, measurements of the translational diffusion coefficient in the presence or in the absence of oxygen yield the same value, indicating that the oxygen—linked conformational changes are not associated to large perturbation in the molecular assembly. Thus oxygenation of ß-hemocyanin under conditions of cooperative binding causes a change of the dynamic behaviour of the internal structure, leading to increased rotational mobility of all tryptophan residues emitting above 370 nm. Concomitantly the lifetime spectrum is shifted to shorter times, as a consequence of dynamic quenching caused by increased rotational mobility. These conformational changes are likely to be a consequence of the ligand linked allosteric transition of hemocyanin and it seems of some interest that the deoxygenated derivative of hemocyanin displays a reduced mobility of some of the internal tryptophans. More detailed studies are in progress in order to single out the population of tryptophan(s) whose dynamic behaviour and internal mobility is affected by oxygen binding to the active site of hemocyanin.

REFERENCES

1) Perutz M.F., Nature 228, 726 (1970).
2) Monod J., Wyman J. and Changeaux J.P., J. Mol. Biol.
   2, 88 (1965).
3) Zolla L., Kuiper H.A., Vecchini P., Antonini E. and
   Brunori M., Eur. J. Biochem. 87, 467 (1978).
4) Brower M., Bonaventura C. and Bonaventura J.,
   Biochemistry 16, 3897 (1977).
5) van Driel and van Bruggen E.F.J. Biochemistry 19, 730
   (1975).
6) van Breenen J.F.L., Ploegnen J.H. and van Bruggen
   E.F.J. Eur. J. Biochem. 100, 61 (1979).
7) Woolery G.L., Powers L., Winkler M., Solomon E.I. and
   Spiro T.G., J. Am. Chem. Soc. 106, 86 (1984).
8) Rigler R., Claesens F. & Lomakka G.: In "Ultrafast Phenomena IV", Eds. D.H. Auston & K.B. Eisenthal, Springer
   Series in Chemical Physics 38, 472 (1984).
9) Ma J.K.H., Luzzi L.A., Ma J.Y.C. & Li N.C. J. Pharm.Sci,
   66, 1684 (1977).

A short term EMBO Followship to Dr. L. Zolla in gratefully
aknoledged.

DETECTION OF CONFORMATIONAL CHANGES IN TARANTULA (EURYPELMA CALIFORNICUM) HEMOCYANIN BY MEANS OF FLUORESCENT PROBES

T. Leidescher and B. Linzen
Zoologisches Institut, Universität München, Luisenstr. 14, 8000 München 2,

Introduction

Tarantula (Eurypelma californicum) hemocyanin (Hc) is a 1.7×10^6 molecular weight protein containing 24 subunits. These are not identical but belong to 7 different types, designated a-g. The quaternary structure of Eurypelma Hc has been revealed several years ago (1). This hemocyanin shows extremely strong positive cooperativity with n_H values running up to 10. However, direct evidence for conformational changes upon oxygenation and deoxygenation of arthropod Hc has not yet been published to our knowledge. In the present paper we report fluorescence spectroscopic studies to detect such changes. The strong intrinsic tryptophan fluorescence cannot be used for this purpose, for it is nearly totally quenched upon oxygenation (2). Therefore we applied various extrinsic probes.

Materials and Methods

A number of fluorescent probes has been examined, both non-covalently binding (ANS, TNS) and covalently binding ones. Among the covalently binding probes, there were several selective for sulfhydryl and amine groups. None of these showed any clear changes in fluorescence properties (wavelength shift or/and change of intensity), except the amine selective label, 7-chloro-4-nitrobenzo-2-oxa-1.3-diazole (NBD-chloride). NBD-chloride (3) (Molecular Probes, Junction City, OR, USA) was dissolved in dimethylformamide and added in 20-fold molar excess to a 4.4-5 mg/ml Hc solution in 0.025 M borate buffer pH 8.1. The reaction was allowed to proceed for 3 h at 25 °C in the dark, and unreacted dye was removed in the dark by gel filtration (Sephadex G 25). The solutions of labelled, purified protein were diluted appropriately. Oxygen binding curves were recorded by the method of Loewe (2). Fluorescence spectra were obtained on a SPEX FLUORO-LOG G 211 fluorescence spectrometer. Oxygenation and deoxygenation of the labelled protein were achieved by bubbling oxygen or argon/nitrogen, respectively, through the solution for 1 h, or by exposing the Hc to appropriate gas mixtures in a tonometer cell until the absorbance at 340 nm remained constant.

Results and Discussion

Among a number of fluorescent labels tested by us, only NBD-chloride responded by changing fluorescence upon oxygenation/deoxygenation of the hemocyanin. Its suitability was further examined by recording oxygen binding curves of the la-

Invertebrate Oxygen Carriers
Ed. by Bernt Linzen
© Springer–Verlag Berlin Heidelberg 1986

belled hemocyanin, as we assumed that the labelling procedure and the multiple binding of probe molecules to the subunits might interfere with the allosteric behaviour of the oligomer. Fig. 1 shows oxygen binding curves of <u>Eurypelma</u> Hc labelled with NBD-chloride and a control (same Hc preparation kept under identical conditions of temperature and pH). Cooperativity was n_H, max = 5.8 in the control and n_H, max = 5.5 in the labelled sample, representing a ca. 5 percent decrease.

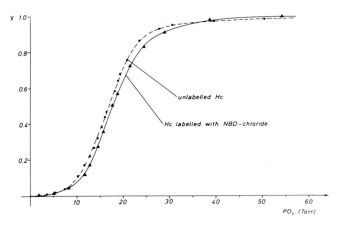

Fig. 1. Oxygen binding curves of NBD-chloride-labelled (——) and unlabelled (---) 37 S Hc of <u>Eurypelma</u> <u>californicum</u>. 0.1 M Tris/HCl buffer pH 7.8; c_{Hc}=1.8 mg/ml; 20 °C; n_H, max=5.8 in the unlabelled sample, n_H, max=5.5 in the NBD-chloride-labelled sample.

The fluorescence emission spectrum of NBD-chloride labelled Hc shows one maximum which is located at 535 nm in the deoxygenated sample, and shifts to 531 nm upon oxygenation (Fig. 2). At the same time, the fluorescence intensity is lowered, possibly due to oxygen quenching. The shift of maximum wavelength is reversible (Fig. 2).

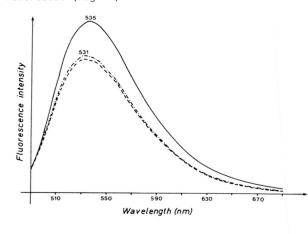

Fig. 2. Fluorescence emission spectrum of 37 S Hc labelled with NBD-chloride. 0.1 M Tris/HCl buffer pH 7.8; c_{Hc}=0.14 mg/ml; 20 °C; excitation wavelength 465 nm. Oxygenated (---), deoxygenated (——) and reoxygenated (·····) hemocyanin.

Deoxygenation and reoxygenation were further examined by recording absorption
spectra of the labelled Hc. The disappearance of the peak at 340 nm showed that
deoxygenation was complete within 1 h of bubbling argon/nitrogen through the solution. 10 min of oxygenation with pure oxygen suffice to restore the peak at
340 nm completely (Fig. 3).

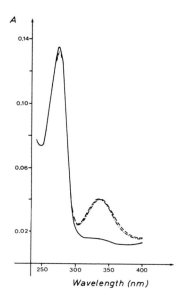

Fig. 3. Absorption spectra
of 37 S Hc labelled with
NBD-chloride. Oxygenated
(———), deoxygenated (---)
and reoxygenated (·····)
sample.

Fig. 4 shows fluorescence emission curves of NBD-chloride-labelled <u>Eurypelma</u> Hc
at four different oxygen concentrations. Clearly the emission maximum shifts
stepwise from 531 nm to 535 nm.

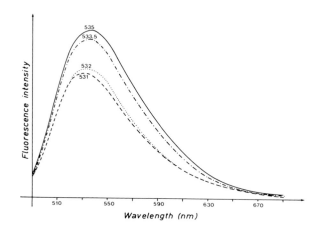

Fig. 4. Fluorescence emission spectra of NBD-chlo-
ride-labelled 37 S Hc
equilibrated with differ-
ent oxygen pressures. 0%
O_2 (———); 1.048% O_2
(·—·—·); 4.19% O_2 (....);
100% O_2 (---). 0.1 M
Tris/HCl buffer pH 7.8;
20 °C; excitation wave-
length 465 nm.

The observed shift in maximum fluorescence emission wavelength (531 nm -> 535 nm) in NBD-chloride-labelled Hc is a strong indication that conformational changes are occurring in the vicinity of the labelled amino acid residues. Probably the environment of the labelled amino acids becomes more polar upon deoxygenation (4).

Acknowledgement. This work was supported by the Deutsche Forschungsgemeinschaft (Li 107/24-6).

1. Markl, J., Kempter, B., Linzen, B., Bijlholt, M.M.C., and van Bruggen, E.F.J., Hoppe-Seyler's Z. Physiol. Chem. 362, 429-437 (1981).
2. Loewe, R., J. Comp. Physiol. 128, 161-168 (1978).
3. Ghosh, P.B., and Whitehouse, M.W., Biochem. J. 108, 155-156 (1968).
4. Brand, L., and Gohlke, J.R., Am. Rev. Biochem. 41, 843-868 (1972).

EFFECTS OF SOLVENT COMPOSITION ON HEMOCYANIN CONFORMATION

M. Beltramini, A. Piazzesi, M. Alviggi, F. Ricchelli° and B. Salvato
Dept. of Biology and °C.N.R. Centre for Hemocyanins and other Metalloproteins, Univ. of Padova, Via Loredan 10, I-35131 Padova, Italy.

The conformation of proteins is due to the balance of the electrostatic and hydrophobic interactions which develop either within aminoacid side chains or between these and the solvent. All chemical and /or physical agents which interfere in this interaction balance act as perturbants of the protein conformation. Their mechanism of action, however, can be quite different: cations as Ca^{++} and Mg^{++} act by binding to charged groups of the protein matrix while some anions (ClO_4^-, SCN^-) and other cations (guanidinium and tetra-alk ylammonium) lower the solution free-energy of aminoacid side chains [1]. Moreover, the effect of aliphatic alcohols can be correlated to changes of the dielectric constant of the medium [1]. The overall effect is either a stabilization or a destabilization of the protein conformation.

In the case of hemocyanin (Hc), SO_4^{--} stabilizes the native conformation against the denaturating effects of urea [2]. Alcohols are known to induce a shift in the equilibrium between R and T conformations of hemoglobin [3].

In this paper, the effects of anions (Br^-, I^-, Cl^-, SCN^-, ClO_4^-, SO_4^{--}, HPO_4^{--}), of Ca^{++} and of alcohols (methanol, ethanol and n-propanol) on the conformation of Hc from Carcinus maenas are reported. Conformational modifications of Hc structure are expected to modify the reactivity of the active site to CN^-. Hence, the kinetics of the reaction between Hc and CN^- (already studied in detail in buffer alone [4]) has been measured in the presence of the various perturbants.

Experimental

The kinetics of the reaction with CN^- has been studied with EDTA-treated Carcinus Hc in Tris/HCl 0.1M pH 8.0 buffer in the presence of Br^-, I^-, SCN^-, ClO_4^-, SO_4^{--}, HPO_4^{--} at anion-to-Cl^-ratios between 0 and 1 (the total salt concentration was kept always 1M), of Ca^{++} up to 14 mM and of alcohols (methanol, ethanol and n-propanol) at cosolvent/water percentages up to 15-20%. The CN^- concentration was 6.0 mM in the case of anions and 7.8 mM in the case of both Ca^{++} and alcohols. Moreover, experiments at constant anion concentration (1M) and at CN^- concentrations up to 12 mM were also carried out. Hc was equilibrated by dialysis overnight against buffer containing either the anion (0.5M)

Invertebrate Oxygen Carriers
Ed. by Bernt Linzen
© Springer–Verlag Berlin Heidelberg 1986

or the alcohol (7.6%) under study.

Results and Discussion

Under all experimental conditions used, CN^- induces a time-dependent decrease of the 337 nm band of oxy-Hc. As for Hc in buffer alone, no site-site interactions are observed in the presence of perturbing agents. Accordingly, the decrease of 337 nm band can be described by a single exponential function. This allows to apply the same model proposed for Hc in absence of perturbing agents [4]; changes of the pseudo-first order rate constant (k_{app}) can be, therefore, correlated to conformational modifications of the Hc structure at the level of the active site.

In fig. 1 the changes of k_{app} against the progressive substitution of Cl^- with other anions are reported. As the anion-to-Cl^- ratio increases, k_{app} decreases. The overall effect follows the order:

$$Br^- < HPO_4^{--} < I^- < ClO_4^- < SO_4^{--} < SCN^-$$

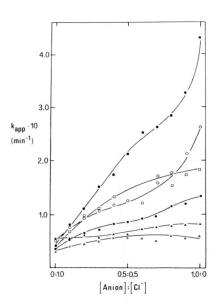

Fig. 1: Effects of anions on the reaction between Hc and CN^-. The pseudo-first order rate constant of the decrease of the absorbance at 337 nm (k_{app}) is plotted versus increasing anions-to-Cl^- ratios. The total salt concentration is kept to 1M and that of CN^- is equal to 6.0 mM. SCN^- (■), SO_4^{--} (□), ClO_4^- (○), I^- (●) Br^- (△), HPO_4^{--} (▲).

This finding, however, is not in agreement with known effects of the same anions on protein conformation [5]. I^-, ClO_4^-, SCN^- are known to have almost the same efficiency in protein denaturation [1]. SO_4^{--} and HPO_4^{--}, known as stabilizing agents [5], induce an increase of k_{app} like I^-, ClO_4^-, SCN^-. These results can be interpreted by assuming that, with the concentrations used, anions induce small modifications of the protein matrix which, however, turn to be more important in the immediate surrounding of the active site. Accordingly, denaturation or stabilization are seen only at anion concentrations higher than 1M. The upward curvature of the plot given by SCN^- and ClO_4^- agrees with the strong denaturing effects of these anions and may reflect the transition of Hc structure to the full denaturation. Experiments at constant anion concentration and at different CN^- concentrations were also carried out. The data, analyzed according to the following equation [4], allow to calculate the ratio of the equilibrium constants of Hc with CN^- and O_2 (K_{CN}/K_{O_2}) and the rate constant for the removal of one copper ion (k_1):

$$\frac{CN^-}{k_{app}} = \frac{1}{k_1} + \frac{1}{k_1} \frac{K_{CN}}{K_{O_2}} \frac{[O_2]}{[CN^-]} \qquad (1)$$

As shown in Tab. I, K_{CN}/K_{O_2} decreases **by** more than two orders of magnitude when going from buffer alone to Cl^- and to SCN^-. k_1 is not affected by the presence of anions. This indicates that the conformational changes induced by anions affect only the accessibility of the active site and not the rate constant of metal removal.

Table I: Effects of anions on the parameters of the reaction between Carcinus Hc and CN^-. For details see the text.

	Buffer° alone	Cl^-	Br^-	I^-	SO_4^{--}	HPO_4^{--}	ClO_4^-	SCN^-
K_{CN}/K_{O_2}	225.4	128	89	69	27	22	10	1.5
k_1^*	1.5	1.2	1.5	1.5	1.3	1.3	1.3	1.5

° data from ref. [4]; * is in $M^{-1}cm^{-1}$

As for the effects of calcium, k_{app} decreases with increasing concentrations of Ca^{++} (fig.2). As known, Ca^{++} interacts with proteins differently from anions. The occurrence of two transitions is in accordance with the existence of two types of Ca^{++} binding sites with different affinities for the ion [6].

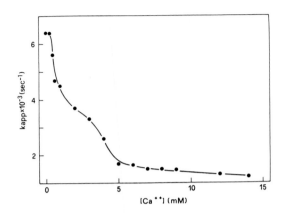

Fig. 2: Effects of Ca^{++} on the kinetics of the reaction between Hc and CN$^-$. k$_{app}$ as in fig. 1.

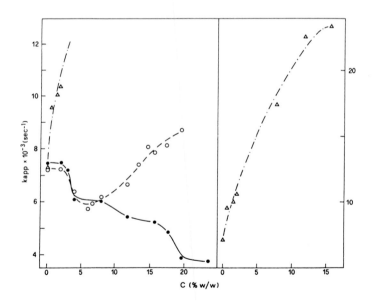

Fig. 3 : Effects of alcohols on the kinetics of the reaction between Hc and CN$^-$. k$_{app}$ (see fig. 1) is plotted _versus_ increasing cosolvent/water percentages. These values are reported as grams of alcohols per 100 grams of solution. Methanol (●), ethanol (○) and n-propanol (△).

The effects of alcohols are reported in Fig. 3. In the range of 0-25% methanol induces several conformational transitions which all imply a decrease of k_{app}. With ethanol, a first transition is superimposed to that of methanol. An opposite transition, however, is induced at high ethanol/water percentages. With n-propanol only the latter effect is observed. Preliminary calculations using eq. 1 have shown that alcohols, like anions, alter the K_{CN}/K_{O_2} ratio without affecting k_1.

Acknowledgements

We thank M. Santamaria for her participation to a part of this work.

References

1. Lapanje, S., Physico-chemical Aspects of Protein Denaturation, J. Wiley and Sons, New York (1978).
2. Ricchelli,F., Filippi, B. and Salvato, B., inInvertebrate Oxygen-Binding Proteins (Lamy, J. and Lamy, J. eds.) pp. 31-39, Marcel Dekker, New York (1981).
3. Cordone, L., Cupane, A., San Biagio, P.L. and Vitrano, E., Biopolymers 20 , 39-51 (1981).
4. Beltramini, M., Ricchelli, F., Tallandini, L. and Salvato, B., Inorg. Chimica Acta 92, 219-227 (1984).
5. Von Hippel, P.H. and Wong K.Y., Science 145, 577-580 (1984).
6. Andersson, T., Chiancone, E. and Forsén, S., Eur. J. Biochem. 125, 103-108 (1982).

The active site of hemocyanins

HALF-APO DERIVATIVES OF HEMOCYANINS

M. Beltramini, A. Piazzesi, M. Alviggi, F. Ricchelli[*] and B. Salvato
Department of Biology and *C.N.R. Centre for Hemocyanin, University
of Padova, Via Loredan 10, I-35131 Padova, Italy

A number of experimental evidences have indicated that the two copper ions in the site of hemocyanin (Hc) are differently shielded by
the protein matrix, thus showing asymmetrical chemical behaviour (1,
2,3). In particular, one metal ion can be selectively removed by CN^-
from the active site of mollusc Hc leading to the formation of an
half-apo derivative still containing one metal ion (3,4). The first
ion has been called "fast reacting copper" and the second "slow reacting copper" (5,6).

Half-apo derivatives of arthropod Hc had not been obtained so far.
However, previous kinetic studies on the reaction with CN^- of Carcinus
maenas Hc (6) indicated that, under anaerobic conditions, half-apo-Hc
could be also prepared from arthropod Hc. As reported, only the removal of the "fast reacting copper" ion shows an hyperbolic dependence
from O_2 concentration (6). Therefore, in absence of O_2, the CN^- concentration can be lowered so as to exclude almost completely the removal of the "slow reacting" copper while the "fast reacting" one is
removed still efficiently. The chemical asymmetry of the two metal
sites suggests that half-apo derivatives could also be prepared by
selective reconstitution of the protein starting from the full apo-Hc.
In this paper the preparation of the half-apo derivative of Carcinus
maenas Hc under mild CN^- treatment is reported. The preparation of
half-apo derivatives by selective reconstitution under controlled
conditions is also described for both arthropod (Carcinus maenas) and
mollusc (Octopus vulgaris) Hcs.

Experimental

Half-apo-Hc was prepared by dialysis at 4°C of deoxy-Hc (7 mg/ml)
against N_2 saturated Tris/HCl buffer I= 0.1 pH 8.0 containing 1 mM
CN^-. At different times aliquots were taken and analyzed for the absorbance ratio A_{337}/A_{280} and the Cu-to-Hc stoichiometry after removal
of CN^- and oxygenation of the sample. Alternatively, a single copper

Invertebrate Oxygen Carriers
Ed. by Bernt Linzen
© Springer-Verlag Berlin Heidelberg 1986

ion was bound to apo-Hc against 20 mM phosphate buffer pH 7.0 con-taining 10 mM $MgSO_4$, 10 mM NH_2OH, 50 mM NaSCN, 0.5 mM $CuSO_4$, 0.05% Tween 80.

Results and Discussion

The absorbance at 337 nm of Hc treated with CN^- in anaerobic con-ditions decreases according to   apparent first-order kinetics (k_{app} 0.7 h^{-1}). After 6 h, the residual absorption is about 3-5% that of the native protein (fig. 1). Copper is lost much more slowly following biphasic kinetics and reaches the limiting value of 45%. The fluor-escence emission of Hc (F) increases up to 4.8 times that of native oxy-Hc (F/F_N= 4.8)(fig. 1, Table 1). These results show that after 6 h about 80% of Carcinus Hc is converted to the half-apo form. The fluorescence emission of this derivative is almost identical to that of apo-Hc (F_{apo}/F_N= 4.9) in spite of the presence of the metal ion still bound to the active site.

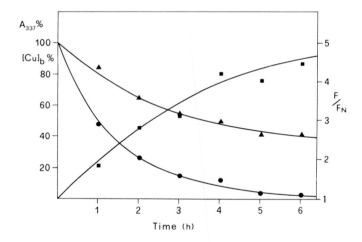

Fig. 1: Copper removal by CN^- from Carcinus Hc under anaerobic condi-tions. Time dependence of the reaction as observed by measuring the absorbance at 337 nm (●) and the copper still bound to Hc (▲). The data are percentages of the corresponding values in oxy-Hc. The time course of the Hc fluorescence intensity (F) is also reported (■). The data are referred to the emission intensity of oxy-Hc (F_N).

As shown in fig. 2 an half-apo derivative of <u>Carcinus</u> Hc is reconstituted from the apoprotein when dialyzed against Cu(I). After 6-8 h
of incubation with Cu(I) this derivative contains 1.1 gr-atoms of Cu
per 75,000 M_r while the absorbance at 337 nm is no more than 5% (fig.
2, Table 1). In contrast to half-apo-Hc obtained by removing the metal,
the binding of a single copper ion per active site causes a strong
fluorescence quenching: the emission intensity is ~ 40% lower than
that of apo-Hc (Table 1). Similar results are given by <u>Octopus</u> apo-Hc.

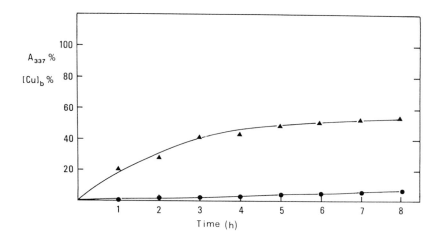

Fig. 2: Time course of the reconstitution reaction of <u>Carcinus</u> apo-Hc
in the presence of Cu(I) as observed by measuring the 337 nm
absorbance (●) and the copper-to-Hc stoichiometry (▲). The
data are reported as percentages of the corresponding values
in oxy-Hc.

As reported also by several authors (7,8) the fluorescence emission
intensity of Hc is strongly affected by the presence of the metal in
the active site. In particular the emission intensity of deoxy-Hc is
about 40% lower than that of apo-Hc. Recent studies on <u>Octopus</u> Hc
have shown that the quenching effect exerted by the binuclear copper
site on the protein fluorescence can be almost completely ascribed to
the "fast reacting copper" ion (8,9) . The results here reported on
<u>Carcinus</u> half-apo-Hc confirm those on mollusc Hc. Moreover, the copper
ion bound to the half-apo-Hc in reconstitution experiments can be
ascribed to the "fast reacting" site since the emission intensity is

quenched to an extent comparable to that of deoxy-Hc. These results emphasize the different reactivity of the two copper sites in Hcs: the "fast reacting" ion can be either selectively removed from the binuclear site or selectively bound to the apo-form. Hence, two half-apo derivatives can be prepared both from mollusc and arthropod Hc: CN^- treatment causes the removal of the "fast reacting copper" leading to an half-apo derivative which still binds the "slow reacting" ion (F-half-apo). On the contrary, reconstitution reaction gives a different half-apo derivative in which the metal ion is bound to the "fast reacting" site leaving the "slow reacting" site empty (S-half-apo).

Table 1: Properties of different Carcinus Hc derivatives.

	native	half-apo (CN^-) (F-half-apo)	half-apo Cu(I) (S-half-apo)	apo
A_{337}/A_{280}	0.21	0.006	0.010	0
Cu/Hc (gr-atoms/mole)	2.0	0.85	1.10	0.01
F/F_N	1.0	4.8	3.0	4.9

References

1. Cox, J.A., and Elliott, F.G., Biochem.Biophys.Acta 371, 392-401 (1974).
2. Symons, M.C.R., and Petersen, R.L., Biochem.Biophys.Acta 535, 247-252 (1978).
3. Salvato, B., and Zatta, P., inStructure and Function of Haemocyanin (Bannister, J.V., ed.), pp. 245-252, Springer Verlag, Berlin (1977).
4. Himmelwright, R.S., Eickman, N.C., and Solomon, E.I., Biochem. Biophys.Res.Comm. 81, 243-247 (1978).
5. Beltramini, M., Ricchelli, F., and Salvato, B., Inorg.Chim.Acta 92, 209-217 (1984).
6. Beltramini, M., Ricchelli, F., Tallandini, L., and Salvato, B., Inorg.Chim.Acta 92, 219-227 (1984).
7. Bannister, W.H., and Wood, E.J., Comp.Biochem.Physiol. 40B, 7-18 (1971).
8. Ricchelli, F., Tealdo, E., and Salvato, B., Life.Chem.Rep., Sup. 1, 301-304 (1982).
9. Beltramini, M., Ricchelli, F., Piazzesi, A., Barel, A., and Salvato, B., Biochem.J. 221, 911-914 (1984).

SPECTROSCOPIC CHARACTERIZATION OF A Co(II) DERIVATIVE OF Carcinus maenas HEMOCYANIN

B. Salvato, M. Beltramini, A. Piazzesi, M. Alviggi and F. Ricchelli*
Department of Biology and *C.N.R. Centre for hemocyanin and other metallo-proteins, Univ. of Padova, Via Loredan 10, I-35131 Padova(Italy)

The substitution of the metal with Co(II) is a useful technique for studying the structure of the active site of metalloproteins and metalloenzymes. As known, Co(II) ions have unique spectroscopic and magnetic features which depend on the nature of the metal ligands and the geometry of the complex (1). The substitution of copper with Co(II) has been reported also in the hemocyanin (Hc) from squid (2), snail (3) and horseshoe crab (4). In our laboratory, a number of Co(II) derivatives of mollusc and arthropod Hc have been prepared where both or only one copper ion is substituted with cobalt (5). In this paper we report the spectroscopic properties of one of these derivatives, namely that of Carcinus Hc which contains 1 Co(II) per 75,000 M_r (1/75 KD Co(II)-Hc).

Experimental

1/75 KD Co(II)-Hc was prepared by dialysis of the apo-Hc (obtained by CN^- treatment (6)) against Tris/HCl I= 0.1 pH 7.0 containing 10 mM $MgSO_4$, 150 mM NaSCN, 0.5 mM $CoCl_2$, 0.05 % Tween 80 for 72 h at 20°C. Excess reagents were removed by further dialysis against buffer plus 20 mM EDTA and finally against 20 mM phosphate buffer pH 7.0 .

Results and Discussion

When Co(II) binds to apo-Hc, a typical absorption spectrum appears with a maximum at 568 nm. The Co(II) uptake follows pseudo-first order kinetics with an half time ($t_{\frac{1}{2}}$) of about 13 h. After 72 h no further increase of absorbance at 568 nm and of the Co-to-Hc ratio is observed. Apo-Hc binds 1.0±0.1 g-atoms of EDTA-resistant Co(II) per 75,000 M_r. Oxy-Hc under the same experimental conditions does not bind Co(II). In fig. 1 the absorption and the circular dichroism (CD) spectra of 1/75 KD Co(II)-Hc are reported. The visible absorption re-

Invertebrate Oxygen Carriers
Ed. by Bernt Linzen
© Springer–Verlag Berlin Heidelberg 1986

gion is dominated by the d-d bands of Co(II) with maxima at 530, 568 and 585 nm. The band position and the molar extinction coefficient at the maximum (ε = 312 M^{-1}cm^{-1}) are both indicative for tetracoordinated Co(II) ions in a pseudo-tetrahedral geometry. As for the spectroscopic properties of the Co(II) complex, <u>Carcinus</u> Hc resembles more squid Hc (ε ~ 300 M^{-1}cm^{-1}) (2) than horseshoe crab Hc (200 $<$ ε $<$ 250 M^{-1}cm^{-1}) (4).

The specific binding of Co(II) is further documented by the relatively intense CD bands displayed by 1/75 KD Co(II)-Hc in the visible region. The optical activity of Co(II)-related transitions would, therefore, arise from distorsion of the metal-protein complex. The CD spectrum is dominated by a rather broad positive band with maximum at 567 nm ($[\Theta]$ = +550) and by a strong negative Cotton band with a maximum at 485 nm ($[\Theta]$ = -680) and with a shoulder at 520 nm. Interestingly, three features are observed in the near-U.V.-visible (λ = 415, $[\Theta]$ = +230; λ = 355 $[\Theta]$ = +900; λ = 327; $[\Theta]$ = +850). They are not re-solved in the absorption spectrum probably because of their low ab-sorptivities as compared to the strong red-edge of the protein absorp-tion band. These features are due to Co(II) binding since they are modified by addition of exogenous Co(II)-ligands.

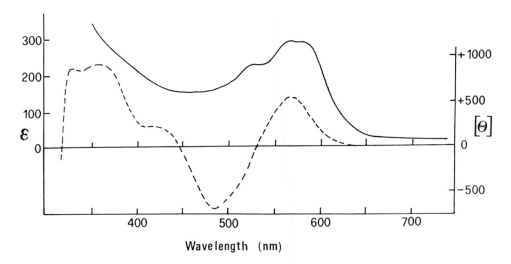

Fig. 1: Absorption (solid line) and circular dichroism (dashed line) spectra of 1/75 KD Co(II)-Hc.

Fluorescence measurements give some information on the localization
of Co(II) within the active site. The fluorescence emission spectrum
of 1/75 KD Co(II)-Hc is shown in fig. 2. Binding of Co(II) causes a
~40% quenching of the Hc fluorescence emission. This is the same both
in deoxy-Hc and in the Co(II) derivative which contains only one metal
ion. It is in agreement with other results on the dependence of Hc
fluorescence versus the metal-to-protein ratio: we observed that the
quenching of Hc fluorescence is completely abolished by the removal
of the "fast reacting" copper ion (7,8,9). This indicates that
Co(II) binds to the same site of the "fast reacting" copper ion
leaving the other site empty.

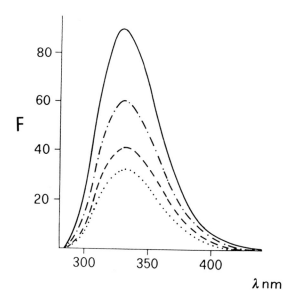

Fig. 2: Fluorescence emission spectrum of 1/75 KD Co(II)-Hc (-.-.-).
The spectra of apo-Hc (———) and of 1/75 KD Co(II)-Hc in the
presence of N_3^- (0.2 M) (----) and SCN^- (0.2 M) (....) are also
shown. All spectra are recorded under excitation wavelength
of 295 nm and are normalized to the same protein concentration.
Fluorescence intensity (F) is in arbitrary units.

Addition of either F^-, I^-, SO_4^{--}, HPO_4^{--}, ClO_4^-, $HCOO^-$, CH_3COO^- does not
cause any spectral change indicating neither binding of the anions
nor the occurrence of conformational modifications on Hc structure.
In contrast, the spectroscopic properties of 1/75 KD Co(II)-Hc are

remarkably modified by ligands such as Cl^- (1 M), Br^- (1 M), N_3^- (0.2 M), SCN^- (0.2 M), CN^- (5 mM) (the concentrations given in parenthesis are required to obtain the maximum of spectral change). A strong hyperchromic effect (ε increases up to $\sim 540\ M^{-1} cm^{-1}$), a 10-20 nm red shift in the absorption maxima and an increase of ellipticity of the Cotton features always occur.

As indicated by titration experiments with the various ligands, in all cases the ligand-to-Co(II) stoichiometry is equal to 1. The apparent stability constants decrease in the order:

$$CN^- (pK=6.0) > N_3^- (pk=1.8) \sim SCN^- (pK=1.7) > Cl^- (pK=1.0) > Br^- (pK=0.8)$$

CN^-, SCN^- and Br^- show anticooperative effects which are more evident in the order: $CN^- > SCN^- > Br^-$.

It appears that, upon binding of exogenous ligand, Co(II) remains tetra-coordinated. This implies that exogenous ligands displace one endogenous ligand. Binding of N_3^- and SCN^- causes a further quenching of 30% and 50% respectively (fig. 2). This can be due to the interaction between the ligand and a tryptophan residue located near the active site: an H-bond between the N- or S- edges of N_3^- and SCN^- and the indole-NH group could be responsible for the observed effect.

References

1. Bertini, I., inCoordination Chemistry of Metalloenzymes (Bertini, I., Drago, R.S., and Luchinat, C., eds.), pp. 1-18, D. Reidel Publ. Co., Dordrecht (1982).
2. Suzuki, S., Kino, J., Kimura, M., Mori, W., and Nakahara, A., Inorg.Chimica Acta 66, 41-47 (1982).
3. Witters, R., and Lontie, R., Life Chem.Rep., Sup. 1, 285-288 (1982).
4. Suzuki, S., Kino, J., and Nakahara, A., Bull.Chem.Soc.Jpn. 55, 212-217 (1982).
5. Salvato, B., Tealdo, E., Beltramini, M., and Peisack, J., Life Chem.Rep., Sup. 1, 291-294 (1984).
6. Salvato, B., Ghiretti-Magaldi, A., and Ghiretti, F., Biochemistry 13, 4778-4782 (1974).
7. Ricchelli, F., Tealdo, E., and Salvato, B., Life Chem.Rep., Sup. 1, 301-304 (1982).
8. Beltramini, M., Ricchelli, F., Piazzesi, A., Barel, A., and Salvato, B., Biochem.J. 221, 911-914 (1984).
9. Beltramini et al., this volume, pp. 437-440 (1986).

PREPARATION OF A METHAEMOCYANIN OF *ASTACUS LEPTODACTYLUS* REGENERABLE WITH HYDROXYLAMINE

R. Witters, D. Van Hoof, W. Deleersnijder, J.-P. Tahon, and R. Lontie
Laboratorium voor Biochemie, Katholieke Universiteit te Leuven
Dekenstraat 6, B-3000 Leuven, Belgium

The classical preparation of arthropodan methaemocyanin is based on the oxidation of deoxyhaemocyanin with a small excess of hydrogen peroxide (1). The methaemocyanin of the crustacean *Astacus leptodactylus* and of the chelicerate *Limulus polyphemus* could not be regenerated with reducing agents like hydroxylamine, hydrazine, nor with hydrogen peroxide. Only the latter could be regenerated markedly with hydrogen sulphide or with hydrogen cyanide (2).

To obtain a methaemocyanin of *A. leptodactylus*, regenerable with hydroxylamine, other methods for the preparation of the derivative were tested.

PREPARATION OF METHAEMOCYANIN

A. leptodactylus haemocyanin was dissolved in 0.1 M acetate buffer pH 5.7. Non-regenerable methaemocyanin was prepared by treatment of deoxyhaemocyanin for 24 h with hydrogen peroxide in a molar ratio of 10 per copper atom in haemocyanin. Regenerable methaemocyanin was prepared by the action of nitrogen monoxide or of nitrite. Nitrogen monoxide was allowed to react with deoxyhaemocyanin at room temperature for 15 min at least; nitrogen was then flushed over the reaction mixture in order to remove the excess of reagent. Nitrite was added to oxy- or to deoxyhaemocyanin in a molar ratio of 100 to copper in haemocyanin and allowed to react for 24 h; the nitrite was then removed by dialysis.

RESULTS AND DISCUSSION

The absence of regeneration of the methaemocyanin, prepared with hydrogen peroxide, by simply reducing the copper atoms - in contrast with molluscan methaemocyanin - may be related to the possible absence of an endogenous bridging ligand between the copper atoms in the active site like found in the haemocyanin of the related crustacean *Panulirus interruptus* (3). This renders a one-electron reduction of hydrogen per-

Invertebrate Oxygen Carriers
Ed. by Bernt Linzen
© Springer–Verlag Berlin Heidelberg 1986

oxide by Cu(I) plausible, whereby the resulting ˙OH radical could
damage the environment of the active site irreversibly. To avoid such
an aspecific reaction *A. leptodactylus* methaemocyanin was prepared in
a strict anaerobic treatment of deoxyhaemocyanin with nitrogen monoxide
according to Verplaetse *et al.* (4). A reaction time of 15 min at room
temperature in 0.1 M acetate buffer pH 5.7 reduced the consecutive
formation of the nitrosyl derivative. Addition of hydroxylamine, in a
molar ratio to copper of 20, to this methaemocyanin (27 mg/ml) in 0.1 M
acetate buffer pH 5.7 or in 0.1 M phosphate buffer pH 6.0 resulted in
a partial regeneration (70 % after 24 h); in 0.1 M borate buffer pH 8.2
no regeneration with hydroxylamine was observed. The regeneration could
be followed by the reappearance of the copper-dioxygen absorption band
at 339 nm, by the partial recovery of the circular dichroic spectrum
in the visible and near ultraviolet, and by the disappearance of the
broad electron-paramagnetic-resonance (EPR) signal of nitrite-methaemo-
cyanin near g = 2 (5).

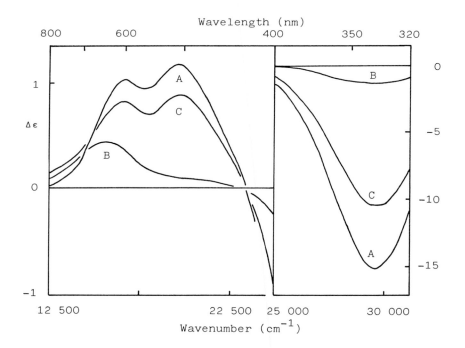

Fig. 1. Molar circular dichroism $\Delta\epsilon$ ($M^{-1}cm^{-1}$), expressed per Cu atom,
of *Astacus leptodactylus* haemocyanin in 0.1 M acetate buffer pH 5.7
measured at 20°C in solutions saturated with air. (A) Oxyhaemocyanin,
(B) deoxyhaemocyanin after 20 h reaction with nitrogen monoxide,
(C) the same solution after regeneration with hydroxylamine in a molar
ratio to Cu of 20 in 0.1 M phosphate buffer pH 6.0.

The binding of dioxygen to the regenerated haemocyanin was reversible as demonstrated by the disappearance of the copper-dioxygen band on de-oxygenation and by the registration of an oxygenation curve under non-cooperative conditions in 0.1 M acetate buffer pH 5.7, yielding a value of one for the Hill coefficient and of 13.3 mmHg for the p_{50}.

The circular dichroic spectrum of the methaemocyanin preparation showed a moderate positive band near 630 nm besides a residual (12 %) negative band at 339 nm (Fig. 1). On regeneration with hydroxylamine this nega-tive band increased in the given conditions to a value of about 70 % of that of the original oxyhaemocyanin, as did the positive bands near 500 and 600 nm.

Arthropodan methaemocyanin in 0.1 M acetate buffer pH 5.7 in the pres-ence of nitrite (in a molar ratio of 100 to copper) showed a broad bi-nuclear EPR signal near $g = 2$ (5). During the regeneration with hydro-xylamine the disappearance of this signal could be followed. Finally only a small signal of mononuclear Cu(II) remained, originating proba-bly from broken copper pairs (Fig. 2).

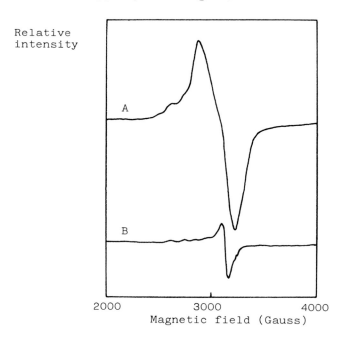

Fig. 2. EPR spectra of *Astacus leptodactylus* haemocyanin (27 mg/ml; 0.6 mM Cu) in 0.1 M acetate buffer pH 5.7 after 20 h reaction of deoxy-haemocyanin with nitrogen monoxide in the presence of 60 mM nitrite (A), and after regeneration with hydroxylamine (12 mM) in 0.1 M phosphate buffer pH 6.0 (B). The spectra were recorded at 100 K with 100 kHz field modulation and at a modulation amplitude of 10 G. Microwave power 30 mW at a frequency of 9.107 GHz. Receiver gain: 3200.

A regenerable methaemocyanin of *A. leptodactylus* could also be obtained by the reaction of deoxyhaemocyanin with nitrite in 0.1 M acetate buffer pH 5.7, which was completed within 4 h. With oxyhaemocyanin the reaction proceeded much more slowly as only deoxyhaemocyanin, in equilibrium with oxyhaemocyanin, seemed to react. Excess nitrite was removed by dialysis. This methaemocyanin showed the same properties as that prepared with nitrogen monoxide.

CONCLUSION

The crustacean methaemocyanin prepared by the reaction of deoxyhaemocyanin with nitrogen monoxide or with nitrite in 0.1 M acetate buffer pH 5.7, in contrast with that obtained with hydrogen peroxide, is regenerable for ≈ 70 % with hydroxylamine.

Acknowledgements

We wish to thank the Nationaal Fonds voor Wetenschappelijk Onderzoek, and the Fonds voor Collectief Fundamenteel Onderzoek for research grants. We wish to express our gratitude to the Instituut tot Aanmoediging van het Wetenschappelijk Onderzoek in Nijverheid en Landbouw for graduate fellowships (D.V.H. and W.D.).

REFERENCES

1. Felsenfeld, G., and Printz, M.P., J. Am. Chem. Soc. *81*, 6259-6264 (1959).
2. Lijnen, H.R., Witters, R., and Lontie, R., Comp. Biochem. Physiol. *63B*, 35-38 (1979).
3. Gaykema, W.P.J., Hol, W.G.J., Vereijken, J.M., Soeter, N.M., Bak, H.J., and Beintema, J.J., Nature *309*, 23-29 (1984).
4. Verplaetse, J., Van Tornout, Ph., Defreyn, Gh., Witters, R., and Lontie, R., Eur. J. Biochem. *95*, 327-331 (1979).
5. Witters, R., Deleersnijder, W., Tahon, J.-P., and Lontie, R., this volume, pp. 449-452 (1986).

BINUCLEAR ELECTRON-PARAMAGNETIC-RESONANCE SIGNALS OF ARTHROPODAN METHAEMOCYANINS IN THE PRESENCE OF NITRITE

R. Witters, W. Deleersnijder, J.-P. Tahon, and R. Lontie
Laboratorium voor Biochemie, Katholieke Universiteit te Leuven
Dekenstraat 6, B-3000 Leuven, Belgium

The deoxy- and oxyderivatives of haemocyanin are diamagnetic; in the oxyderivative the copper(II) ions in the binuclear copper site are strongly magnetically coupled. A strong interaction between the two copper(II) ions in a copper pair is favoured when only very few atoms form a superexchange pathway leading to a coupling of the unpaired spins. The copper(II) ions can also be magnetically coupled as a function of their distance and thus give rise in electron paramagnetic resonance (EPR) to a broad triplet signal near $g = 2$ and a weakly allowed signal at half field near $g = 4$, as observed for *Helix pomatia* methaemocyanin with acetate at pH 5.7 (1, 2).

A similar partial uncoupling of the EPR-undetectable copper pairs, leading to a broad binuclear Cu(II) signal near $g = 2$, was observed by the addition of nitrite to some crustacean and cheliceratan methaemocyanins in 0.1 M acetate buffer pH 5.7.

MATERIALS AND METHODS

The methaemocyanins of the crustacean *Astacus leptodactylus* and of the chelicerate *Limulus polyphemus* were prepared by treating deoxyhaemocyanin with hydrogen peroxide, in a molar ratio to copper of 10, in 0.1 M acetate buffer pH 5.7 according to Felsenfeld and Printz (3). Regenerable methaemocyanin of *A. leptodactylus* was prepared from deoxyhaemocyanin by a treatment with nitrogen monoxide or with nitrite in the same buffer (4).

All EPR spectra were recorded at 133 K with 100 kHz field modulation at a modulation amplitude of 10 G; the microwave power was 30 mW at a frequency of 9.107 GHz. A solution of 0.2 mM $CuSO_4$, 10 mM $HClO_4$ and 2 M $NaClO_4$ was used as a standard in the double integration of the EPR signal near $g = 2$.

RESULTS AND DISCUSSION

The methaemocyanins of *A. leptodactylus* and *L. polyphemus*, prepared by

Invertebrate Oxygen Carriers
Ed. by Bernt Linzen
© Springer–Verlag Berlin Heidelberg 1986

the oxidation of deoxyhaemocyanin with hydrogen peroxide in 0.1 M acetate buffer pH 5.7 or by the reaction of deoxyhaemocyanin with nitrogen
monoxide or with nitrite, normally did not show a distinct EPR signal,
except for a small mononuclear Cu(II) signal near g = 2 (from 5 to 10 %),
likely due to broken copper pairs or to loosely bound copper. The intensity of the mononuclear Cu(II) signal was higher for the crustacean than
for the cheliceratan methaemocyanin, prepared with hydrogen peroxide,
indicating a higher stability of the active site of the latter (Fig. 1a).
In the presence of nitrite at pH 5.7 these methaemocyanins showed a
broad binuclear Cu(II) signal near g = 2 as expected with some uncoupling of the EPR-undetectable copper pairs; in alkaline medium (pH 9.2)
this signal was not observed on addition of nitrite.
L. *polyphemus* methaemocyanin also showed a weak binuclear EPR signal at
half field near g = 4 in the presence of nitrite without, however, the
seven hyperfine lines observed with H. *pomatia* methaemocyanin under the
same conditions (5). At room temperature the broad EPR signal near g = 2
was formed immediately; at 0°C a time dependent formation was observed
for L. *polyphemus* methaemocyanin with a pseudo first-order rate constant
of ≈ 0.25 min^{-1} (Fig. 2). These results confirm the binding of nitrite

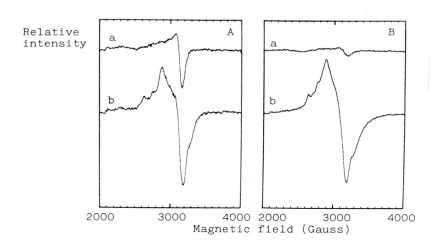

Fig. 1. EPR spectra near g = 2 of methaemocyanin in 0.1 M acetate
buffer pH 5.7, prepared with hydrogen peroxide according to Felsenfeld
and Printz (3), before (a) and 4 h after (b) the addition of nitrite
in a molar ratio to copper of 100 at room temperature.
A. *Astacus leptodactylus* methaemocyanin (23.7 mg/ml; 0.56 mM Cu).
   Receiver gain: 12 500.
B. *Limulus polyphemus* methaemocyanin (38.5 mg/ml; 1.04 mM Cu).
   Receiver gain: 8000.

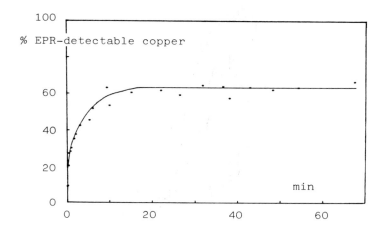

Fig. 2. Percentage EPR-detectable copper as a function of time during the reaction of *Limulus polyphemus* methaemocyanin (prepared with hydrogen peroxide) with nitrite (135 mM) at 0°C in 0.1 M acetate buffer pH 5.0. Protein concentration: 50 mg/ml (1.35 mM Cu).

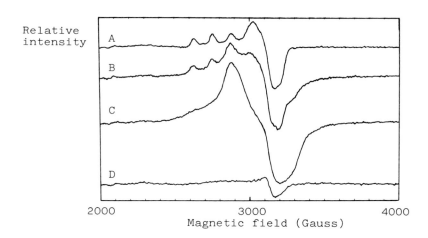

Fig. 3. EPR spectra of *Astacus leptodactylus* methaemocyanin (20 mg/ml; 0.5 mM Cu) prepared in 0.1 M acetate buffer pH 5.7 by reaction of deoxy-haemocyanin with nitrogen monoxide for 30 min (A). The same preparation kept for 15 min (B) and 4 h (C) at room temperature after the addition of nitrite to a final concentration of 47 mM, and after removal of nitrite by dialysis (D). Receiver gain: 10 000.

to methaemocyanin; this binding was reversible as the binuclear Cu(II) signal disappeared on removal of nitrite by dialysis against 0.1 M acetate buffer pH. 5.7.

The addition of nitrite to *A. leptodactylus* nitrosylhaemocyanin, characterized by a mononuclear Cu(II) signal (Fig. 3A), induced a slow transformation of nitrosylhaemocyanin into methaemocyanin, the mononuclear Cu(II) signal being gradually replaced by a broad binuclear signal (Fig. 3B, C), which disappeared on removal of nitrite by dialysis. Besides binding to methaemocyanin at pH 5.7, nitrite can oxidize the copper atoms in deoxyhaemocyanin (4) and expel nitrogen monoxide from nitrosylhaemocyanin as demonstrated in Fig. 3.

CONCLUSION

The copper pairs in arthropodan methaemocyanin, prepared either with hydrogen peroxide or with nitrogen monoxide or nitrite, were EPR-undetectable. They were uncoupled partially on binding nitrite at pH 5.7, resulting in the appearance of a broad triplet signal near $g = 2$, similar to that observed with bovine erythrocyte (Cu,Cu)superoxide dismutase in which an imidazolate serves as a bridging ligand (6, 7).

Acknowledgements

We wish to thank the Nationaal Fonds voor Wetenschappelijk Onderzoek and the Fonds voor Collectief Fundamenteel Onderzoek for research grants. We wish to express our gratitude to the Instituut tot Aanmoediging van het Wetenschappelijk Onderzoek in Nijverheid en Landbouw for a graduate fellowship (W.D.).

REFERENCES

1. Witters, R., De Ley, M., and Lontie, R., *in* Structure and Function of Haemocyanin (Bannister, J.V., ed.), pp. 239-244, Springer Verlag, Berlin (1977).
2. Deleersnijder, W., Witters, R., and Lontie, R., Life Chem. Reports, Suppl. 1, 289-290 (1983).
3. Felsenfeld, G., and Printz, M.P., J. Am. Chem. Soc. *81*, 6259-6264 (1959).
4. Witters, R., Van Hoof, D., Deleersnijder, W., Tahon, J.-P., and Lontie, R., this volume, pp. 445-448 (1986).
5. Verplaetse, J., Van Tornout, Ph., Defreyn, Gh., Witters, R., and Lontie, R., Eur. J. Biochem. *95*, 327-331 (1979).
6. Fee, J.A., and Briggs, R.G., Biochim. Biophys. Acta *400*, 439-450 (1975).
7. Strothkamp, K.G., and Lippard, S.J., Biochemistry *20*, 7488-7493 (1981).

EPR OBSERVATIONS ON THE REACTION OF HEMOCYANIN WITH AZIDE.

B. Salvato, G.M. Giacometti, M. Alviggi and G. Giacometti
Dept. of Biology and   Dept. of Physical Chemistry, University of
Padua, Italy.

INTRODUCTION

The reaction of molluscan hemocyanin with azide, first reported
by Witters and Lontie (1) and further studied by Himmerlwright et
al. (2,3), leads to the formation of a double oxidized form of the
protein active site. This derivative is reported to be EPR
undetectable for about 95% of the total copper and regenerable to
the original native protein by treatment with suitable reducing
agents (1). The small EPR detectable fraction, on the contrary, is
claimed to be not regenerable. The proposed mechanism for this
reaction involves the ligand association to the protein oxy-form,
followed by the displacement of the peroxide, with the formation of
a $[Cu(II) \ Cu(II) \ ] \cdot N_3^-$ complex (4). Such a "ligand displacement"
mechanism would imply a substantial equivalence of the two copper
ions of the active site. This equivalence is, however, questionable
on the basis of a number of evidences; thus, isotopic Cu exchange
(5), photoelectron addition (6) and half-apo preparation (7,8),
point out the different chemical reactivity of the copper ions at
the active site of hemocyanin.
    Working under different experimental conditions, we have
collected a number of evidences concerning the role of azide and
oxygen in promoting the conversion of hemocyanin from its native to
its oxidized forms. The results obtained can be rationalized
assuming an asymmetric chemical behavior for the metal ions.

EXPERIMENTAL

Hemocyanin from _Octopus vulgaris_ was purified according to
Salvato et al. (9). Protein concentration was determined using the
extinction coefficient $E^{0.1\%}$(280) = 1.43. Copper was determined by
atomic absorption using a Perkin Elmer mod. 300 atomic absorption
spectrophotometer. EPR spectra were recorded at 140 °K on a Bruker
ER-200 D 5RC spectrometer. Double integration of the EPR spectra
was performed on a M20 Olivetti micro-computer interfaced to a
digitizer deck. Quantification of the EPR detectable copper was
done by comparison with a 1 mM solution of superoxide dismutase at
pH 7.0 used as a standard. Dialysis procedures were performed using
2 mm in diameter Servapore tubing. All manipulations of the samples
were performed at temperature below or equal to 20 °C. Analytical
determinations of azide and chloride were performed, respectively,
according to (10) and (11).

Invertebrate Oxygen Carriers
Ed. by Bernt Linzen
© Springer–Verlag Berlin Heidelberg 1986

RESULTS AND DISCUSSION

In the absence of oxygen hemocyanin binds azide forming a pale yellow complex which completely reverts to the colorless deoxy form after removal of the ligand. The same ligand, if added in the presence of oxygen, causes irreversible changes at the active site of the protein. Under these conditions, a green-brown color is developed within few hours after the addition of 100 mM azide, giving rise, at equilibrium, to the formation of a derivative characterized by the EPR spectrum reported in fig. 1B. The shape of the spectrum is suggestive of the presence of more than one EPR detectable species, and double integration accounts for approximately 50% of the total copper.

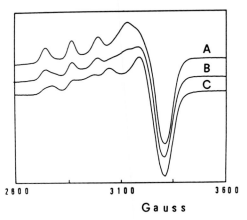

Gauss

Fig. 1
EPR spectra of: (A) green half-met hemocyanin; (B) the product of the direct addition of azide to oxy- or half-met hemocyanin in the presence or absence of oxygen; (C) the product of a dialysis of oxy- or half-met hemocyanin against azide in the presence of oxygen.

The very same derivative, as far as the EPR spectrum is concerned, is obtained by addition of the ligand, both in the presence or absence of oxygen, to the green half-met form of the protein (12) whose EPR spectrum is shown in fig. 1A.
A different situation is observed when the reaction with azide is carried out by dialysis in the presence of oxygen rather than by direct addition. By this procedure, starting both from the native or from the green half-met forms of the protein, a reddish-brown product is obtained whose EPR spectrum is shown in fig. 1C. The double integration of the signal accounts for 64% of the total copper. Removal of the azide by a subsequent dialysis against phosphate buffer at pH 7.0 induces a color change to purple and a modification in shape of the EPR signal with a decrease of the spin concentration to 45% of the total copper (fig. 2B). Analytical

determination gives an upper limit for the azide still present in this sample of about 5% of the active site concentration (which is within the experimental error of the method used). We interpret this form of the protein as a doubly oxidized (met) derivative.

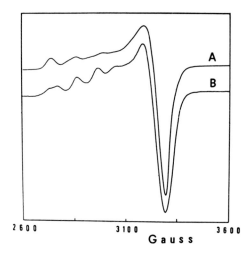

Fig. 2
EPR spectra of: (A) the blue species obtained by treatment of met hemocyanin with 100 mM chloride; (B) met hemocyanin in phosphate buffer pH 7.0.

Fig. 2A shows the EPR spectrum of a blue derivative obtained by treatment of the purple met-form with 50 mM sodium chloride at pH 7.0 and subsequent removal of the reactant by dialysis. No bound chloride was found by analytical determination of this anion after the last dialysis.

These experimental results can be interpreted with reference to the following main points:
i) the product obtained by direct addition of azide (spectrum 1B) is the same either one starts from the oxy- or from the half-met form of the protein. This excludes that the action of azide on hemocyanin involved a ligand displacement reaction with the simultaneous subtraction of two electrons by molecular oxygen.
ii) Spectrum 1B, obtained in the absence of oxygen from the half-met, shows the same spin concentration as the starting material (50%), indicating that under these conditions, azide simply induces interconversion between different EPR detectable species.
iii) a different product is obtained when azide, in the presence of oxygen, is added directly or by dialysis. This implies the occurrence of a dialyzable reaction intermediate which has to be identified in a reduction product of oxygen.

iv) spectrum 1C reflects the presence of an EPR detectable met form in equilibrium with a met EPR undetectable form of the protein.

On this ground we propose the following chemical pathway:

$$[Cu(II)\ Cu(II)]\ O_2 \xrightarrow{N_3^-} [Cu(I)\ Cu(II)]\ +\ O_2^- \qquad (1)$$

$$[Cu(I)\ Cu(II)] \xleftrightarrow{N_3^-} [Cu(II)\ Cu(I)] \qquad (2)$$

$$[Cu(II)\ Cu(I)]\ +\ O_2 \xrightarrow{N_3^-} [Cu(II)\ Cu(II)]\ +\ O_2^- \qquad (3)$$

$$[Cu(II)\ Cu(II)]\ +\ Cl^- \longrightarrow [Cu(II)\ Cu(I)]\ +\ Cl^{\cdot} \qquad (4)$$

Reaction 1 represents a one-electron oxidation, catalyzed by azide, with the production of superoxide and of one of the two half-met species (the known green half-met).

In reaction (2) azide, bridging the copper ions, delocalizes the oxidation and generates an equilibrium between the two possible half met (spectrum 1B).

A further one-electron oxidation by molecular oxygen on the azide half-met complex produces the final doubly oxidized form of the protein. This contains 65% detectable copper in the presence of azide and 45% in its absence. The met derivative so obtained can be completely regenerated to the native oxy-form by treatment with hydrogen peroxide.

The last reaction represents the selective reduction of the external (more accessible) copper of the met by chloride yielding a new blue derivative that we tentatively assign to the other half-met form or "internal half-met".

REFERENCES

1. Witters R. and Lontie R., FEBS Letters, 60 400 (1975).
2. Himmelwright R.S., Eickman N.C. and Solomon E.I., Biophys. Biochem. Res. Commun., 84, 300 (1978).
3. Himmelrwright R.S., Eickman N.C. and Solomon E.I., Biophys. Biochem. Res. Commun., 86, 628 (1979).
4. Solomon E.I., in "Copper Proteins" ed. T.G. Spiro, J. Wiley and Sons, New York, pag. 43 (1981).
5. Cox J.A. and Elliott F.G., Biophys. Biochim. Acta, 371, 392 (1974).
6. Symons M.C.R. and Petersen R.L., Biophys. Biochim. Acta, 535, 247 (1978).
7. Beltramini M., Ricchelli F., Piazzei A., Barel A. and Salvato B., Biochem. J., 221, 911 (1984).
8. Beltramini M. et al., this volume, pp. 437-440 (1986).
9. Salvato B., Ghiretti-Magaldi A. and Ghiretti F., Biochemistry, 13, 4778 (1974).
10. Roberson C. and Austin C.M., Analyt. Chem., 29, 854 (1957).
11. Vogel A.I. "A Textbook of Quantitative Inorganic Analysis" Lowe and Brydone, London (1961).
12. Salvato B. et al. , this volume, pp. 457-462 (1986).

THE REACTION OF HEMOCYANIN FROM OCTOPUS VULGARIS WITH NITRITE. A
REINTERPRETATION OF THE OXIDATION PATHWAYS.

B. Salvato, G.M. Giacometti, M. Alviggi and G. Giacometti

Department of Biology and Department of Physical Chemistry, University of Padua, Italy.

INTRODUCTION

The reaction of Hemocyanin with nitric oxide has been observed
since a long time (1). An interpretation in terms of redox reaction
was first proposed by Schoot Uiterkamp (2) and Schoot Uiterkamp and
Mason (3). Different chemical schemes have been reported since then
to explain this reaction. In all cases NO was assumed to be the oxidant, whereas different interpretation were advanced about the nature and state of ligation of the oxidation product (4,5,6). According to a recent review by Solomon (7), the generally accepted
chemistry of oxidation of molluscan hemocyanin involves the oxidation of the deoxy form by NO to an half-met-NO_2^- intermediate which
can be further oxidized to a "dimer" (i.e. an EPR detectable met)
in the form of a $[Cu(II) \; Cu(II)]N_xO_y$ derivative.

Our preparations of the green half-met hemocyanin from Octopus
vulgaris give a product identical to that reported for other species
(7) as far as the optical and EPR spectra are concerned. However analytical search for the exogenous nitrogenated ligand always yielded
a ratio for the ligand to oxidized copper far below the 1:1 stoichiometric value.

We have therefore undertaken a reinvestigation of the reaction of
hemocyanin with nitrite keeping in mind the complex chemistry of the
oxygenated nitrogen compounds and the possibility of side reactions
on the protein moiety.

EXPERIMENTAL

The hemocyanin from Octopus vulgaris was purified according to

Invertebrate Oxygen Carriers
Ed. by Bernt Linzen
© Springer–Verlag Berlin Heidelberg 1986

Salvato et al. (8). Protein concentrations were determined spectrofo-
tometrically using the value of 1,43 for the $E_{(280nm)}^{0.1\%}$. Copper was
determined by atomic absorption with a Perkin-Elmer mod. 300 spectrophotometer. Dialysis procedures were carried out using 2 mm in diameter tubing to allow the rapid equilibrium of the reactants.

Kinetic experiments were performed on oxy hemocyanin (2 mg/ml)
following the decrease of absorbance at 348 nm.

EPR spectra were recorded at 140 K on a Bruker ER-200D 5RC spectrometer. Double integration of the EPR spectrum was obtained on a M20
Olivetti micro-computer interfaced with a Digitizer deck. Quantification of the detectable spin was done by comparison with a 1 mM solution of superoxide-dismutase at pH 7.0, used as a standard.

Nitrite was determined with the sulfanilic acid method (9).

RESULTS AND DISCUSSION

When Octopus vulgaris hemocyanin is dialysed against 50 mM phosphate
buffer pH 5.5, containing 5 mM $NaNO_2$ and 5 mM ascorbate, the blue native protein turns to a green product within a few hours. Fig. 1
shows the EPR spectrum of this sample recorded after removal of the
reactants by dialysis against phosphate buffer at pH 7.0. Careful
quantitation of the EPR signal accounts for 50% (± 3%) of the total
copper. Analytical determination on the same sample indicates that
the concentration of the exogenous nitrogenated ligand accounts for
no more than 20% of the oxidized copper.

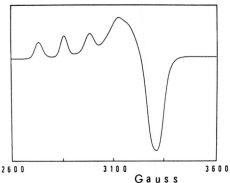

Fig. 1

EPR spectrum of the reaction product of Hc with nitrite. Total Cu
conc. = 3 mM, T = 140 K.

Incubation with sulfamic acid buffered to pH 5.5 strongly decreases the concentration of nitrogenated exogenous ligand without affecting the shape and intensity of the EPR signal. On the other hand, no modification of the EPR spectrum was observed after direct addition of $NaNO_2$ up to 10 mM.

These results demonstrate that the green product of the reaction with nitrite is a ligand free singly oxidized form of the protein. The residual oxigenated nitrogen compound detected by the sulfanilic acid method can therefore be attributed to the formation of nitrous esters from a side reaction of nitrite with alcoholic groups of the protein matrix. This view is further supported by the following observation: when the reaction with nitrite is performed in the presence of TRIS, under the same conditions, the amount of the bound NO_2^- is strongly lowered due to the competition of the -OH groups of the buffer component. In order to get more detailed informations on the mechanism of the hemocyanin oxidation by the "nitrite method", we have measured the rate of disappearance, at pH 5.5, of the 348 nm band of the oxy derivative as a function of the concentration of nitrite and ascorbate.

Fig. 2 shows the dependence of the initial pseudo-first order rate constant on the concentration of the two reactants.

When the nitrite concentration is varied at constant (5 mM) ascorbate (fig. 2, dashed line) a linear dependence starting from the origin is observed. At higher concentration of nitrite (5 mM) an acceleration is observed probably due to the formation of the strong oxidant NO_2. In fact, in the presence of oxygen, the NO species slowly produced by the reaction of nitrite with ascorbate, is very short-lived being rapidly oxidized to NO_2.

This result indicates that the copper oxidation is not driven by NO under these conditions. This has also been directly proved by anaerobic exposure of deoxy-hemocyanin to NO gas at low concentration. No copper oxidation was observed in accordance with the low oxidizing power of this species (10). The oxidation of deoxy-hemocyanin observed by other authors (10) under higher NO pressure, could therefore be due to the NO_2 whose production rapidly increases with the NO concentration (10).

The solid line of fig. 2 is obtained when the concentration of ascorbate is varied at constant nitrite (5 mM).

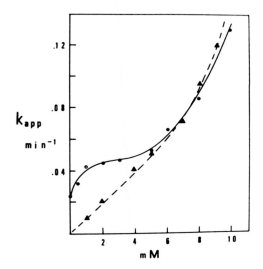

Fig. 2

Rate of disappearance of the copper-oxygen band at 438 nm as a function of the concentration of nitrite at constant 5 mM ascorbate (triangles),or of the concentration of ascorbate at constant 5 mM nitrite (circles). Temperature 20 C.

We first observe that nitrite is effective in oxidizing hemocyanin even in the absence of ascorbate. As the concentration of ascorbate increases, the reaction speeds up to a plateau after which an acceleration follows similar to that observed at high concentration of nitrite. The plateau is obtained at a concentration of ascorbate (1 mM) too low to be responsible for significant production of NO_2. The saturation effect observed at low concentration can be interpreted in terms of the known activity of ascorbate as a scavenger of superoxide (11). In fact, if we assume that O_2^- is produced during the single electron oxidation of hemocyanin, an accumulation of this byproduct can reverse the oxidation giving back the native protein. At higher concentrations of ascorbate the production of

NO_2 becomes the dominant effect.

We have verified the effectivness of NO_2 as an oxidant for hemo-cyanin by anaerobic exposure of the protein to low concentration (0.1 mM) of this gas.

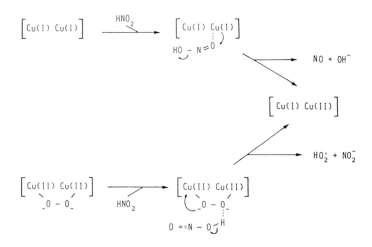

Fig. 3

Reaction scheme for the one-electron oxidation of oxy- and deoxy-hemocyanin by nitrite.

The reaction scheme reported in fig. 3 summarizes the possible chemical pathways for the reactions discussed above. The same kind of reaction mechanism was found effective in explaining the semi-quinone formation during the catechol oxidation by molluscan hemo-cyanin (12 and unpublished results).

References

(1) C. Dhéré, J. Physiol. Pathol. Gen., 18, 503 (1919).

(2) Schoot Uiterkamp, A.J.M., FEBS Lett. 20, 93 (1972).

(3) Schoot Uiterkamp, A.J.M. and Mason, H.S., Proc. Nat. Acad. Sci. U.S.A., 70 993 (1973).

(4) Van der Deen, H. and Hoving, H., Biochemistry 16, 3519 (1977).

(5) Himmelwright, R.S., Eickman, N.C., and Solomon, E.I., Biochem.

Biophys. Res. Commun., 81, 237 (1978).

(6) Verplaetze, J., Van Tornout P., Defrejn G., Witters R., and
Lontie R., Eur. J. Biochem. 95, 327 (1979).

(7) Solomon E.I., in "Copper Proteins", ed. T.S. Spiro, J. Wiley and
Sons, New York, pag. 43 (1981).

(8) Salvato B., Ghiretti Magaldi A. and Ghiretti F., Biochemistry 13,
4778 (1974).

(9) Vogel A.I. "A textbook of Quantitative Inorganic Analysis". Lowe
and Brydone, London (1961).

(10) Cotton F.A. and Wilkinson G., "Advanced Inorganic Chemistry".
Interscience (1972).

(11) Dunford H.B. and Nadezhdin A.D. in "Oxygen and Oxy-Radicals in
Chemistry and Biology", Rodgers M.A.J. and Power E.L., eds., pag.
625, Acad. Press. (1981).

(12) Salvato B., Jori G., Piazzesi A., Ghiretti F., Beltramini M. and
Lerch K., Life Chem. Rep., Supp. 1, 313 (1983).

Achatina fulica Hemocyanin and a Dicopper Oxygen Carrier as Model

S. M. Wang, M. N. Chen, J. R. Tzou, and N. C. Li
Department of Chemistry
National Tsing Hua University
Hsinchu, Taiwan 300
Republic of China

Abstract

Some properties and reactions of Achatina fulica hemocyanin
from Taiwan snails have been compared with those of a synthetic dicopper oxygen carrier, $Cu_2(EDTB)(ClO_4)_2$, where EDTB is N,N,N'N'-
tetrakis(2-benzimidazolylmethyl)-1,2-ethanediamine, serving as an
active-site model. The colorless solution of $Cu_2(EDTB)(ClO_4)_2$ in
dimethylsulfoxide shows an uptake of oxygen to give green solution
with an electronic spectrum characteristic for Cu(II). Addition
of ascorbic acid, AA, results in decolorization of the solution, and
the cycle can be repeated 3-4 times. This is similar to the effect
of AA on hemocyanin, except that oxyhemocyanin is blue instead of
green. The solids of green and colorless forms of the synthetic
compound give ESCA $Cu(2p_{3/2}, 2p_{1/2})$ peaks in the same positions.
However, only the green solid spectrum shows "shake-up" satellites,
indicating Cu(II). By treatment with KCN, copper can be removed
from hemocyanin and $Cu_2(EDTB)(ClO_4)_2$, and copper can be reconstituted
into the protein and the active-site model. Properties and reactions of the reconstituted compounds are compared with those of the
compounds before KCN treatment.

Introduction

Dicopper coordination compounds with formulas $Cu_2(EDTB)(ClO_4)_2$
and $Cu_2(EDTB)Cl_4$ have been prepared (EDTB = N,N,N',N'-tetrakis-(2-
benzimidazolylmethyl)-1,2-ethanediamine). Hendricks et al.[1] have
discussed the importance of $Cu_2(EDTB)^{2+}$ as a potential model for
hemocyanin. The purpose of the present research is to investigate
the reactions of $Cu_2(EDTB)(ClO_4)_2$ with several ligands and compare
these with the reactions of hemocyanin from Taiwan snails (Achatina
fulica[2]) with the same ligands. By treatment with KCN, copper can
be removed from hemocyanin and $Cu_2(EDTB)(ClO_4)_2$, and copper can be
reconstituted into the protein and the active-site model. Properties
and reactions of the reconstituted compounds are then compared with

Invertebrate Oxygen Carriers
Ed. by Bernt Linzen
© Springer-Verlag Berlin Heidelberg 1986

those of the compounds before KCN treatment.

Experimental

EDTB was prepared as described by Hendricks et al.[3] with
final yield of 50%. The dicopper(I) compound, $Cu_2(EDTB)(ClO_4)_2$,
was prepared by mixing under N_2, 20 ml of hot ethanol solution
containing 2 mmol EDTB and 2 ml of hot acetonitrile solution containing $Cu(CH_3CN)_4(ClO_4)_2$. The dicopper(I) EDTB compound was
filtered off under N_2, washed with cold absolute alcohol and diethyl
ether. After drying in vacuo at 50°C, white powders were obtained,
which were not air-sensitive. The copper content was 13.9%, which
agrees with the calculated value, 14.0%, in the formula
$Cu_2C_{34}H_{32}N_{10}(ClO_4)_2$.

Cu(II) and Co(II) compounds of EDTB were prepared by reacting
with $CuCl_2$ and $CoCl_2$, respectively, using M:EDTB ratio of 2:1. A
solution of EDTB in hot ethanol was added to an ethanol solution of
the metal salt. On standing, crystals of the coordination compounds
were formed. These were filtered, washed with ethanol and diethyl
ether, and then dried in vacuo at 60°C. The compounds obtained
were: $Cu_2(EDTB)Cl_4$, yellow, and $Co_2(EDTB)Cl_4$, blue-purple.

By treatment with KCN, Cu can be removed from dicopper compounds of EDTB, and Cu can be reincorporated so as to form the
active-site model compound. We have found that the best yield of
reconstituted $Cu_2(EDTB)(ClO_4)_2$ was obtained by adding KCN/H_2O
dropwise to a $(CH_3)_2SO$ solution of $Co_2(EDTB)Cl_4$ (blue-purple) until
the color changed to slight brownish-yellow. Addition of water
resulted in a white turbid suspension. The aqueous cobalt cyanide
complex formed was filtered off and the precipitate, EDTB, was dried
and dissolved in hot ethanol. This solution was treated with a 2.5-
fold excess of $Cu(CH_3CN)_4ClO_4$ in hot acetonitrile in a N_2 atmosphere.
A white powder formed, which is the reconstituted $Cu_2(EDTB)(ClO_4)_2$.
The powder was filtered off, washed with ethanol and vacuum dried
at 60°C.

For the reconstitution of hemocyanin, apohemocyanin was prepared from oxyhemocyanin by KCN treatment[4] and the apoHc solution
was then treated at pH 5.7 for 24h at 4°C in N_2 atmosphere with an
amount of solid $Cu(CH_3CN)_4ClO_4$ corresponding to twice the copper
concentration of hemocyanin. The reconstituted hemocyanin (R-Hc)
was dialyzed against 0.025 M EDTA in 0.1M acetate buffer, pH 5.7
in N_2 atmosphere and then against the acetate buffer alone. The

native <u>Achatina fulica</u> hemocyanin has a copper content of 0.24%.
After reconstitution, a copper content of 0.21% was found, corresponding to a reconstitution factor of 87.5%.

Results and Discussion

The primary reaction of KCN (0.08-0.2M) with oxyhemocyanin at
pH 5.7 is very fast and rates were about 10 times faster with native
hemocyanin than with R-Hc. The results suggest that in the recon-
situted species, Cu is not bound as tightly to the protein moiety in
the active site as in native hemocyanin, and the effect of KCN is
greater with native hemocyanin. Previous experiments with <u>Busycon</u>
hemocyanin[5] have shown that cyanide causes extensive fragmentation,
which is a subunit dissociation phenomenon. Evidence for strong
binding of cyanide to hemocyanin have been given by Borke et al.[6]
from oxygen-binding and ion-exchange experiments.

The Mn(II) and Cd(II) derivatives of Hc were prepared in a
similar way as the preparation of R-Hc, except that $MnCl_2$ and $CdCl_2$
take the place of $Cu(CH_3CN)_4ClO_4$. The metal contents were 0.12% Cd
in Cd-Hc and 0.15% Mn in Mn-Hc. Since R-Hc contains 0.21% Cu, the
data indicate that the affinity of metals to the active site in Hc
is in the order Cu>Mn>Cd. The Mn and Cd derivatives of Hc do not
bind oxygen.

The dicopper compound, $Cu_2(EDTB)(ClO_4)_2$, chosen as model for
Hc, is colorless. In DMSO solution, there is an uptake of oxygen
to give green solution, with maximum absorption at 690 nm. Addition
of ascorbic acid, AA, results in decolorization of the solution and
the cycle can be repeated 3-4 times. This is similar to the effect
of AA on hemocyanin, except that oxyhemocyanin is blue (max at 345
nm) instead of green. AA exerts similar effects on the reconstituted $Cu_2(EDTB)(ClO_4)_2$ and reconstituted hemocyanin.

The electron spectrum for chemical analysis (ESCA) spectrum of
colorless solid $Cu_2(EDTB)(ClO_4)_2$ showed the following main peaks:
Cl 2p at 206.8 eV; C 1s, 284.0; N 1s, 399.8; O 1s 532.8; Cu $2p_{3/2}$,
934.2, Cu $2p_{1/2}$, 954.2; and Cu 3p, 71 eV. When this colorless
dicopper(I) coordination compound was suspended in a small amount
of absolute ethanol and oxygen bubbled through, the color changed
to green. After filtration, the green solid was dried at 70°C for
6 h. The ESCA Cu($2p_{3/2}$, $2p_{1/2}$) main peaks of the green compound
have binding energies of 934.6 and 954.4 eV. In addition, there
are shake-up satellites at around 945 and 964 eV (higher binding

energies than the corresponding main peaks). Since the shake-up
satellites usually occur with paramagnetic states, our ESCA data
show that part of the Cu in the green compound has been oxidized to
the cupric state. This is confirmed by comparing with the ESCA
spectrum of solid $Cu_2(EDTB)Cl_4$. In this compound, both coppers are
cupric and all the peaks (main peaks at 935.2 and 955.2 for $2p_{3/2}$
and $2p_{1/2}$, as well as the corresponding shake-up satellites at 945
and 963 eV) have higher binding energies. In contrast to
$Cu_2(EDTB)(ClO_4)_2$, $Cu_2(EDTB)Cl_4$ does not absorb oxygen.

Oxyhemocyanin is diamagnetic. On adding KCN the protein solution is still diamagnetic. The model compound, $Cu_2(EDTB)(ClO_4)_2$,
green solution in $(CH_3)_2SO$, is paramagnetic. On addition of KCN
to the green solution, absorbance at 690 nm decreases and is not
increased on bubbling O_2. The Cu(II) ESR signal intensity is also
decreased until above 0.04 M KCN, the solution becomes diamagnetic.
Apparently in the reaction between Cu_2EDTB and KCN, the Cu(II) is
reduced to cuprous cyanide complex, so it becomes diamagnetic and
loses oxygen-carrying ability. Thus, there are certain similarities and dissimilarities between Hc and the dicopper EDTB compound.

References

1. Hendricks, H. M. J., Birker, P. J. M. W. L., van Rijn, J.,
   Verschoor, G. C., Reedijk, J., J. Am. Chem. Soc. 104, 3607-
   3617 (1982).

2. Chen, J. T., Shen, S. T. Chung, C. S., Chang, H., Wang, S. M.,
   and Li, N. C., Biochemistry 18, 3097-3101 (1979).

3. Hendricks, H. M., Ten Bokkel Huinink, W. O., Reedijk, J., Recl.
   Trav. Chim. Pays-Bas, 98, 499-500 (1979).

4. Lontie, R., Blaton, V., Albert, M., Peters, B., Arch. Intern.
   Physiologie Biochimie, 73 (1), 150-152 (1965).

5. Douglas, K. T., Lee, C. H., Li, N. C., in Invertebrate Oxygen
   Proteins (Lamy, J. and Lamy, J., eds.), pp. 603-611, Marcel
   Dekker, N-W Work (1981).

6. Borke, M. L. Lee, B. K., and Li, N. C., Proc. Natl. Sci. Council
   ROC 8 (4), 319-323 (1984).

Hemoglobins and hemocyanins: studies at the levels of DNA and RNA, and biosynthesis

IDENTIFICATION OF mRNA CODING FOR GLOBIN IN THE CHLORAGOGEN CELLS OF LUMBRICUS TERRESTRIS

S.M. Jhiang, J.R. Garey, and A.F. Riggs
Department of Zoology, University of Texas, Austin, TX 78712 USA

Abstract. Chloragogen cells of the outer intestinal walls of Lumbricus terrestris contain RNA which is active in the in vitro synthesis of globin which has been identified immunologically in a Western blot. A Northern blot shows the presence of a poly A^+ RNA fraction of 750–850 bp which hybridizes strongly with a synthetic 17-base oligonucleotide corresponding to amino acid residues 103-108 of chain AIII of Lumbricus globin. A cDNA library in λgt10 is being screened with the oligonucleotide.

Introduction. The hemoglobin of Lumbricus terrestris is a large multi-subunit protein of about 3.9×10^6 Kd (1). The number of kinds of polypeptide chains has been reported to be either 4 (ref. 2) or 6 chains (3). The site of synthesis of the hemoglobin appears to be the chloragogen cells on the outer intestinal wall on the basis of light and electron microscope work (4,5). The goal of the present studies has been to extract and identify the mRNA for the globin from these cells. We hope to clone the cDNA for the globin and to isolate and sequence the globin genes. The structure of the genes and the flanking regions will provide important information on their coordinated expression, the distribution of introns and the evolution of the globin.

Methods. Isolation of mRNA: Chloragogen cells were carefully scraped from the outer intestinal walls of about 1000 worms. They were washed in 0.6% NaCl and extracted with 6M guanidine–HCl as described by Cox (6). Poly A^+ RNA was prepared from the total RNA with an oligo dT column. Production of rabbit antiserum for globin: Whole globin was prepared as described (7). Rabbits were immunized with globin dissolved in 0.85% NaCl mixed with an equal volume of complete Freund's adjuvant and emulsified. The antiserum was processed as described by Garvey et al. (8). Immunodiffusion was used to detect the antibody titer. In vitro translation of RNA: Total RNA was extracted and translated in a wheat germ system. Either ^{35}S-methionine or ^{3}H-leucine was used as label. The translation products were applied to an SDS polyacrylamide gel for electrophoresis (9). Western blot of the translation products: Cold translation products of mRNA in the wheat germ system were applied to a 12.5% SDS-PAGE for electrophoresis, and then transferred to nitrocellulose in Western blot (10). The paper was incubated at 40° C for 30 min with 0.9% NaCl, 10mM Tris, pH 7.4, 5% fraction V bovine serum albumin (BSA). The nitrocellulose was then incubated in Tris–NaCl–BSA buffer containing 20 µl/ml

Invertebrate Oxygen Carriers
Ed. by Bernt Linzen
© Springer–Verlag Berlin Heidelberg 1986

antibody. After washing, the paper was immersed in freshly prepared Tris-NaCl-
BSA with ^{125}I-protein A. The nitrocellulose was washed, dried, and exposed to
X-ray film. Northern blot of mRNA from chloragogen cells: Poly A^{+} RNA (2-3μg)
was applied to a 1% agarose gel containing formamide for electrophoresis (11).
The transfer was done as described (11). The sheet was probed with a ^{32}P-labelled
synthetic oligonucleotide corresponding to amino acid residues 103-108 of chain
AIII (12). After hybridization, the nitrocellulose was washed, dried, and exposed
to X-ray film. Double stranded cDNA synthesis: Poly A^{+} RNA (2μg) was used as a
template for double stranded cDNA synthesis with oligo dT or the synthetic oligonucleotide as primer (13). Reverse transcriptase was used for the first strand
synthesis, and the Klenow fragment of DNA polymerase for the second strand synthesis.
EcoRI linkers were ligated to the cDNA which was then inserted into λgt10 and
cloned.

Results and Discussion. Electrophoresis (Fig 1) of translation products labelled
with ^{35}S-methionine showed three major bands between 16-19 Kd. A prominent band
slightly larger than myoglobin appeared with ^{3}H-leucine as label. Since most globin chains of Lumbricus lack methionine, the band appearing with ^{3}H-leucine probably corresponds to one of the Lumbricus globin chains.

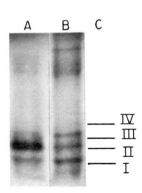

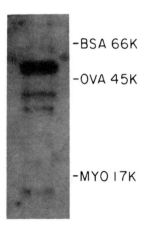

Fig. 1: SDS-PAGE of translation
products. A: fluorogram of
^{3}H-leucine labelled translation
products. B: fluorogram of ^{35}S-methio-
nine labelled translation products.
C: Position of four polypeptides of
reduced Lumbricus globin.

Fig. 2: Western blot of SDS-PAGE
of the cold translation products.

mRNA ↑RNA

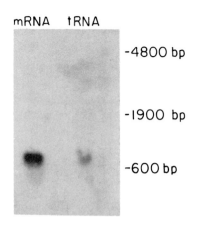

-4800 bp

-1900 bp

-600 bp

Fig. 3: Northern blot of
Lumbricus chloragogen cell RNA.

A Western blot of an acrylamide gel of the cold translation products hybridized
with the anti-globin antibody showed a band of about 52 Kd (Fig. 2). This band
corresponds to the disulfide-linked trimer (1,2). A Northern blot of the RNA
from the chloragogen cells was probed with the synthetic oligonucleotide. The
autoradiogram of this blot showed a single band at 750-850 bp, which should be
the size of the globin message. The combined results of the Northern and Western
blots clearly show that mRNA coding for globin is present in the chloragogen cells
of Lumbricus terrestris. Double stranded cDNA has been prepared from the poly
A^+ RNA. The cDNA primed with oligo dT and the cDNA primed with the synthetic oligonucleotide have been cloned into λgt10. Both cDNA libraries are now being
screened. (Supported by grants from NSF, PCM 8202760, Welch Foundation, F-213,
and NIH, GM28410, to A.F.R.)

References

1. Kapp, O.H., Polidori, G., Mainwaring, M.G., Crewe, A.V., and Vinogradov, S.N.,
   J. Biol. Chem 259, 628-639 (1984).
2. Fushitani, K., Imai, K., and Riggs, A.F., this volume, pp. 77-80 (1986).
3. Vinogradov, S.N., Shlom, J.M., Hall, B.C., Kapp, O.H., and Mizukami, H.,
   Biochim. Biophys. Acta 492, 136-155 (1977).
4. Cooper, E.L., and Stein, E.A., in Invertebrate Blood Cells (Ratcliffe,
   N.A., and Rowley, A.F., eds.), pp. 75-140, Academic Press, New York (1981).
5. Breton-Gorius, J., Ph.D. dissertation, at the Centre National de Transfusion
   Sanguine, Laboratoire de Cytologie du Prof. M. Bessis, Paris.
6. Cox, R.H., Methods in Enzymology 12B, 120-129 (1968).
7. Garlick, R.L., Riggs, A.F., Arch. Biochem. Biophys 208, 563-575 (1981).
8. Garvey, J.S., Cremer, N.E., Sussdorf, D.H., Campbell, D.H., in Methods
   in Immunology (Benjamin, W.A., Inc., eds.), pp. 36-38, Reading, Massachusetts
   (1977).
9. Laemmli, U.K., Nature 277, 680-685 (1970).

472

10. Burnett, W.N., Anal. Biochem. 112, 195-203 (1981).
11. Maniatis, T., Fritsch, E.F., Sambrook, J., in Molecular Cloning, pp. 202-203, Cold Spring Harbor (1982).
12. Garlick, R.L., and Riggs, A.F., J. Biol. Chem. 257, 9005-9015 (1982).
13. Huynh, T.V., Young, R.A., and Davis, R.W., in DNA Cloning: a practical approach (D. Glover, eds.), IRL press, Oxford, pp. 49-78 (1985).

CLONING OF THE cDNA FOR THE GLOBIN FROM THE CLAM, BARBATIA REEVEANA

A.F. Riggs, C.K. Riggs, R.-J. Lin, and H. Domdey
Department of Zoology, University of Texas, Austin, Texas 78712
California Institute of Technology, Pasadena, California 91125

Abstract. The cDNA for the two domain globin from the clam, Barbatia reeveana, has been cloned and largely sequenced. The results show that the amino acid sequences which correspond to the E and F helices in vertebrate globins are 100% identical in the two domains even though other parts of the domains differ substantially. This finding is consistent with the possibility that the E and F helices form important intersubunit contacts in the clam hemoglobin.

Introduction. Grinich and Terwilliger (1) discovered that the clam, Barbatia reeveana, has two intracellular hemoglobins, one with 12-14 polypeptides of 33Kd chains and the other with 16 Kd chains of two kinds which form a tetramer. They showed that the 33Kd chains are composed of two domains which can be cleaved in the native molecule to yield single domain fragments of approximately 16Kd which retain the ability to combine reversibly with oxygen. We report here the isolation of RNA from these cells, the cloning of the cDNA and the sequencing of most of the cDNA which codes for the two domain hemoglobin.

Materials and Methods. Clams were collected from rocks in Bahia Concepcion, Baja California, under a special permit granted by the Mexican government. The red cells were isolated, washed and placed on liquid N_2 until used. RNA was isolated by plunging the frozen pellet into a phenol-SDS solution at 65°C. Poly A^+ RNA was prepared by passage of the total RNA through a column of oligo dT-cellulose. Double-stranded cDNA was prepared with reverse transcriptase and DNA polymerase followed by S1 nuclease treatment. The DNA was blunt-end ligated into the SmaI site of M13 mp8 and cloned into E. coli strain JM103 (2). A second preparation of cDNA was made by the method of Huynh et al. (3) and inserted with EcoR1 linkers into the EcoR1 site of λ gt10 and cloned into the Δ hfl strain of E. coli.

Results and Discussion. The total RNA from the red cells of Barbatia reeveana was added to a rabbit reticulocyte system containing ^{35}S-methionine and shown to be active in protein synthesis. The label was incorporated primarily into a single 32-33Kd protein and two minor proteins of 14Kd and 16Kd, consistent with a two domain hemoglobin and a heterotetramer with 2 kinds of chains.

We have prepared cDNA from the poly A^+ RNA and have cloned it in M13 mp8. One clone, AR12, of the cDNA contained an insert of ≅ 500bp. This insert hybridized strongly with poly A⁺ RNA of ≅ 1200 bases and weakly with RNA of ≅ 700bp. We concluded that the 1200bp RNA probably contained the message for the 33Kd globin and that the 500 bp insert coded for part of it. AR12 was sequenced directly for 250 bases by the Sanger dideoxy procedure (4). The insert was then cut out with EcoR1 and BamH1 and cleaved with ClaI to yield two fragments of approximately 270 and 230bp. These fragments were cloned and sequenced. The translated cDNA sequence of AR12 showed only one open reading frame which coded for the first 129 residues of the first domain of the globin. Proof that AR12 corresponded to the first domain was obtained by a primer extension procedure (5) which showed that only 103 non-coding base pairs are present at the 5' end. A 2nd preparation of cDNA was inserted into λ gt10 and cloned. One clone, J1, contained an insert of ≅ 900bp which hybridized strongly with AR12. The J1 insert was cut out and cloned in M13 mp8 in both orientations, J1G and J1E. J1G was found to correspond exactly to the 5' end of AR12 and began at base 37 of AR12 with respect to the methionyl start codon. A 24-base oligonucleotide corresponding to bases 398-421 of AR12 was synthesized for use as a primer to extend the sequence to about base 600 in J1. The sequencing of J1E in the opposite direction would then complete the sequence. However the dideoxy technique produced an unreadable gel with J1G because the cDNA for the second domain contained an almost identical sequence, so priming occurred simultaneously in the two positions. J1E provided the sequence of the last 3/4 of the cDNA corresponding to the 2nd domain which could then be compared with corresponding parts of the first domain.

The results (Fig.1) show that the first 129 amino acid residues of domain 1 of the Barbatia sequence are 35-40% identical with the corresponding residues in the globin sequences of the related clam Anadara trapezia (6,7). Although the presumed E and F helices of domain 1 are only 19 and 11% identical with their counterparts in human globin chains, they are 57-71% identical with the corresponding sequences of the clam globins. Even more striking is that the presumed E and F helical segments are 100% identical in the two domains. This very high conservation suggests that these segments have a special structural role. The recent x-ray work of Royer et al. (8) indicates that these helices form novel inter-chain contacts in the assembly of a tetramer of hemoglobin in a related clam. We conclude that similar contacts are probably present in the assembly of the two-domain chains in Barbatia. Gel filtration studies (9) suggest that four 2-domain chains form a tetramer of 140 Kd. A trimer of this tetramer would then produce the full size native aggregate of about 420 Kd. [Supported by grants from NSF (PCM 8202760), Welch Foundation (F-213), NIH (GM28410) and University of Texas Research Institute to A.F.R., and NIH grant GM30356 to John Abelson.]

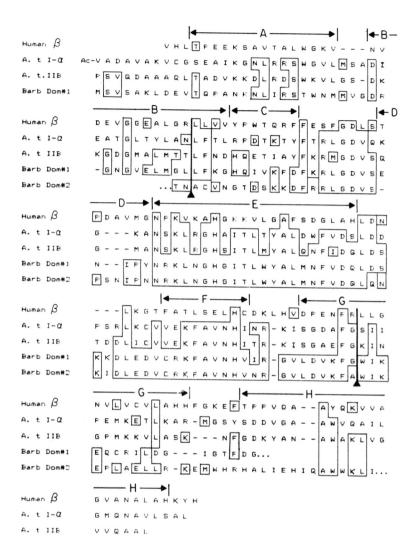

Fig. 1. Comparison of the partial amino acid sequences of the two domains of the 33Kd hemoglobin from Barbatia reeveana with those of Anadara trapezia Iα (6), A. trapezia IIB (5) and the human β chain. The helical designations are those for the human β chain. The black triangles with the dark vertical line show the positions of the introns of the vertebrate globin gene.

References

1. Grinich, N.P., and Terwilliger, R.C., Biochem.J. 189, 1-8 (1980).

476

2. Messing, J., Methods in Enzymology 101, 20-78 (1983).
3. Huynh, T.V., Young, R.A., and Davis, R.W., in DNA Cloning: A Practical Approach (Glover, D., ed.), IRL Press, Oxford, pp. 49-78 (1985).
4. Sanger, F., Nicklen, S., and Coulson, A.R., Proc.Natl.Acad.Sci. USA 74, 5463-5467 (1977).
5. Domdey, H., Apostol, B., Lin, R.-J., Newman, A., Brody, E., and Abelson, J., Cell 39, 611-621 (1984).
6. Fisher, W.K., Gilbert, A.T., and Thompson, E.O.P., Aust.Jour.Biol.Sci. 37, 191-203 (1984).
7. Como, P.F., and Thompson, E.O.P., Aust.Jour.Biol.Sci. 33, 653-664 (1980).
8. Royer, W.E., Love, W.E., and Fenderson, F.F., Nature, in press (1985).
9. Riggs, A.F., unpublished observations (1985).

Identification of a cDNA Transcript Containing the Coding Region for F—I Globin from *Urechis caupo*

James R. Garey and Austen F. Riggs
Department of Zoology, University of Texas
Austin, Texas 78712 U.S.A.

Abstract. A cDNA transcript of 729 bp in length coding for *Urechis caupo* F—I globin has been cloned. The coding region has been completely sequenced. Northern and Southern blot analysis indicate that there may be several genes coding for mRNA transcripts of nearly identical size. This could account for the heterogeneity observed in the hemoglobin.

Introduction. *Urechis caupo* of the phylum Echiura has an intracellular and tetrameric hemoglobin which has non—cooperative oxygen binding and almost no Bohr effect. Each of two major chromatographic fractions have at least five electrophoretic components (1). We have previously shown that mRNA from the red cells translates globin (2). Our goal has been to sequence the cDNA coding for the globins and to determine the structure of the globin genes.

Methods. RNA was isolated from red cells as described (3). Poly A+ RNA was prepared by chromatography on oligo dT cellulose. This was used as a template for double—stranded cDNA synthesis with a mixed oligonucleotide corresponding to residues 11—15 of *Urechis caupo* F—I globin (1). Reverse transcriptase was used for the first strand, and the Klenow fragment of DNA polymerase for the second strand. *Eco*RI linkers were ligated to the cDNA which was then cloned directly into M13 mp8 (4). A second cDNA library was constructed using oligo dT as a primer and cloned into λgt10 (5). Northern and Southern analysis and *in situ* plaque hybridizations were done as described (6). DNA sequencing reactions were carried out using the Sanger dideoxy method (7). Genomic DNA was prepared from sperm as described (8).

Results and Discussion. The synthetic oligonucleotide probe was extended to 60 base pairs by using it as a primer for the synthesis of ds cDNA which was then cloned. One clone, UCG—1, contained 60 bp, 45 of which corresponded exactly to the 15 amino terminal residues of *Urechis caupo* F—I globin.

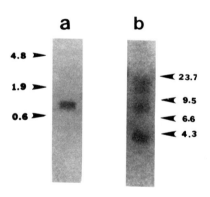

FIGURE 1 (a) Northern blot of *Urechis* red cell poly A+ RNA using UCG—1 as a probe. (b) Southern analysis of *Urechis* genomic DNA using UCG—2 as a probe. Arrows indicate MW markers in kb.

Invertebrate Oxygen Carriers
Ed. by Bernt Linzen
© Springer—Verlag Berlin Heidelberg 1986

UCG−1 DNA was used as a probe in a Northern transfer of the RNA (Fig. 1a) which indicated that the globin message is about 900 bp in size. It was also used to screen the oligo dT primed cDNA library. The clone UCG−2 contained an insert of about 800 bp which hybridized strongly with UCG−1. It was digested with several restriction enzymes: MnII, HincII, AluI, and HinfI). The fragments were isolated, cloned, and sequenced. About 250 bases were sequenced directly from each end of UCG−2. Together with the sequences of the fragments this sufficed to give the structure of the entire coding region (Fig. 2). The sequence of the 3' non−coding region was also determined but parts of it need further confirmation because it has not been sequenced entirely in both directions.

The derived amino acid sequence of F−I globin shows that 141 amino acids are present. Alignment of the sequence with those of other globins show no close similarity to any known globin. The percent of residues identical to those in the mollusc globins of *Anadara trapezia* (9,10) and of two annelid worms, *Lumbricus terrestris* (11), and *Glycera dibranchiata* (12) is 13%, 11%, and 19% respectively. This correspondence is too low to permit any definite conclusion to be drawn concerning phylogenetic relationships. Although *Urechis* hemoglobin is tetrameric, examination of the

```
1              5                 10                 15
Gly Leu Thr Thr Ala Gln Ile Lys Ala Ile Gln Asp His Trp Phe
16             20                25                 30
Leu Asn Ile Lys Gly Cys Leu Gln Ala Ala Ala Asp Ser Ile Phe
31             35                40                 45
Phe Lys Tyr Leu Thr Ala Tyr Pro Gly Asp Leu Ala Phe Phe His
46             50                55                 60
Lys Phe Ser Ser Val Pro Leu Tyr Gly Leu Arg Ser Asn Pro Ala
61             65                70                 75
Tyr Lys Ala Gln Thr Leu Thr Val Ile Asn Tyr Leu Asp Lys Val
76             80                85                 90
Val Asp Ala Leu Gly Gly Asn Ala Gly Ala Leu Met Lys Ala Lys
91             95                100                105
Val Pro Ser His Asp Ala Met Gly Ile Thr Pro Lys His Phe Gly
106            110               115                120
Gln Leu Leu Lys Leu Val Gly Gly Val Phe Gln Glu Glu Phe Ser
121            125               130                135
Ala Asp Pro Thr Thr Val Ala Ala Trp Gly Asp Ala Ala Gly Val
136            140
Leu Val Ala Ala Met Lys
```

FIGURE 2. The complete amino acid sequence of *Urechis caupo* F−1 globin translated from bases 9 through 435 of clone UCG−2.

residue positions corresponding to those in either the $\alpha_1\beta_1$ or $\alpha_1\beta_2$ interfaces in vertebrate globins shows little similarity, so that the subunit contacts must be quite different. The two universally conserved residues, phenylalanine at CD1 and the proximal histidine are present, but glutamine appears in place of the distal histidine. None of the acid groups responsible for the Bohr effect in vertebrate hemoglobins appears to be present.

We have prepared EcoR1 digested genomic DNA and have carried out a Southern hybridization blot with nick−translated UCG−2 cDNA as a probe. The results show the presence of 4 different bands of about 4, 9, 15, and 17 kb (Fig 1b). This indicates that several genes are present. The Northern analysis shows only a single band. Thus, if multiple genes are present, they produce RNA transcripts of nearly identical size. This could explain the electrophoretic heterogeneity which we have observed in F−I globin (1). (Supported by grants from NSF (PCM 8202760), Welch Foundation (F−213) and NIH (GM 28410))

References
1. Garey, J.R. and Riggs, A.F., *Arch. Biochem. Biophys.* **228**, 320−331(1984).
2. Garey, J.R., Brodeur, R.D., and Riggs, A.F., *Life Chem. Reports.* *Suppl.* **1**, 387−391 (1983).

3. Cox, R.H., Methods in Enzymology **12B**, 120–129 (1968).
4. Messing, J., Methods in Enzymology **101**, 20–78 (1983).
5. Huynh, T.V., Young, R.A., and Davis, R.W., *in* DNA Cloning: A Practical Approach (Glover, D., ed.) IRL Press, Oxford, pp. 49-78 (1985).
6. Maniatis, T., Fritsch, E.F. and Sambrook, J., Molecular Cloning, a laboratory manual. Cold Spring Harbor Laboratory. (1982).
7. Sanger, F., Nicklen, S., and Coulson, A.R., *Proc. Natl.Acad. Sci. (USA)* **74**, 5463–5467 (1977).
8. Blin, N. and Stafford, D.W., *Nuc. Acid Res.* **3**, 2303–2308 (1976).
9. Fisher, W.K., Gilbert, A.T., and Thompson, E.O.P., *Aust. J. Biol. Sci.* **37**, 191–203 (1980).
10. Como, P.F., and Thompson, E.O.P., *Aust. J. Biol. Sci.* **33**, 653–664 (1980).
11. Garlick, R.L. and Riggs, A.F., *J. Biol. Chem.* **257**, 9005–9015 (1982).
12. Imamura, T., Baldwin, T.O., and Riggs, A., *J. Biol. Chem.* **247**, 2785–2797 (1972).

A COMPARATIVE STUDY OF ARTHROPOD AND MOLLUSC HEMOCYANIN mRNA

I. Avissar, V. Daniel[1] and E. Daniel
Department of Biochemistry, Tel-Aviv University, Tel-Aviv 69978
[1]Department of Biochemistry, Weizmann Institute of Science, Rehovot 76100
Israel

Introduction

In this communication we sought to examine the relationship between arthropod and mollusc hemocyanins at the mRNA level. For this purpose, we looked for a probe for hemocyanin mRNA. There are indications that the hexaaminoacid stretch His-His-Trp-His-Trp-His found in a number of arthropod hemocyanins (1-4) is involved in the binding of copper (5). An oligodeoxynucleotide complementary to the hexa-aminoacid stretch was synthesized and used in hybridization experiments to probe mRNA that codes for hemocyanin, Hc mRNA.

Results

Hybridization of oligodeoxynucleotide with Hc mRNA. A mixture of 16 deoxynucleo-tide sequences, 18 nucleotide long and complementary to His-His-Trp-His-Trp-His, was synthesized. Poly (A)$^+$mRNA from whole Leiurus, from whole Levantina and from the branchial gland of Sepia were electrophoresed in formaldehyde agarose gel and blotted onto nitrocellulose filter. Hc mRNA was detected by hybridization with ^{32}P-labeled oligodeoxynucleotide probe. The results are presented in Fig. 1. With Leiurus, a single band of Hc mRNA with mobility corresponding to 2.3 kb is observed. The electrophoretic pattern of Hc mRNA from Levantina reveals four bands corresponding to 9.4, 6.7, 4.0 and 1.7 kb. With Sepia, 2-3 bands, with lengths of 9.5, 2.8 and 1.7 kb, were obtained.

A comparison of the thermal stabilities of duplexes of oligodeoxynucleotide probe with Hc mRNA from Leiurus and Levantina was carried out. The duplexes formed between the probe and Hc mRNA from Leiurus and Levantina were found to have similar melting temperatures, 46.5° and 44.5°C respectively.

Hybridization of cDNA from Leiurus with Hc mRNA from Levantina. Poly (A)$^+$mRNA from whole Leiurus was fractionated by sucrose gradient centrifugation. Fractions enriched with Hc mRNA were pooled and used as a template to prepare ^{32}P-labeled cDNA. mRNA from whole Levantina was electrophoresed on formaldehyde agarose gel, blotted onto nitrocellulose filter, and hybridized with the ^{32}P-labeled Leiurus

482

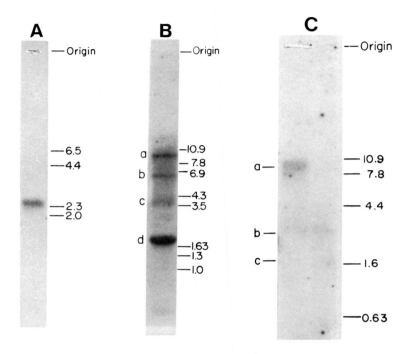

Fig. 1.   Electrophoresis of Hc mRNA.   Poly (A)$^+$mRNA from whole Leiurus (A), whole
Levantina (B) and from 2 preparations of the branchial gland of Sepia (C) were elec-
trophoresed in formaldehyde agarose gel and blotted onto nitrocellulose filter.
Hc mRNA was detected by hybridization with ^{32}P-labeled oligodeoxynucleotide probe.
Nucleotide lengths in kb of DNA markers are indicated.

cDNA.   In the electrophoretic pattern obtained (Fig. 2), bands with mobilities
corresponding to the bands of Levantina Hc mRNA can be distinguished.   The bands
represent hybrids of scorpion hemocyanin cDNA with snail hemocyanin mRNA.

Discussion

In a previous study (6), we have identified Hc mRNA from Leiurus by translation
and immunoprecipitation of the translation products and determined its size as ∿2 kb
by sucrose gradient centrifugation.   In the present communication, we report that
Leiurus Hc mRNA hybridizes with octadecadeoxynucleotide probe corresponding to the
sequence His-His-Trp-His-Trp-His.   The size of Leiurus Hc mRNA determined in this
study, 2.3 kb, is consistent with our earlier determination.   The constituent polypeptide chains of Leiurus hemocyanin have molecular weights about 75 000, corresponding to a message of about 2.1 kb.   The observed size of Hc mRNA is thus consistent
with expectation.

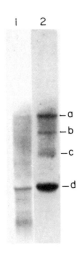

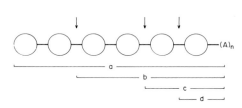

Fig. 2   Fig. 3

Fig. 2. Hybridization of Hc mRNA from Levantina with cDNA transcribed on mRNA from
Leiurus enriched with Hc mRNA (1) and with oligodeoxynucleotide probe (2).

Fig. 3. Possible scheme for nucleolytic cleavage of Hc mRNA from Levantina.
Hc mRNA is assumed to consist of 6 structural units. Nucleolytic cleavage sites
are indicated by arrows. Fragment designations (a,b,c,d) as in Fig. 1B).

The electrophoretic patterns of Hc mRNA from Levantina and from Sepia show a
number of bands. Levantina hemocyanin has polypeptide chains of molecular weight
334 000 (7), corresponding to a message of about 9 kb. The 9.4 kb band of Levantina
Hc mRNA (band a in Fig. 1B), and the 9.5 kb band of Sepia (band a in Fig. 1C), represent therefore authentic Hc mRNA. The remaining Hc mRNA bands probably represent
products of nucleolytic cleavage of authentic Hc mRNA.

Examination of the electrophoretic pattern of Levantina Hc mRNA shows that bands
associated with 6.7, 4.0 and 1.7 kb are relatively narrow. We could hardly have
expected such narrow bands were the authentic Hc mRNA structureless and exposed along
the totality of its stretch to the action of cleaving nucleases. To explain cleavage into well-defined parts, the Hc mRNA chain must be folded in such a way as to
leave only few regions vulnerable to nuclease action (Fig. 3). Thus we are led to
the idea of attributing a multi-unit structure to Hc mRNA from Levantina and probably
from Sepia. The various electrophoretic bands would represent combinations of
structural units, the 1.7 kb band representing a single structural unit (Table 1).
The message contained in a structural unit,$\sim$1.5 kb, corresponds to a functional unit
of hemocyanin of molecular weight $\sim$55 000. The 9.4 kb authentic Hc mRNA is

Table 1. Size relationships among Hc mRNA electrophoretic band components

Species	Band[*]	Length (kb)	No. of structural units	Expected length (kb)[**]
Levantina	a	9.4±0.4	6	9.2
	b	6.7±0.5	4	6.2
	c	4.0±0.4	2	3.2
	d	1.7±0.1	1	1.7
Sepia	a	9.5±0.3	6	9.2
	b	2.8±0.4	2	3.2
	c	1.7±0.1	1	1.7

[*]Band designations as in Fig. 1.   [**]Assuming a message of 1.5 kb for the structural unit and a poly (A) chain of 0.2 kb.

composed of 6 structural units. We thus have a correspondence between the 6-unit structures of the nucleotide chain of Hc mRNA and the polypeptide chain of hemocyanin (7).

Hybridization of the oligodeoxynucleotide probe with Hc mRNA from Leiurus, Levantina and Sepia indicates homology in the coding sequences of the polypeptide chain segments that carry copper in the oxygen binding sites. Hybridization with Levantina Hc mRNA of cDNA transcribed from Leiurus mRNA enriched with Hc mRNA provides further indication of homology between Hc mRNA from the two species, one arthropod and the other mollusc. These results support the hypothesis that hemocyanin from mollusca and arthropoda have evolved from a common ancestral origin.

References

1. Schneider, H.-J., Illig, U., Müller, E., Linzen, B. Lottspeich, F., and Henschen, A., Hoppe Seyler's Physiol. Chem. 363, 487-492 (1982).
2. Schartau, W., Eyerle, F., Reisinger, P., and Geisert, H., Life Chem. Rep. Suppl. 1, 85-88 (1983).
3. Nemoto, T., and Takagi, T., Life Chem. Rep. Suppl. 1, 89-92 (1983).
4. Yokota, E., and Riggs, A.F., J. Biol. Chem. 259, 4739-4749 (1984).
5. Gaykema, W.P.J., Hol, W.G.J., Vereijken, J.M., Soeter, N.M., Bak, H.J., and Beintema, J.J., Nature 309, 23-29 (1984).
6. Avissar, I., Daniel, V., and Daniel, E., Comp. Biochem. Physiol. 75B, 327-330 (1983).
7. Avissar, I., Daniel, E., Banin, D., and Ilan, E., this volume, pp. 249-252 (1986).

HAEMOCYANIN-mRNA-RICH FRACTIONS OF CEPHALOPODAN DECABRACHIA AND OF CRUSTACEA, THEIR *IN VIVO* AND *IN VITRO* TRANSLATION

G. Préaux, A. Vandamme, B. de Béthune, M.-P. Jacobs, and R. Lontie
Laboratorium voor Biochemie, Katholieke Universiteit te Leuven
Dekenstraat 6, B-3000 Leuven, Belgium

The biosynthesis of cephalopodan haemocyanin (Hc) occurs on membrane-bound polyribosomes in the branchial glands (1). Fractions rich in Hc mRNA were isolated by the phenol-oligo(dT)-cellulose method and translated in oocytes of *Xenopus laevis* into immunoprecipitable products comigrating on SDS-PAGE with the subunit of *Sepia officinalis* (2,3), *Loligo vulgaris* (2), and *Loligo forbesi* (3) Hc's.

In order to confirm that the synthesized subunit corresponds to that of Hc and that the mRNA is of a large size, immunocompetition tests with native Hc and centrifugations to isolate active large-size polyribosomes were performed for *S. officinalis*. The influence of the translation system on the size of the synthesized polypeptide chain was also studied.

The arthropodan Hc is synthesized in cyanocytes on free ribosomes (1) in tissues varying with the species: the tissue behind the compound eye for *Limulus polyphemus* (4), the reticular connective tissues surrounding the ophthalmic artery, the gizzard, and the hepatopancreas for *Carcinus maenas* (5) and *Cancer pagurus* (6), the hepatopancreas for *Homarus americanus* (7), and the heart for *Eurypelma californicum* (8).

In order to locate the site of Hc synthesis in the crayfish *Astacus leptodactylus* and to check it further in the crab *C. pagurus*, the four reported tissues were excised from both species and tested for the presence of Hc mRNA's. The SDS-PAGE patterns of the Hc products synthesized in three translation systems were compared with those of the subunits of the native Hc or for the purified 24S component.

The mRNA's were isolated and the translation products identified as described (2,6,9) except that the reticulocyte lysate system was improved as reported (6) and that before immunoprecipitation a preclearing with *Staphylococcus* Cowan I (Pansorbin, Calbiochem) was always performed (6,10).

After microinjection of *X. laevis* oocytes with the total mRNA fraction from the branchial glands, the SDS-PAGE of the immunoprecipitates obtained with antiserum against *S. officinalis* Hc from the homogenate of the oocytes, but in the presence of increasing amounts of native Hc,

Invertebrate Oxygen Carriers
Ed. by Bernt Linzen
© Springer-Verlag Berlin Heidelberg 1986

showed that the intensity of the labelled band with M_r 390 000 decreased, while that of the stained band of added Hc increased (Fig. 1A). This proved definitely their immunological relationship.

From the high M_r of the Hc subunit synthesized in the oocytes a large mRNA was inferred and hence the presence of heavy membrane-bound polyribosomes. To isolate them from the branchial glands, the method described for thyroglobulin (M_r of the subunit 330 000) was applied with slight modifications (11). The homogenization buffer was modified with 200 instead of 50 mM KCl and with 10 mM vanadyl ribonucleoside complexes, prepared as described (12). Translation in X. *laevis* oocytes of the mRNA's isolated from the nuclear pellet and from the cytoplasmic supernatant showed that all active Hc mRNA's were found in the sediment. Therefore this fraction was treated to release the polyribosomes. This was performed with Triton X-100 (2 %) added to the modified homogenization buffer. After separation of the membranes and nuclei, Hc mRNA's were only found in the supernatant. On fractionation of the latter by centrifugation on a sucrose gradient as described (11), the absorbance at 254 nm revealed a peak of very large polyribosomes, sedimenting slightly faster than reported for thyroglobulin, followed by a zone of smaller ones. Only the mRNA fraction from the larger polyribosomes led to the synthesis of Hc, thus proving the large size of the Hc mRNA (10).

In contrast with the results in oocytes, a long acting *in vivo* system, no complete translation of the Hc mRNA was obtained in two *in vitro* systems with a much shorter activity. In the nuclease treated reticulocyte lysates the M_r of the translated polypeptide part could be increased from 50 000 up to 300 000 on replacing the classical system (13) by the modified one (6). In the wheat-germ extract (14) a whole range of Hc polypeptides was obtained with M_r up to 165 000. These differences with the translation systems also point to a long Hc mRNA.

The polyribosomes and the mRNA of S. *officinalis* Hc are thus clearly of a large size.

Only the mRNA fraction from the hepatopancreas of A. *leptodactylus*, when translated in reticulocyte lysate or in X. *laevis* oocytes, yielded products which precipitated with antiserum against the Hc and comigrated with the subunits of the native Hc on SDS-PAGE. The electrophoretic pattern of the Hc from the oocytes was very similar to that of native Hc, that from the reticulocyte lysate differed: the central band was missing and the intensity of the two others was inversed (Fig. 1B). This pointed to the presence of at least two mRNA's and to some possible posttranslational modification (9). Substitution of the phenol (2) by the guanidinium thiocyanate method (15) for the extraction of

A. *S. officinalis* Hc

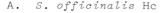
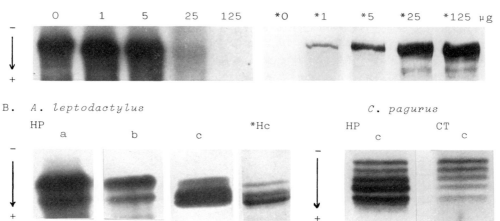

B. *A. leptodactylus*   *C. pagurus*

Fig. 1.   SDS-PAGE on 7.5 % running gels of the Hc immunoprecipitates:
(A) obtained in the presence of 0 to 125 µg of added Hc of *S. offici-
    nalis* per homogenate of 10 oocytes of *X. laevis* microinjected with
    the mRNA from the branchial glands,
(B) of the translation products from the mRNA fraction of the hepatopan-
    creas (HP) of *A. leptodactylus* and of *C. pagurus* and of the reticu-
    lar connective tissues (CT) of *C. pagurus* in reticulocyte lysates
    (a), wheat-germ extracts (b), and oocytes of *X. laevis* (c),
    and of native Hc (*) of *A. leptodactylus*.
Fluorograpny with L-[^{35}S]methionine (SJ.235, Amersham) and/or *staining
with PAGE blue 83 (BDH).

the RNA's also only revealed active Hc mRNA's in the hepatopancreas.
Moreover, as tested for the hepatopancreas and the reticular connective
tissues, it increased the yield of total mRNA about ten times. For the
hepatopancreas it also led to a remarkably more active mRNA fraction.
The Hc translation products from the oocytes and the reticulocyte
lysate showed the same electrophoretic differences as after the phenol
method. Those from a wheat-germ extract and from the reticulocyte lysate
were similar (Fig. 1B). Immunocompetition tests of the translation
products in reticulocyte lysate with native Hc confirmed that the synthesized products corresponded with Hc.
    For *C. pagurus* a fraction rich in Hc mRNA could, besides from the
reticular connective tissues (6), also be isolated from the hepatopancreas. On SDS-PAGE of the translation products, precipitated from the
reticulocyte lysate with antiserum against 24S-Hc, the same four bands
were observed as on translation of the Hc mRNA's from the reticular
connective tissues and as for the native 24S-Hc, but their relative

intensities differed. The translation products in oocytes showed one slower band more for both mRNA preparations (Fig. 1B). By immunocompetition tests with antiserum against 24S-Hc the synthesis of Hc in the reticulocyte lysate was confirmed.

The hepatopancreas is thus the site of synthesis of the *Astacus leptodactylus* Hc and, besides the reticular connective tissues, another one for *C. pagurus* Hc. In both species the heterogeneity of the subunits seems partly due to different mRNA's and partly to some posttranslational modifications.

Acknowledgements. We wish to thank the Fonds voor Collectief Fundamenteel Onderzoek for research grants, the Nationaal Fonds voor Wetenschappelijk Onderzoek (A.V.) and the Instituut tot Aanmoediging van het Wetenschappelijk Onderzoek in Nijverheid en Landbouw (B.d.B., M.-P.J.) for graduate fellowships. We express our gratitude to the Station Marine, Wimereux, France, for the cephalopods and to Dr. W. Peumans for the wheat-germ system.

REFERENCES

1. Préaux, G., and Gielens, C., *in* Copper Proteins and Copper Enzymes (Lontie, R., ed.), vol. 2, pp. 159-205, CRC Press, Boca Raton, FL (1984).
2. Vanderbeke, E., Cleuter, Y., Marbaix, G., Préaux, G., and Lontie, R., Biochem. Intern. *5*, 23-29 (1982).
3. Vanderbeke, E., Marbaix, G., Préaux, G., and Lontie, R., Life Chem. Rep. Suppl. 1, 385-386 (1983).
4. Fahrenbach, W.H., J. Cell Biol. *44*, 445-458 (1970).
5. Ghiretti-Magaldi, A., Milanesi, C., and Tognon, G., Cell Diff. *6*, 167-186 (1977).
6. Jacobs, M.-P., Lontie, R., and Préaux, G., Arch. Intern. Physiol. Biochim. *92*, B30 (1984).
7. Senkbeil, E.G., and Wriston, J.C., Comp. Biochem. Physiol. *68B*, 163-171 (1981).
8. Kempter, B., Naturwissenschaften *70*, 255-256 (1983).
9. de Béthune, B., Cleuter, Y., Marbaix, G., and Préaux, G., Arch. Intern. Physiol. Biochim. *93*, B75-B76 (1985).
10. Vandamme, A., Cleuter, Y., Marbaix, G., Préaux, G., and Lontie, R., Arch. Intern. Physiol. Biochim. *93*, B115-B116 (1985).
11. Davies, E., Dumont, J.E., and Vassart, G., Anal. Biochem. *80*, 289-297 (1977).
12. Berger, S.L., and Birkenmeyer, C.S., Biochemistry *18*, 5143-5149 (1979).
13. Pelham, H.R.B., and Jackson, R.J., Eur. J. Biochem. *67*, 247-256 (1976).
14. Peumans, W.J., Delaey, B.M., Manickam, A., and Carlier, A.R., Planta *150*, 286-290 (1980).
15. McCandliss, R., Sloma, A., and Pestka, S., Meth. Enzymol. *79*, 51-59 (1981).

INTRACELLULAR HEMOCYANIN AND SITE OF BIOSYNTHESIS
IN THE SPIDER
E U R Y P E L M A   C A L I F O R N I C U M

B. Kempter

Zoologisches Institut, Universität München
Luisenstraße 14, 8ooo München 2, F.R.G.

Pressent adress: Physiologisches Institut
Pettenkoferstr. 12, 8ooo München 2, F.R.G.

In several arthropods cells have been detected which contain hemocyanin (Hc) (1,2), or tissues identified with m-RNA coding for Hc
(3,4).
To identify Hc-containing cells in the spider Eurypelma californicum a whole animal was fixed in Bouin´s fluid, embedded in paraffin
and cut into 7 um sections. These were subjected to immuno-histochemistry for Hc according to Sofronief & Schrell (5), the first
antibody directed against whole stripped Eurypelma-Hc, and counterstained with toluene-blue. Endogenous peroxidases were inhibited by
hydroxymethyl-peroxide. Previous attempts to localize Hc by immuno-
fluorescence had failed because of the high background fluorescence.

While the whole hemolymph space was heavily stained, intracellular
staining was only found in hemocytes and oocytes. In these cells
staining for copper according to Okamoto & Utamura (6) was also
successful, where it showed birefringent microcrystals (fig. 1).

Hemocytes with positive peroxidase-staining were seen in many locations in the whole spider´s body. So, in small groups, on the
surface of the pericardium, in the blood sinuses of the lungs and
single cells distributed over the whole hemolymph space. Especially
massive clusters were seen on the inner surface of the heart, from
where they originate (7) (fig. 3).
It must be asked, whether the intracellular Hc of the oocytes (that
has not been noticed before) has the same composition as the Hc

Invertebrate Oxygen Carriers
Ed. by Bernt Linzen
© Springer–Verlag Berlin Heidelberg 1986

dissolved in the hemolymph, or a different one. Serum proteins or proteins derived therefrom have been found to be constituents of the eggs of many species of insects and vertebrates. In some species of Crustacea, eggs have been found to contain Hc (8). In the eggs of Carcinus maenas, the Hc fractions obtained by electrophoresis in 8M urea pH 7.5 showed quantitative relations different from those of the adult animals˙ serum (8). To determine the composition of the Eurypelma oocytan Hc, individual ovaries were homogenized and the clear supernatants subjected to PAGE. A band corresponding to 37 S Hc was observed (fig.2). Two-dimensional immunoelectrophoresis of the alkaline dissociate revealed the typical subunit pattern known from the hemolymph Hc (fig.2).

In order to test whether hemocytes or other tissues have the potential to synthesize Hc, various organs were excised, washed and incubated in a medium similar to Eurypelma blood, omitting proteins and lipids (9), to which 2o uCi of ^{3}H-leucine were added. After 3 hours of incubation the tissues were removed, washed and homogenized and the homogenate centrifuged at 12ooo g. To 1o ul of incubation media and the homogenate supernatants, antiserum against whole dissociated Hc was added. The resulting precipitate was spun down and washed twice. Precipitates, dissolved in SDS, and supernatants were counted. Other aliquots of the samples, treated identically, were subjected to SDS-PAGE.

As already described (1o) cells aspirated from the inner wall of the heart showed the highest degree of labelled Hc. Preparations of heart and lungs also showed some labelled Hc, which may be ascribed to the large count of hemocytes trapped in these tissues. Hepatopancreas and Malpigian tubules showed no labelling of Hc. The radioactivity found in the oocytan preparations ranged in between these two groups (Tab. 1). However, no labelled Hc-bands in SDS-PAGE nor labelled precipitates in two-dimensional immunoelectrophoresis could be found, in contrast to the hemocyte preparations, where massive labelling of the Hc-bands in SDS-PAGE took place. In two dimensional immunoelectrophoresis of the homogenate of the hemocytes from the heart, treated at pH 9.6 to achieve dissociation of Hc, again the radioactivity was found in the precipitated immunocomplexes (fig. 4), giving further proof that the newly synthesized protein was identical to Hc, with a subunit composition, by

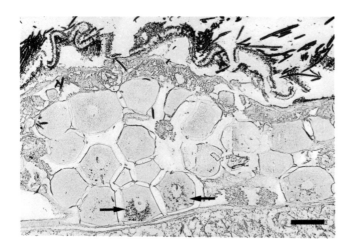

Figure 1

Eurypelma californicum:
7 µm section of oocytes.

Staining for copper according to Okamoto & Utamura (6). Halfmoon shaped or irregular distributed microcrystals of the copper complex are visible around the nucleus (→).

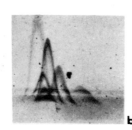

Figure 2

Analysis of the Hc in the homogenate of an ovary of *Eurypelma californicum*.

a) PAGE of the homogenate (H) and 37 S Hc from hemolymph as reference (R).
b) Two-dimensional immunoelectrophoresis of the alkaline dissociate (glycine/OH⁻ pH 9.6) of the homogenate. Antiserum against *Eurypelma* whole stripped Hc. The precipitate pattern is identical to that of the hemolymph Hc treated identically.

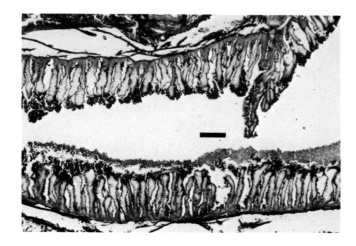

Figure 3

A

Longitudinal section (7 µm) of the heart of *Eurypelma californicum*.

Immunohistochemical staining for hemocyanin, counterstained with toluene blue.
Stained hemocytes are visible on and in between the muscle septa.

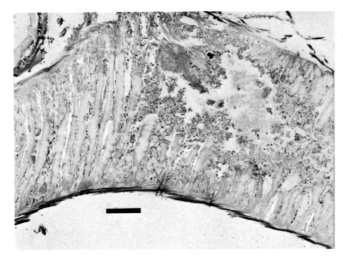

Figure 3

B

Control section(as in Fig. 3 A) reacted with neutral rabbit serum. No staining. The bar represents 2oo μm.

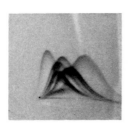

Figure 4

Two-dimensional immunoelectrophoresis of a homogenate of hemocytes from the heart of Eurypelma californicum that have been incubated with ³H-leucine. After 4 weeks of exposure to x-ray film, radioactivity is found in the typical precipitate pattern of Eurypelma Hc subunits.

any means, not very different from the hemolymph protein.

In non-bled animals the main amount of the intracellular Hc was found in the supernatant of the circulating hemocytes and their homogenate. Labelled Hc, however, could only be detected in the supernatant and, to a minor degree, in the cells aspirated from the heart (Tab. 1).

Animals that had been bled previously showed the main amount of Hc in the hemocytes of the heart, and the labelling predominantly in the homogenate of these cells. In the circulating cells Hc was only found in the supernatant, but the degree of labelling was low.

Clearly, bleeding the animals is a strong stimulus for the biosynthesis of Hc as can be seen by the high degree of labelling, but

TABLE 1
Ratio of radioactivity in 10 ul aliquots from the incubation of
several tissues from <u>Eurypelma californicum</u> with^{3}H-leucine. The
original incubation media and the tissue homogenates are listed.
Both were precipitated with antiserum against <u>Eurypelma</u> whole
stripped Hc and the supernatants and immunoprecipitates counted for
radioactivity. From the bled animals hemolymph had been taken on
day 11 and day 3 before the experiment.

A	Sample	R*
	<u>Malpighian tubules</u>	
	Incubation medium	0.1
	Homogenate supernatant	0.1
	<u>Hepatopancreas</u>	
	Incubation medium	0.1
	Homogenate supernatant	0.2
	<u>Ovary</u>	
	Incubation medium	0.1
	Homogenate supernatant	0.6
	<u>Lung sinus</u>	
	Incubation medium	0.2
	Homogenate supernatant	1.1
	<u>Heart</u>	
	Incubation medium	0.2
	Homogenate supernatant	1.3
	<u>Heart hemocytes</u>	
	Incubation medium	0.2
	Homogenate supernatant	21.0
B	<u>Heart hemocytes, bled animals</u>	
	Incubation medium	0.2
	Homogenate supernatant	21.0
	<u>Circulating hemocytes, bled animals</u>	
	Incubation medium	0.3
	Homogenate supernatant	0.5
	<u>Heart hemocytes, non-bled animals</u>	
	Incubation medium	0.2
	Homogenate supernatant	4.3
	<u>Circulating hemocytes, non-bled animals</u>	
	Incubation medium	0.6
	Homogenate supernatant	0.3

$$* \; R = \frac{cpm \; in \; immunoprecipitate}{cpm \; in \; medium \; before \; addition \; of \; antibody}$$

also for the proliferation of hemocytes in the heart as could be observed during the dissection of these organs, where an enormous mass of cells was found inside.

This pattern might be explained by a high fragility of a part of the circulating cells. That leaves us with the question, why just the Hc in the more stable cells was not labelled. Another explanation would be that there are two types or two stages in the life-cycle of the Hc-containing cells: one type that breaks up very soon after completion of Hc biosynthesis and another one that keeps circulating to serve as a Hc reserve.

This hypothesis would be consistent with the observation that after only one bleeding the Hc-level in the hemolymph is quickly restored at least in part. This could occur by depletion of intracellular Hc stores. After a second bleeding, however, the dissolved Hc remains low, and so does the Hc in the circulating cells, so that restoring the oxygen carrying capacity of the blood has to await the proliferation of hemocytes and de novo synthesis of Hc.

I wish to thank Prof. B. Linzen for his support and valuable comments, Prof. Dr. R. Wetzstein and Dr. A. Weindl for the opportunity to work in their laboratory and Dr. U. Schrell for the introduction to immuno-histochemistry. The investigation was partly supported by the DFG (Li 1o7/22-24).

L i t e r a t u r e

1. Ghiretti-Magaldi, A., Milanesi, C., Tognon, G., Cell Diff.,6,167-186 (1977).
2. Schönberger, N., Cox, J.A. & Gabbiani, G., Cell Tissue Res.,205, 397-409 (1980)
3. Senkbeil, E.G. & Wriston, J.C.Jr., Comp. Biochem. Phys.,68B,163-171 (1981).
4. Wood, E.G. & Bonaventura, J., Biochem. J.,196,653-666 (1981).
5. Sofronief, M. & Schrell, U. J. Histochem. Cytochem., 30,504-511 (1982).
6. Okamoto, K. & Utamura, M., Acta Scho. Med. Univ. Imp. Kyoto, 20, 573-581 (1938).
7. Sherman, R.G., Can. J. Zool., 51,1155-1166 (1973).
8. Busselen, P., Comp. Biochem. Physiol., 38A,317-328 (1971).
9. Schartau, W, & Leidescher, T., J. Comp. Physiol., 152,73-77 (1983).
10.Kempter, B., Naturwissenschaften, 5,255 (1983).

CLONING AND SEQUENCING OF EURYPELMA CALIFORNICUM HEMOCYANIN cDNA

R. Voit and H.-J. Schneider

Zoologisches Institut, Universität München, Luisenstr. 14, 8000 München 2, F.R.G.

The hemocyanin (Hc) of the tarantula Eurypelma californicum is a 24-meric macromolecule, built from seven different subunits - a,b,c,d,e,f,g - of molecular
weights ranging from 68 kD to 74 kD. Of these subunits, a, d and e have been sequenced (1,2,3). Recently the site of Hc biosynthesis has been identified in this
tarantula: Hemocytes attached to the inner heart wall and proliferating massively
subsequent to bleeding (4). These findings opened the way to isolation of Hc-mRNA
and to cloning and sequencing of Hc-cDNAs.

Isolation of RNA and in vitro translation

Total RNA was isolated from tarantula heart according to the guanidiniumthiocya-
nate/cesiumchloride method of Chirgwin et al. (5), three to five days after bleeding the animal. 40-60 µg of total RNA could be recovered from one heart of 80-
130 mg fresh weight. Poly(A)-RNA was further enriched on oligo(dT)-cellulose
columns (6).

Total RNA or poly(A)-RNA were translated in vitro using commercially available
(BRL) rabbit reticulocyte lysate (7) and ^{35}S-methionin (1200 Ci/mM, Amersham) for
labelling. The translation products were treated with antiserum against whole
dissociated Eurypelma 37 S Hc, and protein A-Sepharose (8). The immunoprecipitates
were separated on 7.5 % SDS-polyacrylamide gels (9). After fluorography

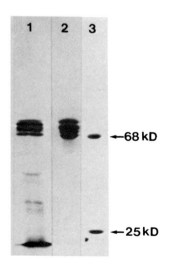

1 2 3

←68 kD

←25 kD

Fig. 1. SDS-PAGE of polypeptides synthesized by in
vitro translation of Eurypelma californicum heart
poly(A)-RNA and precipitated with antibodies
against whole dissociated 37 S hemocyanin. Lane 1,
fluorography of immunoprecipitated in vitro translation products; the upper portion of the gel did
not reveal any bands; lane 2, dissociated Eurypelma 37 S hemocyanin; lane 3, bovine serum albumin and chymotrypsinogen as molecular weight markers.

Invertebrate Oxygen Carriers
Ed. by Bernt Linzen
© Springer-Verlag Berlin Heidelberg 1986

(10) and drying, autoradiographs were made with KODAK X-OMAT X-ray film at -70°C (Fig. 1). About 10 % of the labelled polypeptides could be precipitated with the antiserum indicating preferential synthesis of Hc-mRNA. The immunoprecipitate shows the same electrophoretic band pattern in SDS-polyacrylamide gels as authentic Eurypelma Hc (Fig. 1). Obviously the polypeptides are not altered by post-translational modification.

Synthesis, cloning and sequencing of Hc cDNA

Heart poly(A)-RNA directed cDNA synthesis was performed essentially as described by Gubler and Hoffman (11). Double-stranded cDNA was tailed with dC and annealed with vector plasmid pUC 9, cut with Pst I and tailed with dG. Recombinant molecules were used to transform competent Escherichia coli strain DK1 cells. Positive transformants were selected by their resistance to ampicilline and screened for insertion of Hc sequences using a synthetic oligonucleotide derived from the 'Copper A' binding site - His-His-Trp-His-Trp-His - of chelicerate Hcs (12), and labelled with ^{32}P-ATP at the 5' end. The sequence of the oligonucleotide mixture 5'-CA$^{C}_{T}$CA$^{C}_{T}$TGGCA$^{C}_{T}$TGGCA -3' contains three possible mismatches with Hc cDNA. For colony hybridization and Southern blot analysis standard procedures were applied

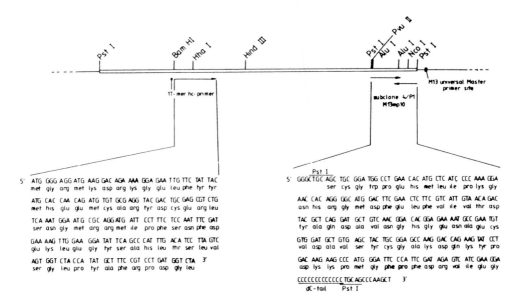

Fig. 2. A physical map of the 1700 base pair insert (═══) of hemocyanin cDNA clone HC4. Preliminary sequence analysis revealed the nucleotide sequences of two fragments.

(13,14). Plasmid DNA was isolated by ethidiumbromide/cesiumchloride-centrifugation or rapid alkaline lysis (15, 16). One clone, HC4, hybridized strongly with the radioactive oligonucleotide probe while a few others gave only weak hybridization signals. HC4 plasmid DNA was found to contain a 1700 base pair cDNA insert. Its restriction map is shown in Fig. 2. The map is accurate in regions already sequenced.

Sequence analysis was done by the method of Sanger (17). A region of about 200 nucleotides was sequenced directly in vector pUC 9 using the 17-mer Hc oligonucleotide as primer. Double-stranded plasmid DNA was denatured either by heat (18) or by alkali treatment (19). The nucleotide and amino acid sequences next to the copper A binding site are shown in Fig. 2. A small Pst I fragment at the 3' end of mRNA was subcloned into the phage M13 mp10 by standard procedures (20). The cloned fragment was sequenced in both directions using the universal M13 master primer (New England Biolabs) (Fig. 2).

```
                190           200          210          220          230          240         250
Eury. a      FGKVKDRKGELFYYMHQQMCARYDCERLSNGLNRMLPFHNFNEPLGGYAAHLTHVASGRHYAGRPDGL

Eury. d      FGKIKDRKGELFYYMHQQMCARYDCERLSVGLQRMLPFQNIDDELEGYSPHLSSLVSGLSYGSRPAG-

Eury. e      MGRMKDRKGELFYYMHQQMCARYDCERLSNGMHRMIPFSNFDEKLEGYSAHLTSLVSGLPYAFRPDGL

clone 4 cDNA MGRMKDRKGELFYYMHQQMCARYDCERLSNGMRRMIPFSNFDEKLEGYSAHLTSLVSGLPYAFRPDGL


                530          540          550          560          570          580          590
Eury. a      SCGKPQHLTVPRGNEKGMQFD LFVMLTDASVDRVQSGDGTPVCADALSYCGVLDQKYPDKRAMGYPFD-
                                   E
Eury. d      SDGKPEHMLVPRGKERGMDFYLFVMLTDYEEDSVQGAGEQTIDQDAVSYCGAKDQKYPDKKAMGYPFD-

Eury. e      SDGKPEHMLIPKGNHRGMDFELFVIVTDYAQDAVDG--ENAECVDAVSYCGAKDQKYPDKKPMGFPFD-

clone 4 cDNA SCGWPEHMLIPKGNHRGMDFELFVIVTDYAQDAVNGHGENAECVDAVSYCGAKDQKYPDKKPMGFPFD-


                600
Eury. a      RKITADTHEE-

Eury. d      RPIQVRTPSQ-

Eury. e      RVIEGLTFEE-

clone 4 cDNA RVIEG  3'end of clone 4 cDNA
```

Fig. 3. Comparison between the amino acid sequences deduced from hemocyanin cDNA clone HC4 and amino acid sequences of subunit a, d and e of Eurypelma californicum hemocyanin determined by Edman degradation. The numbering is from reference (3).

Amino acid sequences obtained so far via sequencing HC4 cDNA show great homology to the primary structure of Eurypelma Hc chain e (Fig. 3) determined by Edman degradation (3). HC4 cDNA does not contain the entire 3' portion of the mRNA and ends about 60 nucleotides before the expected termination codon. Altogether there are eight differences between the chemically determined and the cDNA derived sequences. The ambiguities at pos. 218 (His/Arg) and 533 (Lys/Trp) might be easily explained by ambiguities in the identification of the PTH- and DABITH-derivatives. In the other cases, however, the protein-chemical data appear to be correct;

in particular, between pos. 560 and 570 the sequence runs AVDGENA. This is a variable region (cf. (12)), and the differences probably reflect some microheterogeneity of the protein (see also subunit a in Fig. 3, pos. 248 and 560). The microheterogeneity does usually not appear in sequencing the protein derived from pooled blood.

Acknowledgement. Supported by DFG grants Li 107/24-6, Schn 226/2-4, and Schn 226/2-5. We thank Prof. E.L. Winnacker and his collaborators for their help and advice.

1. Schartau, W., Metzger, W., Sonner, P., and Pysny, W., this volume, p. 177-184 (1986).
2. Schartau, W., Eyerle, F., Reisinger, P., Geisert, H., Storz, H., and Linzen, B., Hoppe Seyler's Z. Physiol. Chem. 364, 1383-1409 (1983).
3. Schneider, H.-J., Drexel, R., Feldmaier, G., Linzen, B., Lottspeich, F., and Henschen, A., Hoppe Seyler's Z. Physiol. Chem. 364, 1357-1381 (1983).
4. Kempter, B., Naturwissenschaften 70, 255-256 (1983).
5. Chirgwin, J.M., Przybyla, A.E., MacDonald, R.J., and Rutter, W.J., Biochemistry 18, 5294-5299 (1979).
6. Aviv, H., and Leder, P., Proc. Natl. Acad. Sci. USA 69, 1408-1412 (1972).
7. Pelham, H.R.B., and Jackson, R.J., Eur. J. Biochem. 67, 247-256 (1976).
8. MacSween, J.M., and Eastwood, S.L., Meth. Enzymol. 73, 459-471 (1981).
9. Laemmli, U.K., Nature 227, 680-685 (1970).
10. Bonner, W.M., and Laskey, R.A., Eur. J. Biochem. 46, 83-88 (1974).
11. Gubler, U., and Hoffman, B.H., Gene 25, 263-269 (1983).
12. Linzen, B., Soeter, N.M., Riggs, A.F., et al., Science 229, 519-524 (1985).
13. Woods, D.E., Markham, A.F., Ricker, A.T., Goldberger, G., and Colten, H.R., Proc. Natl. Acad. Sci. USA 79, 5661-5665 (1982).
14. Southern, E.M., J. Mol. Biol. 98, 503-517 (1975).
15. Maniatis, T., Fritsch, E.F., and Sambrook, J., Molecular Cloning. A Laboratory Manual. Cold Spring Harbor Laboratory (1982).
16. Birnboim, H.C., and Doly, J., Nucl. Acid Res. 7, 1513-1523 (1979).
17. Sanger, F., Nicklen, S., and Coulson, A.R., Proc. Natl. Acad. Sci. USA 74, 5463-5467 (1977).
18. Wallace, R.B., Johnson, M.J., Suggs, S.V., Miyoshi, K., Bhatt, R., and Itakuran, K., Gene 16, 21-26 (1981).
19. Chen, E.Y., and Seeburg, P.H., DNA 4, 165-170 (1985).
20. Messing, J., Meth. Enzymol. 101, 20-78 (1983).

Concluding remarks

CONCLUSIONS

R. Lontie
Laboratorium voor Biochemie, Katholieke Universiteit te Leuven
Dekenstraat 6, B-3000 Leuven, Belgium

Physiology and Biochemistry of Haemocyanins
I - Naples, 30 August - 1 September 1966
F. Ghiretti & A. Ghiretti-Magaldi

Hemocyanins Workshop
NL - Warffum, 20-21 August 1970
M. Gruber & E.F.J. van Bruggen

EMBO Sponsored Haemocyanin Workshop
B - Louvain, 8-10 September 1971
R. Lontie

IV International Workshop on Haemocyanin
I - Padua, 23-25 April 1974
F. Ghiretti & A. Ghiretti-Magaldi

EMBO & University of Malta - V Haemocyanin Meeting
MA - Fort St. Lucian Marine Biological Laboratory, 1-4 August 1976
J.V. Bannister

EMBO & University of Tours - Comparative Study and Recent Knowledge on
Quaternary Structure and Active Sites of Oxygen Carriers and Related
Proteins
F - Tours, 20-24 August 1979
Jean Lamy & Josette Lamy

EMBO Workshop - Structure and Function of Invertebrate Respiratory
Proteins
GB - Leeds, 19-22 July 1982
E.J. Wood

Invertebrate Oxygen Carriers
D - Tutzing, 29 July - 1 August 1985
B. Linzen

Only a few of us have had the privilege to attend all these meetings.
It all started in Naples in 1966 at the initiative of Ghiro Ghiretti,
our mentor. The second meeting, an informal one, took place in Warffum
near Groningen, the third one in Louvain. The problem of the M_r of the
polypeptide chain of molluscan haemocyanins was already present in Padua.
In this respect the meeting in Malta was quite important for the
participants from Benelux, as they reached there the following important
agreement: $(7 + 9)/2 = 8$ for *Helix pomatia* haemocyanin. The meetings
were then extended to invertebrate dioxygen carriers.

Why did Nature develop these three types of dioxygen carriers? Was
it only for providing us with the opportunity to have these pleasant
meetings?

Striking is the similarity in the organization of some invertebrate

Invertebrate Oxygen Carriers
Ed. by Bernt Linzen
© Springer–Verlag Berlin Heidelberg 1986

haemoglobins and of molluscan haemocyanins with their multisite polypeptide chains and between the active sites of haemerythrins and
arthropodan haemocyanins. But the workers on haemocyanins look with
envy at haemerythrins, as their active site seems so well characterized
up to the hydrogen bond, and at some invertebrate haemoglobins, as so
much is still to be learned about their quaternary structure.

In the conclusions in Tours the need was stated for representative
sequences and typical conformations of arthropodan and molluscan
haemocyanins and already now there is a great wealth of arthropodan
sequences with large homologies. Two sequences of molluscan functional
units, a cephalopodan and a gastropodan one, are in sight with a
definite homology.

Striking is the homology of the CuB site of arthropodan haemocyanins
and of *Neurospora crassa* tyrosinase and a Cu-binding site of a gastropodan functional unit, while the CuA site of arthropodan haemocyanins
seems to show only a "scrambled" homology with that of *N. crassa*
tyrosinase.

Many more sequences are, however, needed if we want to understand
the real diversity of the subunits and functional units, their contact
regions in the quaternary structure, the cooperativity of the dioxygen
binding, the Bohr effect, the binding of effectors and modulators like
lactate, the structure of epitopes, the arthropodan subunits (of the
scorpions and one subunit of *Limulus*) which behave with hydrogen peroxide
like molluscan functional units.

Similarly a first amino-acid sequence was presented for a functional
subunit isolated from the polypeptide chain of an invertebrate haemoglobin. Could a non-haem-containing subunit of *Lumbricus terrestris*
e.g. show a sequence homology with the functional subunits?

Isn't it amazing how much more we understand of arthropodan
haemocyanins with the wonderful X-ray data on *Panulirus interruptus*?
How much could also be gained from knowing the conformation of one
molluscan functional unit and of one tyrosinase?

We were also again astonished by the sophisticated methods
developed to analyse quaternary structures: image processing allowing
a choice of enantiomers, immuno-electron microscopy with moreover the
advent of monoclonal antibodies, and the cryo-electron microscopy
announcing a new spring in Groningen.

The structure of dioxygen carriers can now be compared at four
levels: the surface, the conformation, the polypeptide sequence, and
recently at the nucleic acid level.

Although I am tempted to compare it with a museum for the blind, where sensitive fingertips explore the sculptures, the comparison at the surface has already yielded extremely important results in immuno-electrophoresis besides the mentioned immuno-electron microscopy.

The isolation of mRNAs and the preparation of cDNAs are progressing with invertebrate haemoglobins and with several haemocyanins: we saw stimulating results at this symposium. This opens the way to rapid DNA sequencing and to the study of gene expression. In the biosynthesis of haemocyanins and haemerythrins the metal donors have to be identified: Cu-thioneins and 4Fe-4S-proteins respectively are possible candidates. Haemocyanins seem a good model for the study of the biosynthesis of copper proteins, although I am puzzled by the variety of the sites of biosynthesis. Mutagenesis provided already with tyrosinase a striking example for the identification of the ligands of copper. Site-directed mutagenesis offers new possibilities, but I hope that blanks will be performed, as there might be differences in the conformation of a protein when synthesized in a prokaryotic system.

Finally studies are needed at the gene level: the number of genes, their control regions, their sequential expression (larval/adult proteins), the location of the introns. Do the introns show a relation with protein domains, could they provide an explanation for the apparently "scrambled" homologies? Are there occasionally silent genes for haemoglobins and/ or haemocyanins?

As usual there remain nomenclature problems. A unique abbreviation for haemocyanin is desirable: Hc in accordance with Hb for haemoglobin (IUB Nomenclature Rules; Hcy = homocysteine). We need to distinguish subunits with a single active site and functional units (segments of a polypeptide chain with also a single active site). For the latter the term "domain" is less indicated, as it has a different meaning in protein crystallography like already pointed out by Professor H.S. Mason in Tours.

The symbols of the subunits of chelicerates remind me of the early days in histone chemistry. Unambiguous names are needed for the derivatives of the active sites, not to mention linguistic mermaids like "half-met" and "half-apo". They must, moreover, occasionally be able to distinguish between CuA and CuB. Two semi-apo derivatives have namely been obtained with crustacean haemocyanins as shown by interesting fluorescence experiments.

In closing this symposium I would like to ask Bernt Linzen to extend our thanks to the Director and Staff of the Evangelische Akademie for the pleasant and efficient settings, like all of us would like to express our gratitude to him and to all of his collaborators.

AUTHOR INDEX

Abel, S., 403
Alviggi, M., 429,437,441,453,457
Antolini, R., 357,361
Arp, A.J., 129
Ascoli, F., 85,97,101,107
Avissar, I., 249,481

Bak, H.J., 149,153
Banin, D., 249
Beintema, J.J., 149,153,165,169
Bellelli, A., 337
Beltramini, M., 429,437,441
Benoy, C., 223
Beukeveld, G.J.J., 169
Bijlholt, M.M.C., 213
Billiald, P., 185
Bishop, G.A., 389
Boisset, N., 185
Bonaventura, C., 241,407
Bonaventura, J., 241,407
Booy, F., 217
Braunitzer, G., 73
Brenowitz, M., 241
Bridges, C.R., 341,353
Brunori, M., 337,375,421

Cau, A., 337
Chen, M.N., 463
Chiancone, E., 85,97,101,107
Citro, G., 97
Coletta, M., 375
Compin, S., 231
Condo, S.G., 337
Corda, M., 337
Cordone, L., 121

Daniel, E., 37,93,249,481
Daniel, V., 481
David, M.M., 93
de Béthune, B., 485
Decker, H., 383,395
Deleersnijder, W., 445,449
Desideri, A., 101
Di Cera, E., 375
Di Stefano, L., 65,121
Domdey, H., 473
Drexel, R., 255,263

Filippi, B., 235
Fincke, T., 327
Frank, J., 185
Furuta, H., 117
Fushitani, K., 77, 367

Garey, J.R., 469,477
Geelen, D., 81
Ghiretti, F., 3,45
Ghiretti Magaldi, A., 45,53
Giacometti, G., 453,457
Giacometti, G.M., 453,457
Giardina, B., 85,337
Gielens, C., 223,227,255
Gill, S.J., 383,389,395
Gobbo, S., 235
Godette, G., 277
Goettgens, A., 173
Gotoh, T., 69
Graham, R., 333
Grißhammer, R., 173

Haker, J., 217
Hammel, I., 93
Handa, T., 367
Henschen, A., 255
Hol, W.G.J., 135
Huber, M., 265

Igarashi, Y., 89
Ilan, E., 93,249
Imai, K., 77,367

Jacobs, M.-P., 485
Jeffrey, P.D., 207
Jekel, P.A., 149,153,165
Jhiang, S.M., 469

Kajita, A., 89,117
Keegstra, W., 217
Kempter, B., 489
Kihara, H., 367
Kimura, K., 89
Kleinschmidt, T., 73

DATE DUE

APR 1 2 1989		
DEC 1 9 1990		
DEC 7 1990		
MAY 2 3 1993		
DEC 1 5 1994		
FEB 2 6 1997		